The Social and Behavioural Aspects of Climate Change

The Social and Behavioural Aspects of Climate Change

LINKING VULNERABILITY, ADAPTATION AND MITIGATION

EDITED BY PIM MARTENS AND CHIUNG TING CHANG

Published by Greenleaf Publishing Limited
Aizlewood's Mill
Nursery Street
Sheffield S3 8GG
UK
www.greenleaf-publishing.com

Printed in Great Britain on acid-free paper by
Antony Rowe Ltd, Chippenham and Eastbourne

Cover by LaliAbril.com

British Library Cataloguing in Publication Data:
A catalogue record for this book is available from the British Library.

ISBN-13: 9781906093426

Contents

Contents

Preface

Concern about global warming and the influence of human activity can be traced back to the 1980s. Early scientific efforts concentrated on generating knowledge about the potential impacts of a changing climate and how to reduce anthropogenic greenhouse gas emissions. Understanding physical phenomena was at the core of the scientific enterprise in climate research—hence, the dominant position of the natural sciences. The contribution of the social sciences centred on the strategy of mitigation, in a wide sense, and in particular the analysis of reduction measures. Since the IPCC's *Third Assessment Report* in 2001, however, increasing emphasis has been placed on adaptation and the promotion of a risk management approach. Advances on these research frontiers require major contributions from the social sciences. Over the years, social scientists have started to respond to this societal need. On top of individual efforts, dedicated research programmes have started to emerge.

This book presents the results of one of the country's first large research programmes on the social and behavioural aspects of climate change. Funded by the Netherlands Organisation for Scientific Research (NWO), the programme was entitled 'Vulnerability, Adaptation and Mitigation' (VAM), a name that provides a clear hint to the themes that were addressed:

1. **Vulnerability**: the extent to which health, economy and nature and biodiversity will be affected as a result of a certain climatic change

2. **Adaptation**: how to adapt to a changed or changing climate. This covers local, national and global aspects. Adaptation is intended to reduce the vulnerability of systems

3. **Mitigation**: avoidance of climate change. Two strategies can be identified: namely, reducing greenhouse gas emissions and neutralising emitted greenhouse gases, by storing CO_2 underground or by planting forests that absorb CO_2

4. **Adaptation-plus-mitigation**: this denotes two complementary, parallel strategies in climate policy. Mitigation is the only fundamental solution for the climate problem, but adaptation is necessary in order to withstand the inevitable consequences of climate change

The VAM programme explicitly invited the social science disciplines, such as public administration, geography, environmental economics, the socio-cultural sciences, environmental law and psychology, to tender proposals. The call succeeded in attracting proposals for research projects from a wide variety of disciplinary backgrounds, 12 of which were granted. In addition, five experts from the social, legal and behavioural sciences were invited to write essays on the impact of abrupt climate change, a theme that has barely been investigated within the social sciences. An instance of abrupt climate change with particular relevance for the climate of northern Europe is a disturbance of the thermohaline circulation (THC), the phenomenon responsible for the relatively warm climate of the area. A weakening or collapse of the THC might lead to a rapid cooling of the climate of northwestern Europe. The central theme of the essays is which impacts West-European societies can expect when such a sudden climate change event occurs, and how they can respond to it.

VAM's ground-breaking contribution was in providing the social sciences with a structure and a platform for carrying out climate change-related research. At the same time, the openness of the VAM programme to proposals from all social sciences made its scope very broad. Later programmes will be able to build on the foundations provided by VAM, by zooming into promising themes addressed by the VAM projects, or by targeting areas that have so far been left untouched. This book primarily wishes to present an overview of the broadness of the VAM programme and its main conclusions, without attempting to provide a full synthesis of the broad diversity of themes and approaches. In addition, the book will highlight a few promising research themes, and identify a number of important research gaps that could be addressed in future research.

Many people contributed directly or indirectly to this book. First of all, we would like to thank the authors for their valuable input and swift responses. Furthermore, we would like to thank all our colleagues at the International Centre for Integrated Assessment and Sustainable Development (ICIS) at Maastricht University for their help in finalising this book. Special thanks go to Darryn McEvoy for his feedback and assistance in preparing this book. We are also grateful to NWO for funding this synthesis project, the VAM committee members for useful comments, and Ms Mariël Schweizer for administrative support.

Pim Martens and Chiung Ting Chang
Maastricht, April 2010

1

The social and behavioural aspects of climate change

Linking vulnerability, adaptation and mitigation

Chiung Ting Chang and Pim Martens
ICIS, Maastricht University, The Netherlands

Bas Amelung
Amelung Advies, The Netherlands

The VAM programme targets four main themes: vulnerability, adaptation, mitigation, and adaptation-plus-mitigation. In the brochure that accompanied the call in 2004, the programme committee hinted at the political connections between these themes: 'Until recently, social scientific climate research was dominated by the strategy of mitigation, in the wide sense, and in particular the analysis of reduction measures. Yet at COP8—the eighth session of the "Conference of the Parties" in the "United Nations Framework Convention on Climate Change" (UNFCCC) [in 2002]—the developing countries were strongly in favour of adaptation. The two subjects are most likely to be pitted against each other in future negotiations.' This was indeed the case at the following COP sessions, including the one in Copenhagen in December 2009. The emphasis of the developing countries on adaptation is likely to be related to their perceived vulnerability: the extent to which health, economy, and nature and biodiversity will be affected as a result of a certain climatic change. The developing countries are generally considered to be more vulnerable than the industrialised nations.

This chapter's objective is to move beyond an intuitive and political understanding of how the VAM themes are connected. It aims to provide a conceptual framework of the VAM themes and their linkages. Such a framework brings the underlying connections between the (perhaps seemingly disparate) VAM projects to the fore. The chapter also touches on some of the latest trends in climate research in the social sciences, so that the individual projects can also be positioned against that background.

1.1 Vulnerability, adaptation and mitigation: the climate change agenda

Smit *et al.* (1999) propose a schematic representation of the links between vulnerability, adaptation and mitigation. The scheme was later adapted by Martens *et al.* (2009), and further revised here (see Fig. 1.1). The scheme can be read as follows. Human interference with the global carbon cycle and other natural processes leads to changes in the climate system. These changes can manifest themselves in many different ways, such as temperature increases, sea level rise, reduced precipitation, and more intense rainfall. Such manifestations vary widely in both space and time, and their relevance differs strongly between economic sectors and regions. Exposure to climate change therefore not only depends on the physical realities of climate change, but also on the characteristics of the system (e.g. region, country, company or economic sector) exposed. How exposure translates into initial impacts depends on the sensitivity of the system: for example, a slight change in precipitation patterns may be sufficient to make the cultivation of a certain crop unprofitable, whereas large increases in summer temperatures may hardly affect the popularity of certain beach resorts.

Typically, systems resist change; they respond to the initial impacts of climate change by means of 'autonomous adaptation'. For example, a farmer may decide to replace a crop that cannot cope with the changes in climate conditions by a crop thriving on these same changes. It is a matter of dispute what exactly differentiates 'autonomous adaptation' from 'planned adaptation', which also features in the scheme. According to Smit and Pilifosova (2001), 'planned adaptation' involves some form of public coordination through government intervention, whereas 'autonomous adaptation' occurs through private actors. Another key distinction is that autonomous adaptation is always reactive, whereas planned adaptation may be proactive as well. The residual or net impacts that result after autonomous adaptation are usually smaller than the initial impacts. In some cases, however, positive feedback effects may be at work that reinforce rather than dampen the initial effects.

Governments and other actors can respond to (projected) impacts by taking policy measures. One type of response is to mitigate, i.e. to reduce greenhouse gas emissions or to capture and store emissions. The other type of response is to adapt

Figure 1.1 **The analytical framework for vulnerability–adaptation–mitigation research in the context of climate change**

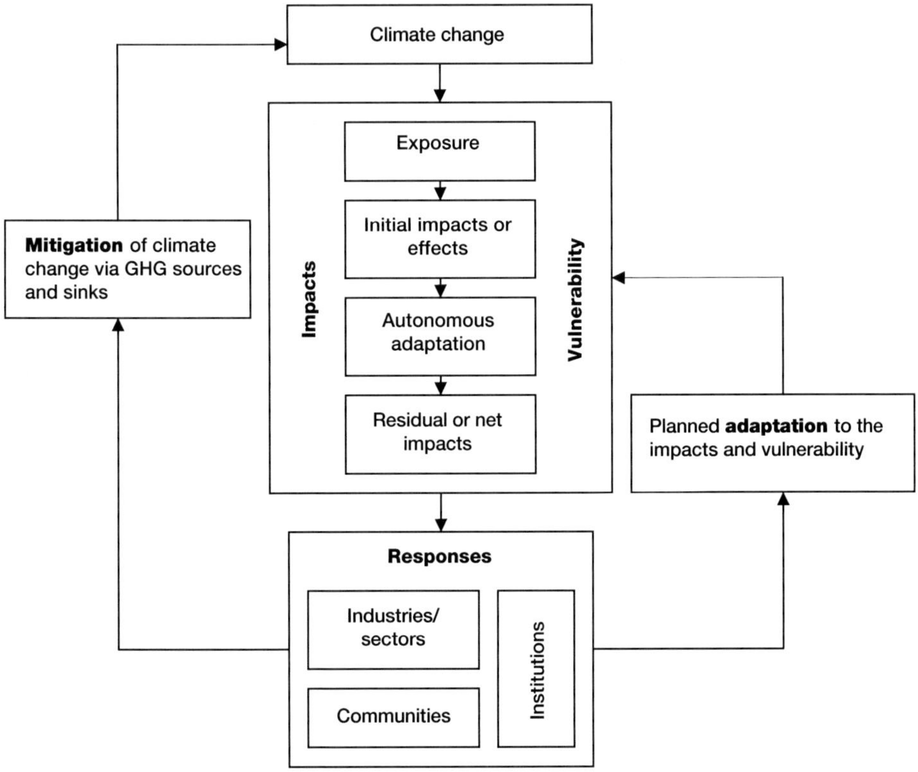

Source: adapted from Martens *et al*. 2009

to the (projected) impacts in order to reduce any negative effects and foster any positive effects, or to reduce vulnerabilities. The Intergovernmental Panel on Climate Change (IPCC) (2001) defines vulnerability as 'the degree to which a system is susceptible to, or unable to cope with, adverse effects of climate change, including climate variability and extremes'. The IPCC conceptualises vulnerability as being composed of three elements: exposure, sensitivity and adaptive capacity. Policy responses aimed at reducing vulnerability can operate on any of these three components. Exposure can be reduced (e.g. by changing the sectoral composition of the economy), sensitivity can be reduced (e.g. by making operational adjustments), and the adaptive capacity can be increased (e.g. by making contingency plans).

Since 1999, when Figure 1.1 was devised, climate change research has changed. Martens *et al.* (2009) summarise four key developments. The first is an increase in scientific consensus concerning climate change. Through the IPCC, the collaborative efforts of scientists have concluded that climate change is happening and, importantly, that human activity is making a discernible contribution to this change. The second development is a shift of focus from impacts to risk management. This means going beyond mere consideration of climate-related hazards, to more explicit considerations of issues surrounding the vulnerability and exposure of different elements at risk, as well as addressing conditions of uncertainty. The third development is the consideration of non-climate stressors, acknowledging that the climate is not the only driver of change in our societies. Finally, policy and research agendas have become less dominated by mitigation. There is an increasing awareness that actors also need to be preparing for changes that are unavoidable. This has resulted in a greater consideration of vulnerability and adaptation, and recognition of the need for better interdisciplinary cooperation. Improved linkages between natural and social scientists will be crucial in order to effectively address the complexities of climate change.

Klein *et al.* (2007) deduced four possible connections between mitigation and adaptation: adaptation actions that have consequences for mitigation; mitigation actions that have consequences for adaptation; decisions that include trade-offs or synergies between adaptation and mitigation; and processes that have consequences for both adaptation and mitigation. Obvious as this list may look, connecting the policy and research efforts regarding mitigation and vulnerability/adaptation is a formidable challenge. Martens *et al.* (2009) explored the potential and pitfalls, which are introduced below. Before doing so, however, it is perhaps good to note that few, if any, of these challenges are restricted to the endeavour of connecting the domains of mitigation and adaptation. Perhaps they reach their maximum acuteness in connecting these domains, but for the most part the challenges are common features of climate research at large.

First, a common link between mitigation and adaptation is the capacity of a system to respond. For example, adaptive capacity can be simply defined as the ability of a system to adjust to climate change; this is thought to be determined by a range of factors including technological options, economic resources, human and social capital, and governance. Mitigation has similar determinants—in particular, the availability and penetration of new technology. Although technological solutions have a role to play in both mitigation and adaptation, it should be recognised that 'soft engineering' has a particularly important role in adapting to climate change. The willingness and capacity of society to change is also critical. Information and awareness-raising can be useful tools to stimulate individual and collective climate action (McEvoy *et al.* 2006).

Second, an integrated response is challenging as there is a mismatch between mitigation and adaptation in terms of scale, both spatially and temporally. Mitigation efforts are typically driven by national initiatives operating within the context of international obligations, whereas adaptation to climate change and variability tends to be much more local in nature, often in the realm of regional economies, communities, land managers and individuals. Besides the spatial element, there are also differences in the timing of effects. As greenhouse gases have long residence times in the atmosphere, the results of mitigation action will be seen only in the long term. Adaptation, on the other hand, has a stronger element of immediacy. Regional differences and the dynamic features of vulnerability and averting behaviour should therefore be taken into account in both theoretical and practical analyses.

Third, disconnection in space and time can make it difficult for people to link the consequences of their activity with long-term environmental consequences. It also raises the question of environmental equity: that is, who are the likely beneficiaries of the different types of response? Mitigation, being an action targeting the long term, means attaching value to the interests of future generations and to some extent can be considered an altruistic response by society. Conversely, the impacts of climate change are felt more immediately by individuals in society and adaptation is typically viewed as obeying the everyday 'self-interests' of individuals. As such, studies on risk perception by individuals, industries and organisations will be critical to understand its influence on the acceptability and ultimate effectiveness of different responses.

Fourth, mitigation and adaptation have different distributional effects: in particular, who pays and who gains, and whether there is a willingness to invest if the benefits of adaptation are perceived to be private. It is also important to note discrepancies in that those responsible for the majority of emissions (i.e. developed countries) also have the highest adaptive capacity, while the poorest countries, producing the lowest emissions, are most vulnerable to the impacts of a changing climate. Consequently, the urgency that different countries attach to any mitigation response varies widely. The same phenomenon holds true within national territories where uninsured, unaware and relatively immobile populations living in poorer-quality accommodation are often the hardest hit. This means that, in reality, those most vulnerable to climate change are often those already socioeconomically disadvantaged in society. Thus, not merely geo-physical vulnerability but also socioeconomic vulnerability should be taken into account. Studies on distributional effects and cost–benefit analysis of mitigation, adaptation and mitigation-plus-adaptation measures will contribute to the analyses of trade-offs and avoidance of unwanted side effects,

Finally, another important difference between mitigation and adaptation relates to those who are directly involved. Mitigation policy is primarily focused on decarbonisation and involves interaction among the large emitting sectors such as energy and transport, or else targets efficiency improvements according to specific end-users—commercial and residential. The limited number of key personnel, and their experience in dealing with long-term investment decisions, means that the mitigation agenda can be considered more sharply defined. In contrast, the many actors involved in the adaptation agenda, in contrast to mitigation policy, come from a wide variety of sectors that are sensitive to the impacts of climate change and operate across a wide range of spatial scales. As a result, the implementation of adaptation measures is likely to encounter greater institutional complexity than the implementation of mitigation policies. A relevant research question is therefore how formal and informal institutional conditions affect socioeconomic vulnerability, adaptive capacity and mitigation choices.

1.2 Outline of the book

The above-mentioned developments and challenges underscore the relevance and timeliness of the VAM programme. Below is an outline of the structure of the book and a brief introduction to the various chapters. The diversity of projects made it difficult to cluster the projects around a small number of themes, although one option available to us was to organise the book according to the four VAM themes. The portfolio of VAM projects is reasonably balanced in terms of representation of these themes. Vulnerability and adaptation (which are often highly interlinked within projects) are the focus of six out of 12 projects, as well as the collection of essays. Four chapters have an emphasis on mitigation, while the two remaining chapters address the interaction between adaptation and mitigation. While consistent with the VAM scheme, a structure based on the main themes provided little coherence within the resulting sections.

Other dimensions for classification included geographical scale, temporal scale, and scientific discipline, each with its own advantages and disadvantages. A workable structure for the book resulted from a hybrid framework for classification, made up of three sections: industry/sector, community, and institution. This classification has an element of 'nesting': an industry may be part of a community, and a community in its turn belongs to a society with institutions. It also contains an element of increasing institutional complexity, with trade-offs being clearer or more easy to resolve for industries than for communities, and the level of society being even more complex. These ideas were used only as ordering principles; it is fully acknowledged that they do not always do justice to the levels of complexity found in sectors and communities.

Chapters on inland navigation, tourism, partnerships, and energy conservation in housing make up the section on industries. The communities section contains chapters on the impact of Hurricane Mitch on Nicaragua, local adaptation in Mozambique and the Netherlands, and 3D simulation of flood events. Finally, the section on institutions is made up of chapters on white certificates, EU climate change law, the legal challenges of applying the precautionary principle, and the explicit incorporation of adaptation into Integrated Assessment models. Each of the chapters is introduced in somewhat more detail below.

1.3 Industries

Industries contribute substantially to the emission of CO_2 and other greenhouse gases. At the same time, they face significant economic losses as a result of climate change. Some industries are particularly vulnerable to climatic changes. Inland navigation for instance, discussed in Chapter 2 ('Climate change and inland navigation between the Netherlands and Germany: an economic analysis'), is subject to changing water levels and is susceptible to hydrological extremes. The sector is expected to suffer both in winter and in summer. In winter, more navigation delays on the River Rhine are likely to occur due to higher water levels, whereas in summer transport capacity will be reduced owing to lower water levels. Both transport volumes and speed are negatively affected. Chapter 2 estimates economic losses and analyses how these are distributed between the upstream and downstream countries of Germany and the Netherlands. In addition, the chapter explores the possible adaptive measures by examining the relationships between transport carriers and their clients.

Tourism is another industry in Europe that is particularly vulnerable to changing climatic conditions, as well as climate policy. Many tourism activities require favourable conditions for their success, and tourism is increasingly reliant on energy-intensive forms of transport that are likely to be affected by mitigation measures. The economic stakes are high. In 2008, Europe earned US$435 billion from international tourism alone (UN World Tourism Organization 2009), a significant share of which is climate-driven. Chapter 3 ('Climate change impacts: the vulnerability of tourism in coastal Europe') adds to the scarcely available knowledge on the climate preferences of tourists. Furthermore, it analyses the likely consequences of climate change for the spatial and temporal distribution of future tourism patterns in Europe. The chapter also provides a vulnerability assessment framework specifically tailored to the tourism sector.

Next to sectors that face predominantly negative consequences (inland navigation) or mixed effects (tourism), there are also industries that are set to benefit from new opportunities, perhaps including those that offer products and services responding to or supporting people's adaptive behaviour. Chapter 4 ('Corporate responses to climate change: the role of partnerships') studies the existing and potential contributions of businesses to climate change mitigation and adaptation. The chapter reports on the results of an empirical study on partnerships for climate change—forms of cooperation with governmental and societal actors. The motives and strategic benefits for companies to engage in partnerships are explored. The chapter also shows how the findings can be taken into account in the policy-making process.

Improving energy efficiency in order to cut down CO_2 emissions has been one of the major strategies to counteract climate change. Options for energy saving within the housing sector are studied in Chapter 5 ('Energy conservation in Dutch housing renovation projects'). The chapter explores factors that explain energy conservation of the housing stocks in 11 selected sites and provides an estimate of the amount of energy conservation that is realistically attainable. The chapter comments on the Dutch policies, projects and subsidies on CO_2 reduction measures in the built environment, and analyses the role of housing associations in promoting energy efficiency.

1.4 Local communities

Vulnerability is location-specific and case-dependent, and so is the willingness of communities to support mitigation policies and take adaptation measures. The knowledge, creativity and organisation of local communities are vital for their adaptability and success in coping with climatic hazards. Social innovation is essential, and this process may be stimulated by exchanging experiences among different parts of the world. The effects of floods are a recurrent theme in the four chapters in this section.

Nicaragua's local responses to the devastation caused by Hurricane Mitch in 1998 are explored in Chapter 6 ('Natural hazards, poverty traps and adaptive livelihoods in Nicaragua'). A key question in the chapter is whether disasters have long-term effects on people's preferences and survival strategies as well as the short-term effects of damaging their resource base. The chapter analyses whether Mitch pushed people into chronic poverty, and tests for differences in risk aversion between affected farmers and their unaffected counterparts.

Chapter 7 ('Climate change adaptation in Mozambique') also reports on post-flood/disaster research, but this time in Africa and from the perspective of sustainable livelihoods. The chapter examines the aftermath of the 2000 flood in the southern region of Mozambique which claimed more than 700 lives and resulted in damages of over US$600 million. The role of local knowledge and experiences in the development of adaptation strategies is explored, and the impact of disaster relief efforts on local adaptive capacity assessed.

The drivers of local adaptation to flooding in the Netherlands are assessed in Chapter 8 ('Adaptation to climate change induced flooding in Dutch municipalities'). The chapter provides a systematic study of local community initiatives to improve adaptation capacities by assessing Dutch municipalities in the context of multi-level governance. Two possible drivers of adaptation are examined in detail: historical experience with flooding impacts and the projected risk of new climate change impacts. However, local contextual factors, the quality of climate change information and other potentially influential elements are also taken into account.

Chapter 9 ('Human responses to climate change: flooding experiences in the Netherlands') reports on experiments with a virtual environment in which people can 'experience' floods in a laboratory setting, with the aim of detecting changes in the perception and understanding of climate risks. The chapter discusses if and how such techniques can be used to enhance local adaptive capacity.

1.5 Institutions

The sections on industries and local communities emphasise the issue of 'autonomous adaptation': how individual people, businesses and communities respond to the impacts of climate events. Occasional or even frequent reference may have been made to possible deliberate social arrangements to respond to climate change, but these were not central to these chapters. Conversely, institutions are at the heart of each of the chapters in this third and final section.

The effectiveness of a particular policy measure may be significantly affected by the interference of policy instruments that are already in place. Chapter 10 ('Interactions between white certificates for energy efficiency and other energy and climate policy instruments') examines how the effectiveness of White Certificates—a market-based instrument for energy efficiency improvement—is influenced by the interaction with similar policy instruments, such as green certificates and carbon taxes. Cost savings are estimated under various scenarios.

Chapter 11 ('Distributional choices in EU climate change policy seen through the lens of legal principles') examines distributional choices in EU climate law and policy from the perspective of legal principles of EU law. These choices relate to the distribution of climate change impacts and financial burdens between generations, countries and sectors. The chapter identifies a number of important tensions, and provides recommendations to reduce them.

Persisting scientific uncertainties are among the defining features of the climate change phenomenon. Chapter 12 ('Climate change liability and the application of the precautionary principle') analyses how they affect climate change liability. Major challenges are identified that may inhibit the adoption of enhanced mitigation and adaptation measures. The precautionary principle is proposed as a potential instrument to deal with these challenges, and its merits are subsequently scrutinised against legal principles.

Linking the issues related to mitigation and adaptation is the key objective of Chapter 13 ('Incentives for international cooperation on adaptation and mitigation'). It presents a methodology for adding adaptation as an explicit decision variable to integrated assessment modelling frameworks (IAMs). By doing so, it becomes possible to analyse regional differences in the costs of adaptation. The chapter applies the methodology to the RICE model, creating the AD-RICE model, which is subsequently used to investigate the optimal mitigation paths for various scenarios of cooperation between countries.

Chapter 14 ('Imagining the unimaginable: synthesis of essays on abrupt and extreme climate change') is a synthesis of five essays on the potential implications of a shutdown or slowdown of the thermohaline circulation (THC). Such an event would cause an abrupt decrease in temperature in northwestern Europe, with potentially serious social and economic consequences. The essays are written from a legislative, institutional, sectoral, multi-sectoral and economic perspective respectively. The legal essay discusses which actors should be politically and legally responsible for the impacts of a climate-related extreme event. The institutional essay analyses the risk of rapid cooling for urban and rural areas, as well as the water sector in the Netherlands, and considers their management by existing and potentially new institutional structures. The sectoral essay considers the implications of abrupt change for both human health (focusing on heat-related and cold-related mortality, and infectious diseases) and tourism (focusing on thermal comfort, and the likelihood of Elfstedentocht skating events). The multi-sectoral essay considers three additional weather- and non-weather-related events that could potentially lead to rapid cooling in Europe, in addition to a shutdown of the THC. These analogies are used to better understand some of the potential socioeconomic implications and to begin to explore what adaptation options are likely to be necessary to cope with a rapid change of the climatic regime. The economic essay addresses decision-making under uncertainty. Central to the discussion is the issue of timing of response, i.e. do we need to take action now and what can be done or delay the decision until more information is available?

Chapter 15 concludes the book by synthesising the major insights generated by the studies reported in the various chapters and identifying directions of further research. The synthesis is organised around the developments and challenges in climate research that have been identified in this introductory chapter.

References

Hulme, M., and H. Neufeldt (eds.) (2010) *Making Climate Change Work for Us: European Perspectives on Adaptation and Mitigation Strategies* (Cambridge, UK: Cambridge University Press).

Klein R.J.T., S. Huq, F. Denton, T.E. Downing, R.G. Richels, J.B. Robinson and F.L. Toth (2007) 'Inter-relationships between Adaptation and Mitigation', in M.L. Parry, O.F. Canziani, J.P. Palutikof, P.J. van der Linden and C.E. Hanson (eds.), *Climate Change 2007: Impacts, Adaptation and Vulnerability: Contribution of Working Group II to the Fourth Assessment Report of the Intergovernmental Panel on Climate Change* (Cambridge, UK: Cambridge University Press): 745-77.

Martens, P., D. McEvoy and C. Chang (2009) 'The Climate Change Challenge: Linking Vulnerability, Adaptation, and Mitigation', *Current Opinion in Environmental Sustainability* 1: 1-5.

McEvoy, D., S. Lindley and J. Handley (2006) Adaptation and Mitigation in Urban Areas: Synergies and Conflicts. Proceedings of the Institution of Civil Engineers, *Municipal Engineer* 159: 185-191.

Ostrom, E. (1990) *Governing the Commons: The Evolution of Institutions for Collective Action* (New York: Cambridge University Press).

Smit, B., and O. Pilifosova (2001) 'Adaptation to Climate Change in the Context of Sustainable Development and Equity', in J.J. McCarthy *et al.* (eds.), *Climate Change 2001: Impacts, Adaptation and Vulnerability. Contribution of Working Group II to the Third Assessment Report of the Intergovernmental Panel on Climate Change* (Cambridge, UK: Cambridge University Press): 876-912.

——, I. Burton, R.J.T. Klein and R. Street (1999) 'The Science of Adaptation: A Framework for Assessment', *Mitigation and Adaptation Strategies for Global Change* 4: 199-213.

UNWTO (United Nations World Tourism Organization) (2009), *UNWTO World Tourism Barometer* 7.2; unwto.org/facts/eng/pdf/barometer/UNWTO_Barom09_2_en_excerpt.pdf, accessed 3 March 2010.

2

Climate change and inland navigation between the Netherlands and Germany

An economic analysis

Erhan Demirel, Jos van Ommeren and Piet Rietveld
Department of Spatial Economics, VU University, The Netherlands

In this chapter we study two aspects of inland navigation between the Netherlands and Germany, where an important part, about 50%, of freight transport takes place along the river Rhine. The first aspect is that the rivers on which this transport moves are influenced by climate change. Climate change may well lead to increases in the length of low-water periods, which decreases the capacity of barges[1] and increases the cost of transport. Furthermore, the river may have less-predictable water levels as a result of heavier rainfalls in certain parts of the year.

The other aspect is that inland navigation in this region is characterised by an imbalance in demand for transport. Carriers often find it easy to ship freight from the Netherlands to Germany, but have difficulty to find freight for the return trips. This is known in the transport literature as the 'backhaul problem'.

The imbalance aspect is important for an economic analysis of climate change. The point is that climate change leading to long periods with low water levels implies higher costs per tonne for the barge operators. In a competitive market, the additional burden will ultimately not be borne by the barge operators themselves,

1 Because a minimum distance between the barge and the riverbed must be maintained, loads may need to be reduced.

but by their clients (referred to as '[transport-demanding] customers'). We will demonstrate that the way the cost increases are distributed between the various shippers depends strongly on the imbalances in the transport market. The exploration of this issue is also the main aim of this chapter.

This research is relevant for the climate change adaptation theme of the book, since it sheds light on the motivation of stakeholders to be involved in adaptation strategies. The motivation of a government to spend money for adaptation policies will depend on whether or not households or firms within its territory will benefit from them. Given the supra-regional and international character of barge operations, it is by no means clear where the main beneficiaries of adaptation measures will be located, and hence whether or not public authorities are well motivated to be involved in adaptation policies. Thus, this chapter is also relevant for the distributional effects of adaptation. Another reason why this research has high policy relevance is the strong link with EU transport policies that are aimed at developing alternatives for road transport, which is the dominant transport alternative for freight. Inland waterway transport is considered to be one of the promising alternative transport modes, on account of, among other things, its favourable performance from the viewpoint of energy efficiency. Adaptation measures to keep inland waterway transport viable are therefore also important from a mitigation perspective in order to reduce the reliance on energy-intensive road transport. In this chapter we use a theoretical model to focus on the interaction between climate change and imbalance in the demand for transport between regions, and study the effect of both on freight prices and transported volumes. This theoretical analysis is based on Demirel *et al.* (2010), which is a part of the Vulnerability, Adaptation and Mitigation (VAM) project.[2] We compare these findings with empirical evidence from Jonkeren (2009).

2.1 Inland navigation and climate change

In sub-section 2.1.1 we explain the importance of inland navigation as a mode of transport between the Netherlands and Germany. In sub-section 2.1.2 we describe the effects of climate change on inland navigation in Western Europe. We end this section with sub-section 2.1.3 where we briefly discuss the imbalance theme in freight transport.

2 The VAM project investigates the effects of climate change on society from a social science perspective. The VAM project is financed by the Netherlands Organisation for Scientific Research (NWO).

2.1.1 Importance of inland navigation between the Netherlands and Germany

The importance of inland navigation as a transport mode for countries in Western Europe can best be understood from the figures on modal split. From Table 2.1 we can see that within the Netherlands, for both domestic and international transport, one-third of all tonne-kilometres are transported by inland navigation. For Germany this is approximately one-seventh.

Table 2.1 **Modal split in tonne-kilometres for Western European countries (in percentages) in 2007**

Country	Road	Rail	Inland navigation
The Netherlands	61.8	5.0	33.2
Belgium	71.1	13.2	15.7
Luxemburg	92.5	4.1	3.3
Germany	65.7	21.9	12.4
France	81.4	15.2	3.4

Source: Eurostat 2008[3]

To get an idea of the volume of transport between the Netherlands and Germany, in 2005, 127 million tonnes were transported from the Netherlands to Germany by road, rail and inland navigation. From Germany to the Netherlands the quantity was 73 million tonnes. Of these amounts, transport by inland navigation accounted for 58% of all transport, measured in tonnes, from the Netherlands to Germany. From Germany to the Netherlands, this was 41%.[4] The reason that there is such an imbalance in the volume of tonnes transported in both directions is that the port of Rotterdam is the main port of entry for bulk materials such as coal, iron ore and agricultural products moving into its hinterland. The flow of primary products of this type in the opposite direction is considerably smaller.

Transport by inland navigation between the Netherlands and Germany occurs mainly along the Rhine (see Fig. 2.1 for a map of the river Rhine and the Rhine area). For example, in 2006, the Rhine corridor accounted for 63% of total transport by inland navigation in Europe (CCNR 2007). The vulnerability of inland navigation in this area is explained in the next sub-section.

3 epp.eurostat.ec.europa.eu, accessed 4 March 2010.

4 Figures from TLN (2007) and Central Bureau of Statistics (2008), The Hague; statline.cbs.nl/statweb, accessed 4 March 2010.

Figure 2.1 **Map of the river Rhine and the Rhine area in Western Europe**

Note: The thick line represents the Rhine.

2.1.2 Effects of climate change on inland navigation

Inland navigation between the Netherlands and Germany mainly takes place along the Rhine. Some properties of the Rhine that have, until now, made it attractive for transport, such as stability of water levels and sufficient depth, seem to be vulnerable to climate change.

Low water levels have the effect that transport is only possible at low capacities. The summer of 2003 was an example of how water levels could drop so much that navigation was only possible at low freight capacities for a long period of time, and the transport cost per tonne increased accordingly. High water levels increase flood risk and cause difficulties with bridges and motorways, especially for container transport. In cases of extreme high water levels, navigation may be halted completely.

Climate change scenarios have been developed to address fundamental uncertainties about future climate conditions. Table 2.2 shows scenarios for the Netherlands in terms of expected temperature change in 2050 provided by Koninklijk Nederlands Meteorologisch Instituut (Royal Netherlands Meteorological Institute—KNMI) in 2006.

Table 2.2 **Values for the steering parameters of the KNMI 2006 climate scenarios for 2050 relative to 1990**

Scenario	Global temperature increase in 2050	Change of atmospheric circulation
M	+1°C	Weak
M+	+1°C	Strong
W	+2°C	Weak
W+	+2°C	Strong

Note: M = 'medium' (temperature increase); W = 'warm'.

Source: KNMI 2006

Possible future water levels on the Rhine can be derived on the basis of these climate scenarios. In Figure 2.2 we present predicted water discharge (in cubic metres per second) within one year in 2050 at Lobith, which is a place on the border between the Netherlands and Germany. Four of the lines correspond to the four scenarios presented in Table 2.2, and the solid black line corresponds to the current situation.

Figure 2.2 **Predicted water discharges on the Rhine at Lobith (on the border between the Netherlands and Germany) in 2050, over one year for different scenarios**

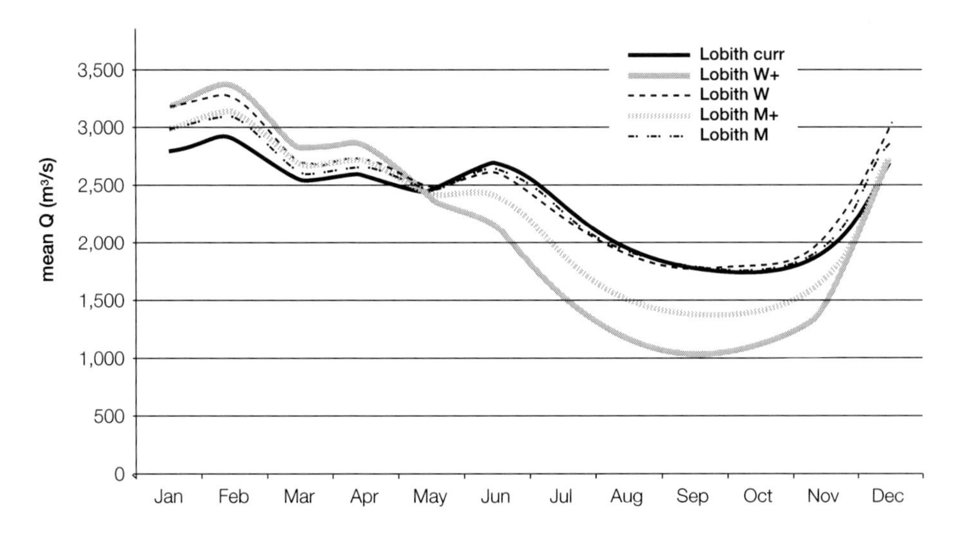

Source: Te Linde 2007

It can be seen that especially the M+ and W+ scenarios predict low water levels between May and December. All scenarios predict higher water levels between January and May.

As mentioned above, low water levels cause low freight capacities, lower speeds and increased freight prices per tonne. High water levels may cause navigation halts due to flood risk and for safety reasons. To illustrate the impact of water levels on the transport cost per tonne, Figure 2.3 presents the negative relation between freight prices per tonne (which are paid by customers)[5] and water levels observed for the years 2003–2005.

Figure 2.3 **Freight price per tonne and water levels at Kaub (Germany) for the years 2003–2005**

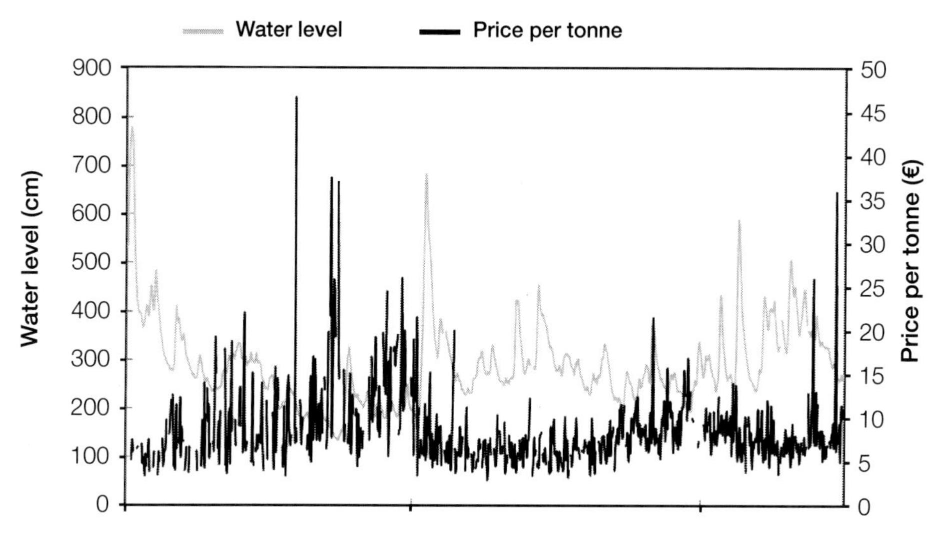

Source: Jonkeren 2009. Freight prices from Vaart!Vrachtindicator 2003–2005;[6] water levels from iidesk.nl 2003–2005[7]

Both extreme low water levels and extreme high water levels are harmful in terms of economic welfare. A welfare analysis of the negative effect of low water levels on welfare was studied by Jonkeren *et al.* (2007). Jonkeren (2009) estimates a welfare loss of €227 million following an increase of the transport cost caused by low water levels for the Rhine area. For the remainder of this chapter we focus on the effects of low water levels.

5 For convenience, transport-demanding customers are simply called 'customers' in the text.

6 Inland waterway transport portal Vaart! at www.vaart.nl.

7 Database design, building and maintenance organisation at www.iidesk.com/water.

2.1.3 Imbalance in freight transport

In freight transport, freight prices and transported volumes are to some extent determined by the degree of imbalance in demand for transport between regions. Different degrees of imbalance have different implications for economic welfare when the setting is changed as a result of changes in policy or environmental changes such as climate change. In this chapter we focus on freight prices, transported volumes and the number of carriers.

Freight prices do not only depend on costs, but are also determined by direction-specific demand. In addition, demand for goods is also determined to a certain extent by freight prices. For example, freight prices influence trade and therefore regional and international transport demand (Krugman 1991). In economic terms, we can say that freight prices are endogenous with respect to transport demand.

While demand for transport is direction-specific, a large proportion of the transport cost is shared. As carriers often have incentives to return to the location of origin, this means that the cost for a return trip is made together with the fronthaul trip (see Pigou 1913). This situation, where carriers have difficulty in finding backhaul (or passengers) for the return trip because of low(er) demand for transport on the return journey, is frequently referred to as 'the backhaul problem'.

The backhaul problem is a well-known phenomenon in transport economics, both in freight and passenger transport studies. It arises in situations where the volume of transported goods or persons is not in balance between two (or more) locations, which means that transport flows are mainly in one (or more) dominant direction(s).

One might suppose that backhaul factor makes the issue of low water levels less problematic since unloaded ships have less difficulty in moving through low river water. It is indeed correct that navigating with empty ships in general does not pose specific problems during low water periods. Thus the backhaul problem leads to a waste of resources that is not affected by water levels as such. An exception may occur in some specific situations such as narrow rivers where one-way traffic may be imposed during low-water periods. In that case empty ships sailing in one direction could cause waiting times for loaded ships sailing in the opposite direction.

To model the inland navigation market between the Netherlands and Germany, we are interested in competitive transport markets because, in many countries, the trucking as well as the inland navigation markets are characterised by a high level of competition. In these markets there are a large number of suppliers, so that the market power due to the size of the supplier can be discounted.[8] In these markets, there is also free entry and low capital costs, which enhances competition. For example, in the inland navigation market in Western Europe, thousands of carriers are active, most of which are sole proprietorships.

We observe that market frictions, such as lack of information on the time and

8 The Rhine fleet consisted, on 31 December 2006, of 9,796 ships (CCNR 2007).

location of demand and supply, play an important role in freight price formation in this transport market. Therefore, we deviate from the standard competitive model that is used in textbooks (see Boyer 1998; Felton 1981) and adopt imperfect information by means of a matching model.

This study, which combines climate change and freight imbalance, sheds light on policy questions, such as how to share the burden of a climate-related infrastructure improvement in a transnational context. A common practice is for each country to pay for the infrastructure costs on its own territory. However, in the case of backhaul problems the benefits of the improvement are distributed unevenly: one country will receive a much larger share of the benefits than the other one. Knowledge of how the benefits are spread between the two locations (countries) may help us to arrive at a proper division of infrastructure costs. One case with high political relevance concerns the distribution of costs between Belgium and the Netherlands for dredging in the Western Scheldt estuary, which is located on Dutch territory, but where the Belgian port of Antwerp is the main beneficiary. In this case, it has been decided that Belgium will pay for the dredging, even though the estuary is situated on Dutch territory. Similar discussions, but more implicit, are taking place between other pairs of countries, along the Rhine and the Danube, where water management costs to improve navigability have to be shared between countries and where benefits are unevenly spread because of the backhaul problem.

We now introduce in Section 2.2 a two-location transport framework, which incorporates a matching model to study the issues presented above. In Section 2.3 we present a numerical outcome of the theoretical model, for which we use input values chosen from the inland navigation market on the Rhine in Western Europe. In particular, we investigate the effect on the freight price of (anticipated) changes in transport costs due to climate change, in the context of imbalance. In Section 2.4 we compare the numerical outcome of Section 2.3 with empirical evidence for a similar setting from Jonkeren (2009). Section 2.5 concludes.

2.2 Theoretical model

In this section we present a theoretical model to study the inland navigation market between the Netherlands and Germany. In sub-section 2.2.1 we introduce the two-location network that we used and explain the economic assumptions. In sub-section 2.2.2 we present theoretical outcomes for the two different equilibria that are possible. Sub-section 2.2.3 is an introduction to the numerical climate-related sensitivity analysis of sub-section 2.3.2.

2.2.1 Theoretical freight transport framework

In order to model the inland navigation market between the Netherlands and Germany, we abstract from reality by considering transport between two locations: one location, labelled *H*, with a high demand for transport; and one location, labelled *L*, with a low demand. In this setting, the Netherlands, with Rotterdam as the main port, can be considered as a location with high demand for transport, and Germany, as hinterland, can be viewed as a location with low demand for transport. This abstraction can be represented by the two-node network in Figure 2.4.

Figure 2.4 **A two-node network representing transport between a location with high demand for transport (H) and a location with low demand for transport (L)**

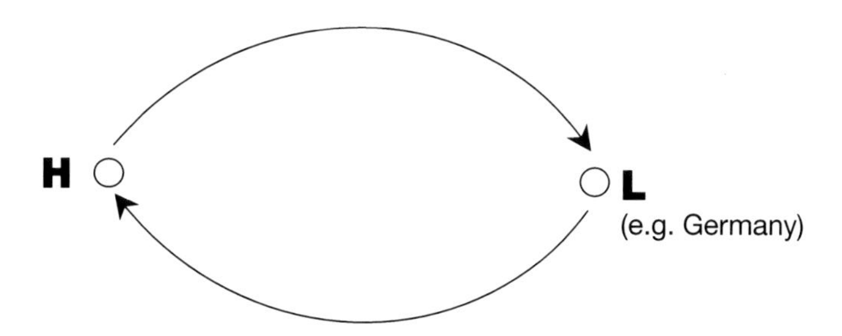

In the economic literature, the market structure with **standard perfect competition** is often used to model the backhaul problem (see e.g. Boyer 1998). In a market structure with standard perfect competition, the sum of the fronthaul and backhaul prices equals the marginal cost of a round trip. As we have already observed, these costs are affected by water levels, and hence by climate conditions. In many transport markets we observe an unbalanced distribution of marginal costs between the fronthaul and the backhaul trip. The imbalance in freight prices reflects the imbalance in transport flows. In theory, the imbalance could be so strong that the freight price for the fronthaul trip is equal to the round trip cost of transport, and the freight price for the backhaul trip is zero, which means that transport is essentially free for the return trip

The above situation can be shown to arise in situations where some carriers return from the low-demand location to the high-demand location without freight. If temporal aspects are disregarded, we conclude that when some carriers have no freight (and therefore no revenue) on the return trip, they prefer to transport at any price than not to transport at all. This type of price competition decreases freight prices on backhaul trips to zero. This means that the transport cost from one location to the other is fully borne by the high-demand location's customers. This cost is equal to the round-trip transport cost. The type of equilibrium where the quantities transported are not the same in both directions will be called throughout this chapter the **unbalanced equilibrium**.

Another equilibrium may arise in the standard perfect competition case, which we will call throughout this chapter the **balanced equilibrium**. In this type of equilibrium, the quantities transported are exactly the same in both directions and *all* carriers prefer to return with backhaul. The fronthaul prices will be higher than the backhaul price owing to the difference in demand, but the backhaul prices will be positive.

The inland navigation market in Western Europe could serve as an example of unbalanced equilibrium. Jonkeren *et al.* (2007) show that backhaul prices (from Germany to the Netherlands) are 76% of fronthaul prices although the quantity transported from the fronthaul location (the Netherlands) exceeds the quantity transported from the backhaul location (Germany) by at least 50%.[9] Another example is the trucking industry in the USA (see Felton 1981).

The market structure of standard perfect competition assumes that customers and carriers possess complete information about each other's (future) location and all other characteristics to form a match. Customers and carriers find each other immediately. This implies the absence of market friction and search time for carriers, as well as for customers. In many competitive transport markets, search time for customers is an obvious cost component, not only in the taxi market, where taxis either cruise or passively wait for customers (see Arnott 1996), but also in the inland shipping market (see Meelker 2006).

To describe a transport market where no perfect information is available about each other, we use a market structure of 'perfect competition with imperfect information'. In order to model this type of market, we use a matching model to model the information imperfection, and where matching occurs in both locations. The model developed is also consistent with another fact for inland shipping. It has been documented that many carriers regularly spend time waiting and or searching for new customers having arrived at a destination (see e.g. Meelker 2006). In this chapter we also aim to determine the waiting time endogenously, which is left unexplained in competitive models with perfect information.

Nowadays, matching models are a popular way to analyse markets where agents are searching for each other and face a difficulty in finding each other and forming a match. This kind of model is now standard in the labour market economics literature (see e.g. Pissarides 2000) and is also used in housing economics (see e.g. Wheaton 1990). Examples of matching models applied to the taxi market can be found in Lagos 1996 and Arnott 1996. The matching model is also applicable in many freight transport markets, as carriers and customers search for each other and have difficulty in finding each other. As argued above, this difficulty is due to a combination of spatial and temporal variation in demand and supply.

We assume that a fixed number of identical risk-neutral customers with a demand for transport are located in locations H and L. The numbers of customers located at each side are denoted by N_H and N_L, respectively, with $N_H > N_L$. For convenience, other exogenous parameters are assumed to be identical for both locations.

Each customer aims to have one good transported by identical risk-neutral carriers to the opposite side. We assume that a customer withdraws from the market after it has found a carrier that is willing to move the good to the other side and is immediately replaced by a new customer, so the number of customers remains con-

9 Central Bureau of Statistics (2008), The Hague; statline.cbs.nl/statweb.

stant over time. A carrier may at most move one good, so it is either full or empty.

We assume that customers and carriers have to search for each other because they have imperfect information about each other. It is assumed that, when they search for one another, customers and carriers contact each other according to a well-defined contact function. Carriers and customers are only able to contact each other when they are in the same location L or H. The contact function in location i then specifies the number of contacts taking place in location i during a time-period as a function of the number of searching carriers and customers in that location.

Having made contact, the customer and the carrier bargain about the freight price. If they agree on a price, which can be shown as always occurring, the customer and carrier are matched. Hence, the matching function is identical to the contact function. When a carrier and a customer are matched, the customer pays the carrier the agreed freight price, the cargo is loaded, and the carrier is obliged to move the good towards the other location.

The carrier's cost of moving the good is assumed to be independent of whether or not the carrier moves the freight and is proportional to the time it takes to move the good. We define mc as the cost of moving per unit of time. The duration of the trip is stochastically determined. We assume that the average duration of a *round* trip is distributed with a mean equal to $1 / \lambda$. So λ can be interpreted as the average speed at which a carrier makes round trips (measured in round trips per time unit). As we have assumed that customers and carriers are risk neutral, all decisions made by carriers will *not* depend on any other properties of the distribution except its mean. Hence, for convenience, we will assume that the duration of a round trip is *exponentially* distributed. In this case, the duration of a single trip is exponentially distributed, with the mean equal to $1/(2\lambda)$. Consequently, 2λ can also be interpreted as the average speed of arriving at the other location (measured over the course of a single trip).

2.2.2 Two types of equilibria

When we make some further assumptions (see Demirel *et al.* (2010) for a full overview), it can be shown that, as in the situation of standard perfect competition, either a balanced equilibrium or an unbalanced equilibrium will occur.

In the extreme case that $N_H = N_L$, it appears that a balanced equilibrium will occur. When $N_H > N_L$, one may have either balanced or unbalanced equilibrium. Intuitively, when N_H is close to N_L, a balanced equilibrium will occur, and, when N_H is much greater than N_L, then an unbalanced equilibrium will arise.

For a balanced equilibrium the following pair of price equations are obtained:

> p_H = expected search cost in $H + \alpha$ round-trip transport cost, and
> p_L = expected search cost in $L + (1 - \alpha)$ round-trip transport cost,
> where α can be written in other (endogenous and exogenous) variables and $0.5 < \alpha < 1$.

For an unbalanced equilibrium the following pair of price equations are obtained:

p_H = expected search cost in H + round-trip transport cost, and
p_L = expected search cost in L.

We conclude that, as search costs differ in both locations, freight prices in both directions have another differing component other than the unbalanced split of the transport costs due to demand imbalance. This has an impact on the numerical outcomes of the climate change analysis in Section 2.3.

2.2.3 Sensitivity analysis for climate change

The parameter of transport speed λ can be used as a parameter to study the costs of climate change. Climate change may cause low water levels in summer, which decreases the capacity of barges and therefore reduces the speed[10] of transport, which corresponds to a lower λ. Climate change may also cause extreme high water levels, which may lead to navigation-halts on rivers, for example due to flood risk. Furthermore, in this situation, transport speed decreases, which also corresponds with a lower λ. With λ as an instrument, we can perform a sensitivity analysis on our model to explore the impacts of climate change on the equilibrium in transport markets.

In the next section we present the numerical outcome of this theoretical model and perform a sensitivity analysis with respect to a reduction in transport speed due to climate change.

2.3 Numerical outcome of the theoretical model

In sub-section 2.3.1 we present a subset of chosen values for the exogenous parameters (see Demirel *et al.* 2010). We call this initial setting the standard case. We then present the numerical outcome for the standard case. In sub-section 2.3.2 we present the sensitivity analysis with respect to a reduction in transport speed due to climate change.

2.3.1 Standard case

In Table 2.3 the chosen values for a subset of the exogenous parameters are shown.

10 This reduction in speed could be regarded as a more abstract form of a reduction in speed of delivery measured, e.g. in tonnes per unit of time.

Table 2.3 **Subset of exogenous parameters**

Parameter	**Value**	**Description**
λ	50	Annual trip speed measured in round trips per year. We assume a return trip duration of *c*. 1 week
N_H	400	Number of customers located in *H*
N_L	100	Number of customers located in *L*

Using the model of Section 2.2, the equilibrium outcome is given in Table 2.4 with a description of every variable.

Table 2.4 **Equilibrium outcome for the standard case of the matching model**

Variable	**Value**	**Description**
p_H	€5,830	Price from *H* to *L*
p_L	€837	Price from *L* to *H*
z_L	0.251	Fraction of arriving carriers that search in *L* for customers
C	383.7	Total number of carriers
Carrier's search time in *H*	2.030 days	–
Carrier's' search time in *L*	2.040 days	–
Single-trip moving time	3.653 days	–
Single-trip moving cost	€2,499	–
Carrier's search cost in *H*	€834	–
Carrier's search cost in *L*	€837	–

As z_L, which is defined as the fraction of carriers that search for customers and return with backhaul from *L*, is smaller than 1, it can be seen that the unbalanced equilibrium arises.[11] It is clear that the freight price for the return trip, p_L, is much lower than p_H. p_L is seen to be 14% of p_H, even though 75% of carriers return empty to *H* ($z_L = 0.25$). p_L is positive due to a positive search time, which is 2.04 days in *L*.

In the next sub-section, we present a case where we study a decrease in speed λ and compare this with the standard case of Table 2.4.

11 In our model it appears that the balanced equilibrium only arises if N_H is extremely close to N_L.

2.3.2. Decrease in transport speed due to climate change

As a result of climate change, it is expected that more navigation halts on the Rhine will occur owing to extreme high water levels in winter, while the capacity of transport will be reduced owing to extreme low water levels in summer. This phenomenon has received considerable attention in the media in discussions on the impacts of climate change on the future of the transport sector. Both navigation stops and decreases in capacity lead to a decrease in the speed of transport (or delivery of the whole load). In the model developed above, the effects of a reduction in speed can be analysed by a decrease in λ. We decrease λ from 50 (normal speed) to 40 (low speed), so speed falls by 20%. It appears that the type of equilibrium (unbalanced equilibrium) does not change. The equilibrium outcome can be found in Table 2.5.

Table 2.5 **Effects of variations in the round-trip rate λ**

Variable	Normal speed	Low speed
λ (exogenous)	50	40
p_H	€5,830	€7077
p_L	€837	€837
z_L	0.251	0.251
C	383.7	454.3
Carrier's search time in *H*	2.030 days	2.028 days
Carrier's search time in *L*	2.040 days	2.040 days
Single-trip moving time	3.653 days	4.566 days
Single-trip moving cost	€2,499	€3,123
Carrier's search cost in *H*	€834	€833
Carrier's search cost in *L*	€837	€837

The results of a lower speed, captured by λ, implies a substantially longer expected moving time per trip and a higher moving cost, but it appears that the carrier's search time is hardly affected. However, the increase in moving cost is borne entirely by customers in *H*, which pay a (21%) higher freight price.

This has important implications for the Western European inland navigation market, which covers a large share of the overall freight market because of the presence of a number of well-known rivers, such as the Rhine. In this market, the majority of goods are moved from the Netherlands (in particular, from the sea harbour in Rotterdam) to industrial areas in Germany, whereas many fewer goods are moved from Germany to the Netherlands. Our results show that, under low-water conditions German firms importing goods from the Netherlands will be confronted with higher prices. However, Dutch firms importing goods from Germany will hardly be affected in the freight price that they pay. The latter result can be attributed to the observation that search times are hardly affected, and that, since both the 'normal speed' and the 'low speed' cases are in an unbalanced equilibrium, the moving cost comes in the accounts of the H to L direction. Moreover, customers from overseas using the port of Rotterdam as a transfer point will also hardly be affected. Hence, the welfare losses of changes in anticipated water levels will be borne almost entirely by the German firms importing goods from the Netherlands. This strongly suggests that, in this context, new infrastructure to improve navigation on the Rhine would mainly benefit German firms.

In this section we have studied the numerical outcomes of the theoretical framework and performed a sensitivity analysis with respect to climate change. In the next section, we compare these findings with the empirical analysis of Jonkeren (2009), which also studies the inland navigation market between the Netherlands and Germany.

2.4 Empirical evidence

Low water levels have a negative effect on available freight capacity for individual barges in inland navigation because of a minimum required distance between the barge and riverbed. As tonnages decrease, and costs hardly do, lower water levels are reflected in higher freight prices per tonne. In Jonkeren *et al.* (2007) the effect of low water levels on freight prices is empirically estimated for data from the Vaart!Vrachtindicator.[12] In Jonkeren (2009) an overall marginal effect of low water levels in centimetres of –0.68% on freight prices per tonne is found. This means that a 1 cm decrease in water levels causes an increase in freight price per tonne of 0.68%. In Jonkeren (2009) an empirical study is also conducted on the interactive effect of imbalance and low water levels on freight prices per tonne. The latter part of the study enables a comparison with our theoretical framework and its results.

Jonkeren (2009) divides the inland waterway network of Western Europe into 20 regions and assigns to every region an imbalance variable, which is based on the number of incoming trips and outgoing trips in a region. This imbalance is weighted[13] for the distance to, and imbalances of, other regions in the network (for all details, see Jonkeren 2009). As a last transformation, the logarithm is taken of this imbalance variable. The regional imbalances (and their logs) are given in Table 2.6 for the 20 regions.

12 www.vaart.nl.

13 Data based on Vaart!Vrachtindicator 2003–2005; www.vaart.nl.

Table 2.6 **Imbalance by region (I_i), and its logarithm**

Region	Ii	ln*(Ii)*	Region	Ii	ln*(Ii)*
Rotterdam port area (NL)	1.811	0.594	Upper Rhine area (D, F, CH)	1.002	0.002
Amsterdam port area (NL)	1.649	0.500	Main and Danube (D, H)	0.960	-0.041
Netherlands, South (NL)	1.626	0.486	North German Canals (D)	0.923	-0.080
Northern France (F)	1.523	0.421	Ruhr area (D)	0.829	-0.187
Antwerp port area (B)	1.409	0.343	Netherlands, East (NL)	0.811	-0.210
Flanders (B)	1.230	0.207	Middle Rhine area (D)	0.808	-0.213
Netherlands, Centre (NL)	1.154	0.143	Lower Rhine area (D)	0.761	-0.273
Wallonia (B)	1.103	0.098	West German Canals (D)	0.746	-0.293
Netherlands, North (NL)	1.060	0.058	Moselle and Saar area (D, F)	0.742	-0.299
Meuse area (NL, B)	1.050	0.049	Neckar area (D)	0.656	-0.422

Note: NL = the Netherlands; B = Belgium; D = Germany; F = France; CH = Switzerland; H = Hungary.

Source: Jonkeren 2009

An imbalance of 1.811 for the Rotterdam port area (NL) can be roughly interpreted as an imbalance where the number of outgoing loaded trips is two times that of incoming loaded trips (while also weighting for the surrounding regions is taken into account). A region with $I_i > 1$ or $\ln(I_i) > 0$ is therefore a good place for barge operators to find freight. These types of region may be compared with the H location in the theoretical framework of Section 2.1. Regions with $I_i < 1$ or $\ln(I_i) < 0$ are candidates for *L* locations.

Jonkeren (2009) also studies the interactive effect of low water levels and imbalance on freight prices. The findings for two routes (both fronthaul and return trips) are presented in Table 2.7.

The difference in the logarithm of the regional imbalance for the route from the Rotterdam port area to the Moselle/Saar area is calculated as ln(0.742) – ln(1.811) = – 0.299 – 0.594 = – 0.893. The row 'Marginal effect of an increase of water level of 1 cm' corresponds to the change (increase) in price per tonne when water levels change (decrease) by 1 cm. We see that this marginal effect of water levels on freight prices is stronger on routes from high-imbalance regions (or, in the framework of Section 2.1, high-demand locations) to low-imbalance regions (low-demand locations). For example, on the route from the Rotterdam port area to the Moselle/Saar area the marginal effect of a water-level decrease of 1 cm on price per tonne is 0.733%, while for the return trip this effect is 0.433%, which is lower by a factor of almost 2. We may conclude from these empirical findings that the effect of low water levels on freight prices, which is 0.68% without taking imbalance into account, is increased for trips from high- to low-imbalance regions and decreased for trips in the opposite direction.

While not exactly comparable quantitatively, these findings are quite in line with the findings from our theoretical framework. In the numerical outcome of

Table 2.7 **The relative change of the freight price when the water level increases by 1 cm**

Route	**Rotterdam port area–Moselle/ Saar area**	**Upper Rhine area–North German canals**	**North German canals–Upper Rhine area**	**Moselle/Saar area–Rotterdam port area**
Difference in (logarithm of) regional imbalance between origin and destination	-0.893	-0.082	0.082	0.893
Marginal effect of an increase in water level of 1 cm	-0.733%	-0.619%	-0.597%	-0.433%

Note: By construction, the difference in regional imbalance for the trip 'Rotterdam port area–Moselle/ Saar area' is the negative of the difference in regional imbalance for the trip 'Moselle/Saar area–Rotterdam port area'. The difference in regional imbalance for the trip 'Upper Rhine area–North German canals' is the negative of the difference in regional imbalance for the trip 'North German canals–Upper Rhine area'.

Source: Jonkeren 2009

Section 2.3 we found that water-level decreases, modelled through a decrease in λ, increased the freight price from the high-demand to the low-demand location. A decrease of 20% in speed (or capacity) caused a 21% increase in fronthaul prices and hardly any increase in backhaul prices. While our theoretical framework gives more extreme results, the main idea, that capacity reductions due to climate change makes especially transport from high-demand to low-demand locations more expensive, is valid both empirically and theoretically. For transport by inland navigation between the Netherlands and Germany, this again means that transport in the direction from the Netherlands to Germany is expected to become relatively more expensive as a result of climate change than transport in the opposite direction.

2.5 Conclusion

In this chapter we have stressed the importance of inland navigation as a transport mode between the Netherlands and Germany, and have studied the effect of climate change in this market taking into account that imbalance exists in transport flow between these countries.

Climate change affects inland navigation on the Rhine by causing lower water levels in summer and higher water levels in winter. Our focus has been on low water levels. From Jonkeren *et al.* (2007) we know the effect of low water levels on freight prices per tonne. Low water levels decrease the available capacity of individual barges and increase freight prices per tonne, which has a negative effect on economic welfare.

When studying the effect of climate change through low water levels, we took the imbalance in transport flows between the two countries into account. We presented a theoretical framework that contains a two-node network which describes transport between a location with high demand for transport to a location with low demand for transport. Climate change, or low water levels, was in this framework modelled by a reduction in transport speed (or capacity).

The numerical outcome of this theoretical framework showed that climate change increased the freight price from locations with high demand for transport (e.g. the Netherlands) to locations with low demand for transport (e.g. Germany), while in return directions the freight price hardly increased.

We also compared these theoretical findings with empirical evidence from Jonkeren (2009). Jonkeren shows a similar result: that is, the effect of decreases in water levels is increased when trips are more of the type 'from regions with high demand for transport to regions with low demand for transport'.

From both theoretical and empirical perspectives, we may therefore conclude that climate change will make transport by inland navigation from the Netherlands to Germany relatively more expensive than transport from Germany to the Netherlands. This may affect policy-related issues, such as infrastructural decisions on canalisation, dredging and the construction of barrages. As, under climate change, importing firms in Germany are faced with relatively higher freight prices than firms importing from Germany, Germany can be considered to benefit most from infrastructural investments to combat climate change. This result is important for discussions on optimal adaptation strategies. A difficult point may be that most of the transport on the Rhine takes place by Dutch barge operators. This might reduce political support for infrastructure measures in Germany (for the German part of the Rhine) even though the main beneficiaries of adaptive measures in the German part of the Rhine would be German consumers. These results underline the relevance of distributional issues in this domain of adaptation to climate change. Other adaptation policies to be considered concern changes in the size and design of ships, and these would of course be the responsibility of barge owners.

References

Arnott, R. (1996) 'Taxi Travel Should Be Subsidized', *Journal of Urban Economics* 40.3: 316-33.

Boyer, K.D. (1998) *Principles of Transportation Economics* (Reading, MA: Addison-Wesley): 252-57.

CCNR (Central Commission for Navigation on the Rhine) (2007) 'Market Observation for Inland Navigation in Europe, 2007–1' (Strasbourg: CCNR, European Commission).

Demirel, E., J. Van Ommeren and P. Rietveld (2010) 'A Matching Model for the Backhaul Problem', *Transportation Research: Part B* 44.4: 549-61.

Felton, J.R. (1981) 'The Impact of Rate Regulation upon ICC-Regulated Truck Back Hauls', *Journal of Transport Economics and Policy* 15.3: 253-67.

Jonkeren, O.E. (2009) 'Adaptation to Climate Change in Inland Waterway Transport' (PhD thesis, Vrije University, Amsterdam).

——, P. Rietveld and J.N. van Ommeren, J.N. (2007) 'Climate Change and Inland Waterway Transport: Welfare Effects of Low Water Levels on the River Rhine', *Journal of Transport Economics and Policy* 41.3: 387-411.

KNMI (Koninklijk Nederlands Meteorologisch Instituut) (2006) 'KNMI Climate Change Scenarios 2006 for the Netherlands' (KNMI Scientific Report WR 2006-01, De Bilt, Netherlands: KNMI).

Krugman, P. (1991) 'Increasing Returns and Economic Geography', *Journal of Political Economy* 99.3: 483-99.

Lagos, R.A. (1996) 'An Alternative Approach to Market Frictions: An Application to the Market for Taxicab Rides' (CARESS Working Paper 96-09; University of Pennsylvania, November 1996).

Meelker, C. (2006) 'The Dutch Inland Waterway Transport Market: The Backhaul Problem' (Master's thesis, Vrije University, Amsterdam).

Pigou, A.C. (1913) 'Railway Rates and Joint Costs', *Quarterly Journal of Economics* 27.2: 687-94.

Pissarides, C.A. (2000) *Equilibrium Unemployment Theory* (Cambridge, MA: MIT Press).

Te Linde, A.H. (2007) *Effect of Climate Change on the Rivers Rhine and Meuse* (WL | Delft Hydraulics report prepared for Rijkswaterstaat; Delft, Netherlands).

TLN (Transport en Logistiek Nederland) (2007) *Transport in Cijfers, editie 2007* (Zoetermeer, Netherlands: TLN).

Wheaton, W.C. (1990) 'Vacancy, Search, and Prices in a Housing Market Matching Model', *Journal of Political Economy* 98.6: 1270-92.

3

Climate change impacts

The vulnerability of tourism in coastal Europe

Alvaro Moreno
ICIS, Maastricht University, The Netherlands

The right to leisure was recognised by the Universal Declaration of Human Rights (UN 1948: Article 24). As a result of this, the right to tourism was proclaimed universal in the *Global Code of Ethics for Tourism* of the United Nations World Tourism Organization (UNWTO 1999). Although there are very important disparities worldwide in the access to and practice of these rights, it is unquestionable that recreation and tourism are key phenomena of our society and that they have global effects. The contribution of the tourism sector to global GDP and employment is estimated to be around 10%. The same percentage also applies to GDP and employment in the European Union; between 7.3 and 20.6 million jobs are directly and indirectly related to tourism (Leidner 2004). The number of tourists travelling internationally in 2008 reached 922 million (UNWTO 2009) and, according to the projections of the UNWTO (2001), this number is expected to reach 1.6 billion by 2020. In other words, assuming international tourism started only in the 1950s, projections of future growth in the sector suggest that international tourism arrivals will experience the same growth in 15 years as it had in the last 55 years (the 800 million benchmark was reached in 2005).

International tourist flows are unevenly distributed between the different world regions. Europe is currently the world's leading destination, with a total of 487.9 million international arrivals (a market share of 53%) in 2008. Despite the growth of cultural and rural tourism in recent years, coastal areas remain the preferred space for recreation in Europe. To a certain extent most types of leisure activity depend on weather and climate conditions but coastal and marine recreation in general, and 'sun, sea and sand' tourism in particular, are especially exposed and highly sensitive to weather.

Our understanding of the relationship between climate and coastal tourism is still limited. The need to improve our knowledge about this link is strongly intensified by the increasing evidence that the climate is changing. With a potential impact on all the actors directly and indirectly involved in or related to tourism (from demand to offer, from tourism providers to other sectors such as agriculture), climate change is seen as one of the major challenges the tourism sector will have to face in the coming years.

This chapter summarises some of the key elements that characterise the relationship between climate (change) and tourism in Europe. More specifically the chapter addresses the sensitivity of key actors in tourism to weather and climate (change). The focus is on coastal and marine tourism because recreation in these environments is arguably the most climate-sensitive. Moreover, coastal and marine areas are known to be disproportionately exposed to the effects of climate change. The vulnerability of the coastal tourism sector is, therefore, at the core of this chapter.

The chapter is structured around four main sections, which summarise the main findings of a four-year project on climate change and coastal tourism in Europe. The first section presents the state-of-the-art knowledge on climate (change) and tourism, paying special attention to coastal and marine environments. Due to the importance of sun, sea and sand recreation in Europe, especially in the Mediterranean, the rest of the chapter focuses on this type of recreation. Section 3.2 explores the way weather conditions determine participation in sun, sea and sand recreation. In this context, sun, sea and sand tourism (also referred as 3S tourism) refers to light activities commonly associated with sunbathing by the seashore and therefore implying a high exposure of the body to the weather elements. In addition, Section 3.3 summarises the effects climate change will have on the climatic suitability of destinations for 3S tourism as compared to current conditions. The chapter continues in Section 3.4 with an overview of the most relevant issues in relation to the vulnerability of coastal tourism to climate change. Finally, the main knowledge gaps are identified and some suggestions for future research are presented.

3.1 Climate change and tourism: knowledge and implications for coastal and marine environments

Our knowledge and understanding of the ways in which climate change is manifesting itself and will affect the environment and society have improved substantially in recent years. As is evident from the Fourth Assessment Report (AR4) published by the Intergovernmental Panel on Climate Change (IPCC 2007)—the most authoritative source of information on the science, impacts, adaptation and mitigation of climate change—increasing attention is being paid to the relationship between climate change and tourism. Our research analysed how tourism is incorporated into the AR4 and compared it to previous IPCC reports (see Amelung *et al.* 2008 for more information). The findings show that the AR4 pays much more attention to the impacts of climate change on environments such as coastal and mountain areas, which are at the centre of many tourism activities and likely to be highly affected by climate change. Geographically, the published research is highly unbalanced: Europe, North America, Australia, New Zealand and Small Island States dominate the literature on tourism impact assessment. On the other side of the spectrum, Africa, Latin America, Asia and the Polar regions are virtually unexplored, suggesting that these are priority areas for future research (Table 3.1). The projected increase of tourism in many of these regions (e.g. Polar regions) adds to the urgency. Adaptation strategies for destinations and tourists are poorly accounted for in the AR4.

Coastal areas have been identified as being highly vulnerable to climate change impacts. In addition to the changes in temperature and precipitation patterns experienced by all regions, coastal areas face specific impacts such as acidification (severely threatening coral reefs), sea-level rise and coastal erosion. These regions are also more exposed to extreme events such as hurricanes and storm surges. The high exposure of coastal environments to these impacts combined with the importance of these environments for recreation makes coastal tourism one of the most vulnerable tourism segments. Given this vulnerability, this project revealed that there is surprisingly little known about the weather requirements of coastal and marine tourism activities. The findings, published in the *Journal of Coastal Research* (Moreno and Amelung 2009a), show that beach tourism has received most of the attention due to its high dependency on weather conditions and its popularity. However, even for this segment, many uncertainties about weather requirements for 3S activities still exist. As for other coastal leisure activities, they have been virtually unexplored so far. As the knowledge about physical impacts of climate change on certain marine ecosystems and species improves, it is expected that more attention will be paid to how these impacts will affect the tourism industry that makes use of those resources. For example, knowledge of the impacts of climate change on coral reefs has increased significantly over the years and now some studies are available that translate those ecosystem impacts into impacts on the diving tourism industry (Cesar 2000; Ngazy *et al.* 2005; Andersson 2007).

Studies addressing greenhouse gas emissions from coastal and marine activities are very limited, although a few studies have been made available in the last years. The lack of tourism data with the required resolution (e.g. activity-specific data) is a major factor hindering the assessment of emissions from tourism in general and from coastal activities specifically. This lack of information influences how the IPCC Working Group III approaches the mitigation and tourism link, which is

Table 3.1 **Word search results by IPCC AR4 chapters**

Chapter	Word search
Working Group II: Impacts, Adaptation and Vulnerability	
1 Assessment of observed changes and responses in natural and managed systems	17
2 New assessment methods and the characterisation of future conditions	0
3 Fresh water resources and their management	2
4 Ecosystems, their properties, goods and services	9
5 Food, fibre and forest products	3
6 Coastal systems and low-lying areas	30
7 Industry, settlements and society	60
8 Human health	1
9 Africa	25
10 Asia	7
11 Australia and New Zealand	29
12 Europe	37
13 Latin America	11
14 North America	22
15 Polar regions (Arctic and Antarctic)	7
16 Small islands	48
17 Assessment of adaptation practices, options, constraints and capacity	9
18 Interrelationships between adaptation and mitigation	5
19 Assessing key vulnerabilities and the risk from climate change	6
20 Perspectives on climate change and sustainability	0
Working Group III: Mitigation of Climate Change	
All chapters together (occurrences in five separate chapters)	8

Source: Based on Amelung *et al.* 2008

seen as a one-way relationship (effect of climate change mitigation on tourism). However, the important role of tourism as a source of greenhouse gases has been acknowledged by the UNWTO (UNWTO *et al.* 2008), the tourism industry (WTTC 2009) and the overall scientific community. In the IPCC AR4, tourism is seen either as a victim of climate change mitigation policies, mainly due to the potential effect of taxation on air transport, or as a beneficiary of mitigation strategies due to, for example, the protection of forests and their use for recreation. This is a very incomplete, one-sided view that does not reflect the scientific, political and social consensus about the role of tourism—and more specifically aviation—as a source of greenhouse gases. Clearly, emissions and mitigation related to tourism are issues that need to be addressed in the next IPCC report.

3.2 The influence of weather and climate on 3S tourism

Detailed statistics on its magnitude are lacking, but 3S tourism is unquestionably one of the most important forms of recreation in Europe, and it is likely to retain this position in the future. Due to its magnitude and its high sensitivity to weather conditions, it is also projected to be one of the tourism segments most affected by climate change impacts. In order to assess these impacts, baseline information is required about the exact role of different weather attributes in beach tourism. Existing studies have commonly emphasised the role of temperature as a key determinant (and limiting factor) of beach tourism (e.g. Perry 2005, 2006). The results from these and similar studies depict a Mediterranean region with deteriorating climatic conditions during the summer, resulting in decreasing attractiveness for tourism. The increase in temperatures causing the Mediterranean region to be too hot for tourism would make climate conditions in northern Europe more suitable and comfortable for this same tourism. The consequence of this would be more northern European tourists spending their holidays in higher latitudes instead of travelling to the Mediterranean (Amelung *et al.* 2007; Amelung and Viner 2006; Nicholls 2006).

The validity of this and similar assumptions have been assessed in recent years in different studies (see Table 3.2). In our research project two assessments have been carried out using different methodologies: observing tourist behaviour and comparing it to weather conditions (Moreno *et al.* 2008) and using questionnaires to elucidate weather preferences and the relative importance of certain weather parameters (Moreno 2009). The results from the studies complement each other. Responses to the questionnaires highlighted the indisputable importance of climate as a key attribute of Mediterranean destinations. This is the case even for people who do not travel with beach recreation as the main activity during their holiday period. Overall, the lack of precipitation is the most important factor for beach recreation, followed by comfortable temperatures. When looking into the weather conditions that hinder beach recreation, precipitation appears as the number one factor, which is consistent with previous studies. Temperature, however, is less important when it comes to making a day unfavourable, especially when high temperature is considered. Overall, the ideal weather conditions respondents associated with beach tourism are a mean temperature of 28.3°C, a light breeze, between six and ten hours of sun and no cloud. From the information presented in Table 3.2 it can be seen that different nationalities seem to perceive differently the importance of weather components. It can also be seen that the assumption that temperature plays a leading role might not be applied as a general rule for 3S tourism.

Table 3.2 **Overview of some recent studies on priorities for climatic parameters for beach tourism**

Authors	**Respondents**		**Type of tourism**	**Sample**	**Ranking of weather parameters**			
	Number	**Nationality**			**Precip.**	**Temp.**	**Wind**	**Sun.**
Moreno (2009)	115	Belgian & Dutch	3S	Tourists	1	2	4	3
de Freitas *et al*. (2008)	331	Canadian	3S	Students	2	1	3	1
Scott *et al*. (2008)	333	Canadian	3S	Students	3	2	4	1
	207	New Zealand	3S	Students	1	2	4	3
	291	Swedish	3S	Students	3	2	4	1
Morgan *et al*. (2000)	1,354	North Europeans	3S	Tourists	1	4	3	2
Mieczkowski (1985)	Based on literature	n.a.	Sight-seeing	n.a.	2	1	3	2

Source: Based on Moreno 2009

The study that employed observation of tourist behaviour as method and was also carried out as part of this research (Moreno *et al*. 2008) reinforced the findings of the questionnaires. Precipitation, even in small quantities or for short periods of time, had a clear effect on participation, overriding other favourable weather conditions. Contrary to what existing literature suggests, high temperatures seemed to stimulate rather than deter participation in beach activities. These findings provide new insights about the role of weather for 3S recreation. The results also have important implications for the tourism and climate change literature, urging a reassessment of the impacts that reflects the specific weather requirements of beach tourism as discussed below.

3.3 Climate change and 3S tourism in Europe

Climate change is shifting the main weather parameters relevant to tourism and putting pressure on the environments where tourism activities take place (Fig. 3.1). These factors make the assessment of climate change impacts on tourism a major necessity. Existing studies on the impacts of climate change on Europe's tourism have commonly used composite indices such as the Tourism Climate Index (TCI) (Mieczkowski 1985) to assess destinations' climate suitability for recreation.

Figure 3.1 **Changes in average summer temperature and precipitation in Europe between the baseline and the 2060s, according to the integration of two climate models (HadCM3, CSIRO2) and (left) four scenarios (SRES A1FI, A2, B1, B2) and (right) four European regions**

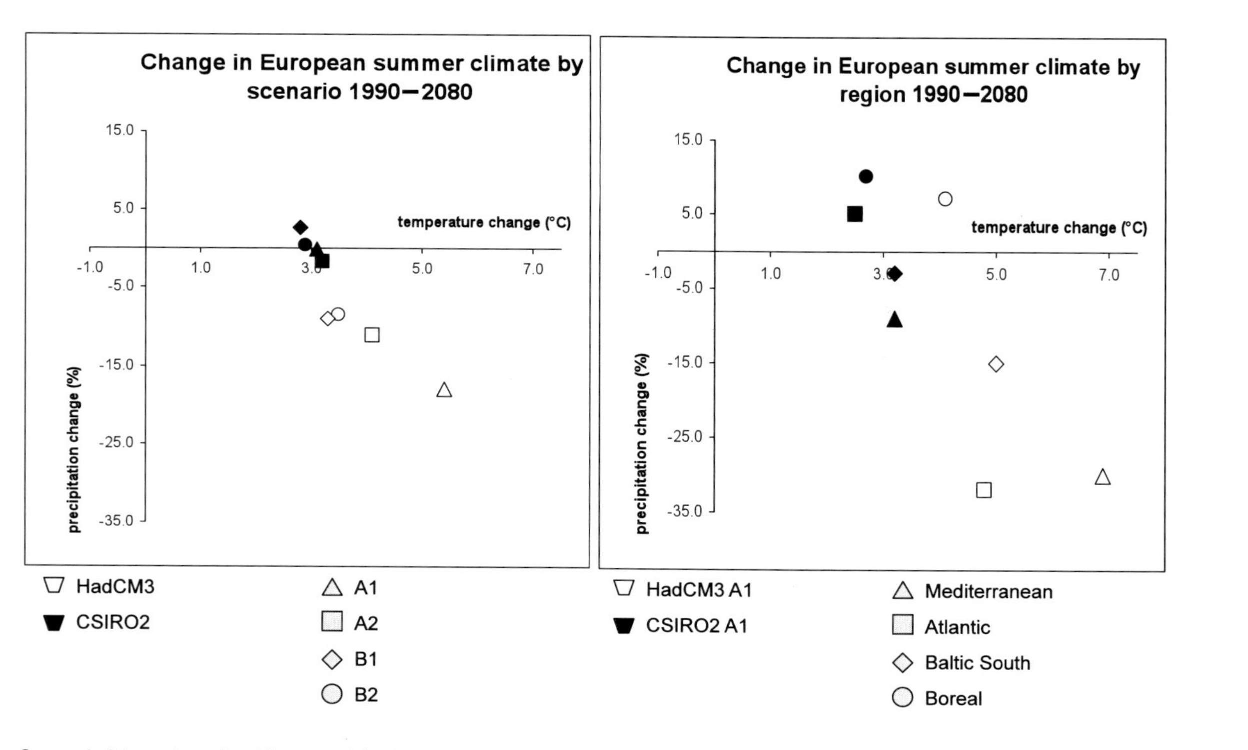

Source: Left image based on Moreno and Amelung 2009b

These types of study, however, were not specifically designed for beach tourism and therefore overemphasised the negative role of high temperature compared to other weather aspects. The findings presented above confirmed the importance of climate for beach tourism, but revealed that the role of high temperature is less important than for other forms of tourism. Another limitation of the TCI is that the weighting of the weather parameters is rooted in theoretical knowledge rather than empirical evidences.

This research reassessed the impacts of climate change on beach tourism using an adapted index, the Beach Climate Index (BCI), which addressed both limitations of the TCI (see Moreno and Amelung 2009b). The index is based in research carried out by Morgan *et al.* (2000) with a broad sample of north European beach users. Using baseline climate data and future projections of the main climate variables, the resulting index offered a view of the main changes in beach tourism climate comfort in Europe during the summer season. Conditions during the baseline period (1961–90) depicted a Mediterranean climate ranging from very good to excellent for beach tourism; the lack of precipitation, high sunshine, moderate wind breeze and comfortable temperature contribute to this high climatic comfort (Fig. 3.2).

Index values were subsequently calculated for two models of climate change in the period 2051–80 (2060s): the more extreme HadCM3 (Hadley Centre Coupled Model, version 3) and the moderate Commonwealth Scientific and Research Organisation CSIRO2 General Circulation Models (GCMs). The selected scenario for both models was the A1FI scenario, a high emissions scenario presenting the fastest emission growth and the greatest potential change for tourism in Europe of all IPCC Special Report on Emissions Scenarios (SRES). The results of this analysis (see Fig. 3.3) indicated a trend of deteriorating conditions in the Mediterranean, similar to that of previous studies (e.g. Amelung and Viner 2006), although the magnitude and speed of change is significantly different. Future conditions reflected by both models show deteriorating comfort in some areas of the Mediterranean, mainly in the south of Spain in the HadCM3 model. Climatic comfort extends towards northern latitudes and the value of the beach climate index improves in the north of Spain and south-west France. But the similarities with previous assessments end here. As a contrast to existing studies, in this research the majority of the Mediterranean retains very good or excellent climatic comfort. This is especially the case for most of the east coast of Spain, the south of Italy, the coasts around the Aegean Sea and all the islands, where even the extreme HadCM3 model does not bring the beach climate index to values under 70 (very good). Absolute 'winners', that is, the regions that will benefit by even a moderate climate change scenario (CSIRO2 model), are mainly the eastern coast of the Adriatic Sea, areas in the north-west of the Iberian Peninsula and the regions of Poitou-Charentes and Aquitaine in south-west France. Areas that currently enjoy very good or better conditions but that will see their climate resources deteriorate to levels under 70 even if the moderate CSIRO2–A1FI scenario materialises ('losers') include various regions in the south of the Iberian Peninsula (Algarve, Andalucía, Valencia) and a few other

Figure 3.2 **Beach tourism climate index for baseline summer conditions**

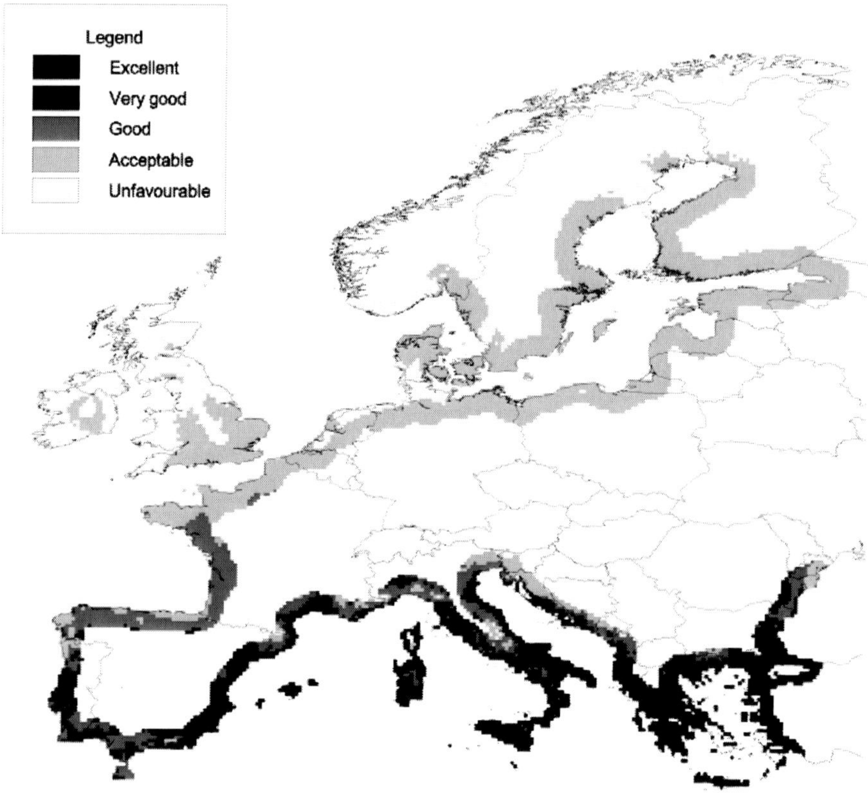

Source: Based on Moreno and Amelung 2009b

areas in the Mediterranean (see Moreno and Amelung 2009b for a detailed description of the results).

These findings confirm the importance of climate as a destination attribute, but they challenge some of the major statements about climate change impacts on European tourism and more specifically on the Mediterranean. Climatic comfort for beach recreation may shift to northern latitudes and the suitability of the Mediterranean may decrease, but these changes are not likely to be significant at least in the coming 50 years and most likely not before the end of the century (Fig. 3.4).

The existing projections of climate change impacts on Europe's tourism industry have typically focused on high temperatures as the main driver of change. An increased frequency and intensity of heatwaves have therefore received most of the attention of all climate change impacts. The findings of this project and presented above, however, show that high temperature does not deter beach recreation, suggesting that the effect of heatwaves may have been overestimated. This

Figure 3.3 **Beach tourism climate index during summer in the 2060s (left) according to the HadCM3–A1FI and (right) according to the CSIRO2–A1FI**

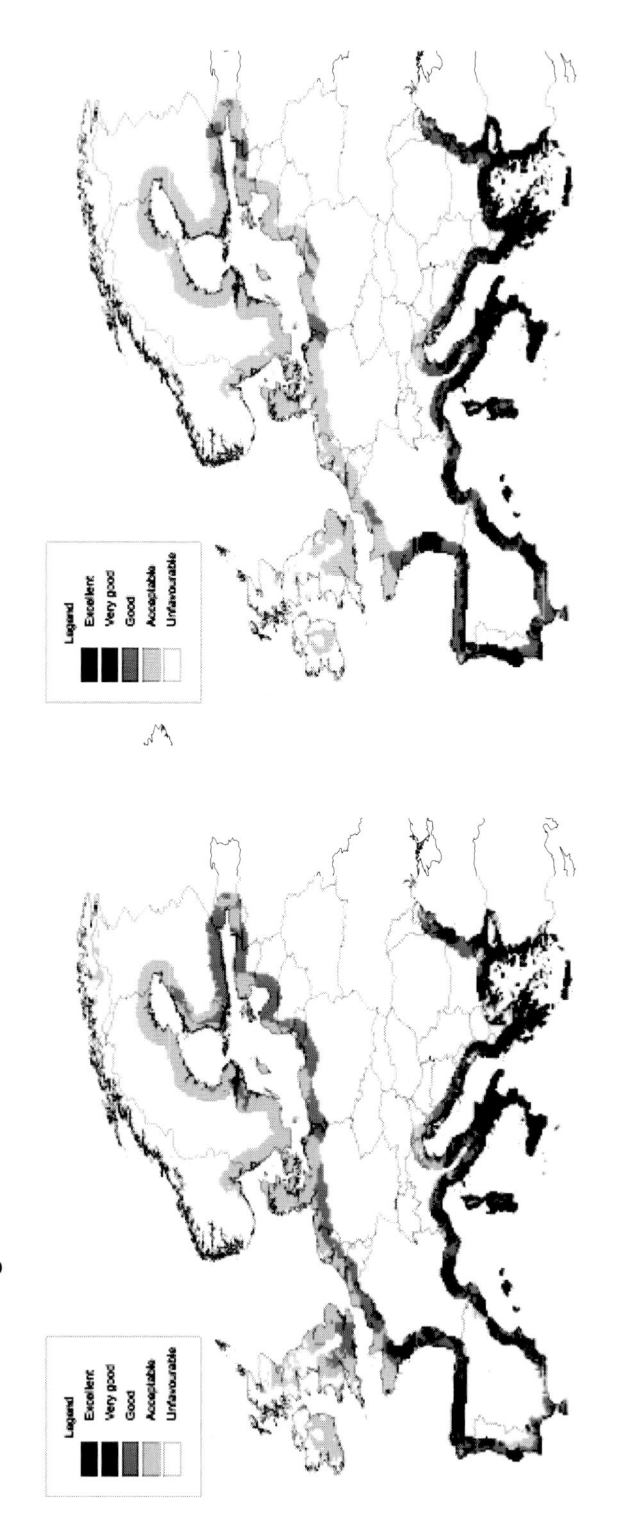

Source: Based on Moreno and Amelung 2009b

Figure 3.4 **Change in BCI in different European coastal regions (including only the first 100 km inland from the shoreline)**

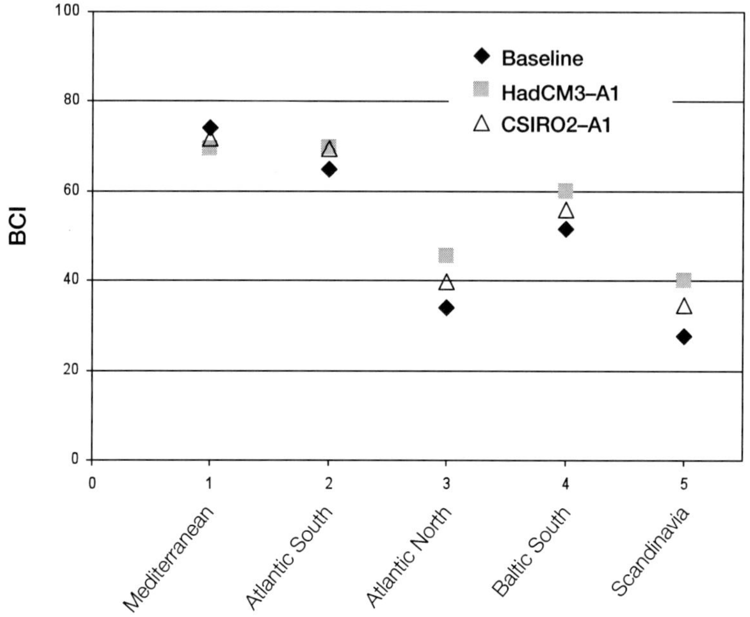

Source: Based on Moreno and Amelung 2009b

hypothesis was tested in this research (see Moreno 2009). Using a questionnaire, this study explored the relative importance for tourists of five projected impacts of climate change in the Mediterranean. The results confirmed the hypothesis: of the five impacts, heatwaves were seen as the least negative of all. This indicates that, together with studies looking into tourists' comfort, there is a pressing need to assess the importance of other climate change impacts, such as the risk of diseases or water availability, as these types of impact assessment have been overlooked so far by the tourism and climate change research community.

3.4 Tourism vulnerability to climate change: a framework

Although initially applied to fields such as risk-hazard studies, the concept of vulnerability is now widely used in climate change science (Adger 2006). The IPCC (2007) defines it as 'the degree to which a system is susceptible to, and unable to cope with, adverse effects of climate change, including climate variability and extremes'. Determined by the nature, magnitude and rate of climate variation, there are three main components that define the vulnerability of a system: its exposure, its sensitivity and its capacity to adapt to the specific change (IPCC 2007).

Coastal tourism destinations are highly vulnerable to climate change, and this is exacerbated by existing pressures on these, often highly populated, areas. Coastal areas are not only highly exposed and sensitive to climate change, but in many cases their adaptive capacity is low, especially for Small Island Developing States because of limited financial resources and technical knowledge (UNWTO *et al.* 2008). This high vulnerability jeopardises the sustainability of many tourism destinations, putting at risk the economy of these locations and threatening many of the environmental assets used by tourists.

The quantification of vulnerability to climate change requires consistent and structured methodologies (Adger 2006). Different examples of assessment tools are available for economic sectors such as agriculture (e.g. Fraser *et al.* 2005; Luers *et al.* 2003) or for certain communities in exposed locations (e.g. Adger 1999; Ford *et al.* 2007). Although general guidelines for performing vulnerability assessments exist (Füssel 2007; Snover *et al.* 2007), frameworks addressing specifically the needs of the tourism sector were not available until recently. This research proposed a five-step methodology framework to analyse the vulnerability of the sector to climate change in a consistent, structured and replicable way (Fig. 3.5) (see Moreno and Becken 2009 for more information).

Step 1 involves a detailed analysis of the destination, including the identification and involvement of the main stakeholders, other relevant economic sectors related to tourism and the main tourism activities. The link between potential climate hazards and key tourism activities is made in Step 2. In this step, a set of activity-hazard sub-systems is created; the destination is disaggregated into smaller units to facilitate their analysis and because not all activities have the same sensitivity to climate the vulnerability of each of them is likely to be different. The decision about which sub-systems will be included in following steps of the analysis is based on stakeholder input and it depends on the elements that these actors consider to be a priority.

After deciding the sub-systems to be analysed, Step 3 involves the identification of quantitative indicators to estimate the vulnerability of the sub-system. To structure this step, each sub-system is analysed based on the three dimensions of vulnerability: exposure, sensitivity and adaptive capacity. To assist participants in the process of identifying indicators, the Vulnerability Scoping Diagram (VSD) proposed by Polsky *et al.* (2007) is suggested. The use of indicators not only facilitates the assessment of the present vulnerability of the sub-system, but it also allows for a monitoring of the change over time and therefore any progress or regress in the path towards reducing vulnerability. Although in the research it is acknowledged that the use of indicators to measure the state or change of a certain parameter is a common practice and it attempts to be as objective as possible, the interpretation of the indicators and the identification of 'threshold(s) of risk, danger or harm'

Figure 3.5 **A framework to analyse tourism vulnerability to climate change**

Step 1. System analysis
- Economic, environmental and social context of destination
- Identification and characterisation of tourism activities
- Prioritisation according to their importance

Step 2. Climate
- Characterisation of climate conditions and identification of key hazards
- Creation of activity-hazard sub-systems
- Selection of sub-systems for the analysis

Step 3. Vulnerability
- Identification of vulnerability indicators
- Fine-tuning of components and indicators (with the help of VSD)
- Operationalisation of vulnerability
- Validation of steps 1–3

Step 4. Integration of individual vulnerability assessment
- Scenario construction
- Analysis of interdependencies and feedback loops
- Discussion of results and overall vulnerability assessment

Step 5. Communication of results
- Communication to broader audience (e.g. use of visualisation tools like VSD)

Source: Based on Moreno and Becken 2009

(Adger 2006) can be highly subjective. Therefore this step requires stakeholder involvement in order to include the values and perceptions of those directly interested in and affected by in the assessment.

In Step 4 the individual sub-system analyses are put together and the overall vulnerability of the destination to climate change is analysed. Assisted by tools such as scenario analysis, stakeholders can explore future developments in the tourism sector as well as the effects of unexpected shocks affecting the system. The complexity that characterises the tourism sector indicates that all the different elements (service providers and activities) are not independent entities but are connected to each other. Therefore, linkages between different sub-systems and how they affect the overall vulnerability of the destination are investigated in this step. It is also in this step that different adaptation options can be proposed, a process that will also depend on stakeholders' views and attitudes towards risk. Finally, the aim of Step 5 is the dissemination of the results to those that were not involved directly in the assessment process and, ideally, to incorporate them into the tourism management and planning of the destination.

Although the methodology has not been empirically tested, it provides a useful first step towards the elaboration of vulnerability assessments specifically design for tourism. The potential to explore both current vulnerability as well as the vulnerability of future activities reflects the dynamic nature of tourism. Moreover, the disaggregation of the analysis into activity-hazard sub-systems makes the analysis more manageable and easier to perform. Another advantage of this approach is that it reflects the different nature of climate change impacts depending on the specific destination and activity, a distinction neglected by previous studies. By incorporating stakeholders into the process, the assessment is likely to gain in credibility and social relevance. It is also the recognition that the process is, eventually, a result of stakeholder dialogue and decision-making, increasing the likelihood of it being successfully implemented. However, and as is indicated in the original paper, the validity of the methodology ultimately depends on 'its capacity to improve the understanding of vulnerability of the particular destination (short term), and, to the extent that adaptation measures are put in place, to reduce climate change vulnerability (long term)' (Moreno and Becken 2009).

3.5 Conclusions

This chapter presents the main findings of the four-year project on the impacts on climate change on coastal tourism and the vulnerability of the sector. The research demonstrates the importance of weather and climate for coastal and marine recreation in Europe and how climate change will affect tourism in the coastal zone. Despite increasing attention on the tourism and climate change relationship since the 1990s and recognition of the potential disruptive role that climate change might have on the sector, the field is still relatively unexplored. In fact, it has been noted that tourism is lagging a decade behind other sectors with regard to impact assessments and, more significantly, adaptation and mitigation (Ceron and Scott 2007). The important role of human behaviour and preferences makes tourism a complex research field as compared to other, more physical, studies such as agriculture.

Climate change will impact many of the qualities that make a coastal destination successful, including the attractions and environment where tourism activities take place and the climate that facilitates, and in certain cases allows or hinders, these activities to take place. The need to assess the vulnerability of destinations to changes in climate comfort and, very importantly, to other indirect impacts of climate change is therefore of paramount importance. These other impacts include sea-level rise, saltwater intrusion, health issues, coastal erosion and ecosystem changes, to mention a few. These issues will be especially relevant for destinations in the Mediterranean and should be high on research and policy agendas in the coming years. Although certain non-Mediterranean parts of Europe might increase their potential for specific types of coastal tourism, adaptation to climate change impacts should also be included in any planning or development programmes in order to minimise negative impacts. To this end, integrated planning and management approaches for sustainable coastal tourism development can significantly benefit from existing knowledge and tools such as the handbook recently published by the United Nations Environment Programme (UNEP 2009).

Effective policy implementation for sustainable tourism practices will play an important role on facilitating the adaptation of the tourist sector to climate change (Scott *et al.* 2008). The United Nations Development Programme (Lim *et al.* 2004) proposed four principles to guide adaptation: consider adaptation as part of the broader context of a region's sustainable development policies; build on past and present adaptive experiences; acknowledge that the implementation of adaptation measures will primarily be at the local or the destination level; and recognise that adaptation is an ongoing and reflexive process that needs to be monitored and adjusted over time. Such strategies can be supported by vulnerability assessments, identifying key hazards for specific tourism activities and regions and potential adaptation options. In this sense, vulnerability assessments can be considered the first step towards adaptation.

Although the tourism sector is showing an increasing interest in climate change, basic information about the influence that weather conditions have on key tourism stakeholders—from tourists to governments—is still missing. Therefore, consistent, structured and integrated approaches to explore the climate change and tourism relationship lack a solid ground. A number of recommendations can be put forward to improve this situation and to expand the existing knowledge on the coastal tourism and climate change relationship. As a fundamental step, more information is needed on the weather requirements of different activities, as well as data on the number of users and temporal and spatial distribution of activities. This is a necessary step before any attempt is made to assess the vulnerability of a destination. In terms of impacts, current experiences with extreme events could be used as analogues for tourist responses to future climate conditions, a technique applied already to ski tourism but which has been scarcely used with other tourism segments. Innovative approaches to exploring tourism vulnerability may be inspired by progress in other disciplines such as ecology that have a longer tradition in climate change impact research. Direct collaboration with other disciplines such as psychology, sociology and marketing would also contribute greatly to the study of the relationship between climate change and tourism, as they can assist in building knowledge in aspects such as the decision-making process of tourists. Stakeholders in general, and local businesses and tourism planners in particular, need to be more aware of the vulnerability of their destinations and the potential adaptation measures to be implemented if the impacts of climate change are to be minimised.

Sun, sand and sea activities are especially vulnerable to climate change due to their high sensitivity to weather parameters. Although the results of this research for beach tourism in the Mediterranean are promising in terms of tourism comfort, there are important caveats in knowledge that need to be addressed. Most research on climate change and tourism has focused on tourists' comfort. However, climatic comfort does not necessarily determine the success of a destination on its own as other characteristics may, in the future, override climate as the key attribute. Mitigation policies related to air transport may play a critical role in the future, especially for islands and long-haul destinations, and therefore needs further attention. Besides the uncertainties intrinsic to any social system such as tourism, the magnitude of climate change impacts is also uncertain; the use of available General Circulation Models to project climate change impacts does not mean that more dramatic and unexpected changes could not occur. Therefore, the exploration of impacts related to more extreme scenarios should not be disregarded. Finally, in addition to the already mentioned need to pay attention to air transport mitigation policies and to tourists' preferences and market trends, other climate change impacts such as coastal erosion, water availability or health-related impacts are also important issues that require consideration as they may become critical in many coastal destinations.

References

Adger, W.N. (1999) 'Social Vulnerability to Climate Change and Extremes in Coastal Vietnam', *World Development* 27.2: 249-69.

—— (2006) 'Vulnerability', *Global Environmental Change* 16.3: 268-81.

Amelung, B., and D. Viner (2006) 'Mediterranean Tourism: Exploring the Future with the Tourism Climate Index', *Journal of Sustainable Tourism* 14: 349-66.

——, S. Nicholls and D. Viner (2007) 'Implications of Global Climate Change for Tourism Flows and Seasonality', *Journal of Travel Research* 45: 285-96.

——, A. Moreno and D. Scott (2008) 'The Place of Tourism in the IPCC Fourth Assessment Report: A Review', *Tourism Review International* 12: 5-12.

Andersson, J.E.C. (2007) 'The Recreational Cost of Coral Bleaching: A Stated Revealed Preference Study of International Tourists', *Ecological Economics* 62: 704-715.

Ceron, J.P., and D. Scott (2007) 'Overview of Issues Regarding Impacts of, and Adaptation to Climate Change', paper at the *E-CLAT Technical Seminar: Policy Dialogue on Tourism Transport and Climate Change: Stakeholders Meet Researchers*, Unesco, Paris, 15 March 2007.

Cesar, H. (2000) 'Impacts of the 1998 Coral Bleaching Event on Tourism in El Nido, Philippines: Report for Coastal Resource Center, Coral Bleaching Initiative' (Narragansett, RI: University of Rhode Island).

De Freitas, C.R., D. Scott and G. McBoyle (2008) 'A Second Generation Climate Index for Tourism (CIT): Specification and Verification', *International Journal of Biometeorology* 52: 399-407.

Ford, J., T. Pearce, B. Smit, J. Wandel, M. Allurut, K. Shappa, H. Ittusujurat and K. Qrunnut (2007) 'Reducing Vulnerability to Climate Change in the Arctic: The Case of Nunavut, Canada', *Arctic* 60.2: 150-66.

Fraser, E.D.G., W. Mabee and F. Figge (2005) 'A Framework for Assessing the Vulnerability of Food Systems to Future Shocks', *Futures* 37: 465-79.

Füssel, H.M. (2007) 'Vulnerability: A Generally Applicable Conceptual Framework for Climate Change Research', *Global Environmental Change* 17: 155-67.

IPCC (Intergovernmental Panel on Climate Change) (2007) *Climate Change 2007: Synthesis Report* (Geneva: IPCC).

Leidner, R. (2004) 'The European Tourism Industry: A Multi-sector with Dynamic Markets' (Luxemburg: European Commission).

Lim, B., E. Spanger-Siegfried, I. Burton, E. Malone and S. Huq (eds.) (2004) *Adaptation Policy Frameworks for Climate Change: Developing Strategies, Policies and Measures* (Cambridge, UK: Cambridge University Press/UNDP; www.undp.org/climatechange/adapt/apf.html).

Luers, A.L., D.B. Lobell, L.A. Sklar, C.L. Addams and P.A. Matson (2003) 'A Method for Quantifying Vulnerability, Applied to the Agricultural System of the Yaqui Valley, Mexico', *Global Environmental Change* 13: 255-67.

Mieczkowski, Z. (1985) 'The Tourism Climate Index: A Method of Evaluating World Climates for Tourism', *The Canadian Geographer* 29.3: 220-33.

Moreno, A. (2009) 'Mediterranean Tourism and Climate (Change): A Survey-Based Study', paper for *7th International Symposium on Tourism and Sustainability*, Eastbourne, UK, 8–10 July 2009.

—— and B. Amelung (2009a) 'Climate Change and Coastal & Marine Tourism: Review and Analysis', *Journal of Coastal Research* SI 56: 1,140-44.

—— and B. Amelung (2009b) 'Climate Change and Tourist Comfort on Europe's Beaches in Summer: A Reassessment', *Coastal Management* 37.6: 550-68.

—— and S. Becken (2009) 'A Climate Change Vulnerability Assessment Methodology for Coastal Tourism', *Journal of Sustainable Tourism* 17.4: 473-88.

——, B. Amelung and L. Santamarta (2008) 'Linking Beach Recreation to Weather Conditions: A Case Study in Zandvoort, Netherlands', *Tourism in Marine Environments* 5.2–3: 111-19.

Morgan, R., E. Gatell, R. Junyent, A. Micallef, E. Ozhan and A.T. Williams (2000) 'An Improved User-Based Beach Climate Index', *Journal of Coastal Conservation* 6: 41-50.

Ngazy, Z., N. Jiddawi and H. Cesar (2005) 'Coral Bleaching and the Demand for Coral Reefs: A Marine Recreation Case in Zanzibar', in M. Ahmed (ed.), *Economic Valuation and Policy Priorities for Sustainable Management of Coral Reefs* (Penang, Malaysia: WorldFish Center).

Nicholls, S. (2006) 'Climate Change, Tourism and Outdoor Recreation in Europe', *Managing Leisure* 11: 151-63.

Perry, A. (2005) 'The Mediterranean: How Can the World's Most Popular and Successful Tourist Destination Adapt to a Changing Climate?', in M. Hall and J. Higham (eds.), *Tourism, Recreation and Climate Change* (Clevedon, UK: Channel View Publications): 86-97.

—— (2006) 'Will Predicted Climate Change Compromise the Sustainability of Mediterranean Tourism?', *Journal of Sustainable Tourism* 14.4: 367-75.

Polsky, C., R. Neff and B. Yarnal (2007) 'Building Comparable Global Change Vulnerability Assessments: The Vulnerability Scoping Diagram', *Global Environmental Change* 17.3–4: 472-85.

Scott, D., S. Gossling and C.R. De Freitas (2008) 'Preferred Climates for Tourism: Case Studies from Canada, New Zealand and Sweden', *Climate Research* 38.1: 61-73.

Snover, A.K., L. Whitely Binder, J. Lopez, E. Willmott, J. Kay, D. Howell and J. Simmonds in association with ICLEI (2007) *Preparing for Climate Change: A Guidebook for Local, Regional, and State Governments* (Oakland, CA: ICLEI Local Governments for Sustainability).

UN (United Nations) (1948) 'The Universal Declaration of Human Rights'; www.un.org/en/documents/udhr/index.shtml#a13, accessed 8 March 2010.

UNEP (United Nations Environment Programme) (2009) *Sustainable Coastal Tourism: An Integrated Planning and Management Approach* (Paris: UNEP).
UNWTO (United Nations World Tourism Organization) (1999) 'Global Code of Ethics for Tourism' (Madrid: UNWTO).
—— (2001) *Tourism 2020 Vision: Global Forecast and Profiles of Market Segments* (Madrid: UNWTO).
—— (2009) UNWTO World Tourism Barometer 7.2; unwto.org/facts/eng/pdf/barometer/UNWTO_Barom09_2_en_excerpt.pdf, accessed 3 March 2010.
——, UNEP (United Nations Environment Programme) and WMO (World Meteorological Organization) (2008) *Climate Change and Tourism: Responding to Global Challenges* (Madrid: UNWTO).
WTTC (World Travel and Tourism Council) (2009) *Leading the Challenge on Climate Change* (London: WTTC).

4

Corporate responses to climate change

The role of partnerships

Ans Kolk and Jonatan Pinkse
University of Amsterdam Business School, The Netherlands

Lia Hull Van Houten
Van Houten Communications, USA

This chapter relates to the Netherlands Organisation for Scientific Research (NWO)-funded research project '"Getting down to business": Economic responses to climate change', which involved the first two authors.[1] It studied the (potential) contribution of business to climate change mitigation and adaptation, and how the realities of business can be taken into account in policy-making to help further common objectives. It built on earlier research that found that the position of mainstream oil, car and electricity-producing companies towards climate change has shifted considerably over the last decade (Kolk 2008; Kolk and Levy 2004). When government support for an international agreement in Kyoto turned out to be more widespread than initially expected, an increasing number of companies in these sectors stopped their opposition to measures for dealing with climate

1 The NWO project was carried out by Ans Kolk (project leader) and Jonatan Pinkse (post doc). In the specific study on partnerships for climate change reported in this chapter, Lia Hull Van Houten also participated, as part of her MBA internship and thesis, in the broader framework of a collaborative project by the research team at the University of Amsterdam Business School with Pleon, as reported on in the Pleon *Climate Change Stakeholder Report* (Pleon 2007).

change. What is more, even companies in sectors that do not depend on carbon-intensive fossil fuels to the same extent have also embraced climate change as a business issue, and there is now broad-based support for the position that climate change (policy) will substantially influence business operations. This set in motion a wave of corporate activities and initiatives to reduce emissions, through product and process improvements, exchange of technologies and expertise, and the exploration of new modes of governance such as emissions trading and cross-sectoral partnerships: that is, cooperation with other companies, government agencies and non-governmental organisations (NGOs) (Kolk and Pinkse 2004, 2005, 2007; Pinkse and Kolk 2009).

Although there are many positive signs regarding business and climate change, all these initiatives do not necessarily answer the question of what the corporate contribution is to climate change mitigation and adaptation. On the basis of our research we can say that there are many positive indicators regarding business and climate change, but that all the initiatives taken by companies do not (yet) translate into mitigation and adaptation as they are defined in the climate policy debate. Our conclusion is that for substantial organisational adaptation to occur (and for existing business activities to translate into real mitigation) concerted efforts are required. These efforts include policy steps and behavioural changes that relate to the sector- and firm-specific business realities that we have explored in our project. Challenges that arise include the fact that the impact of corporate mitigation efforts is difficult to assess and that business understands the concept of adaptation in a different way from how it is framed in the climate policy debate.

To start with mitigation, most efforts are directed at the reduction of greenhouse gases, particularly carbon dioxide. In the past few years, there has been great development in the implementation of a whole set of business practices such as emissions inventories, emissions reduction targets and carbon accounting for tracking and disclosing climate change-related information. However, owing to a lack of standardisation of these practices and the many options that companies have in choosing an approach that best fits their situation, it is unclear to what extent this leads to reliable and comparable information about the corporate impact on climate change (Kolk, Levy and Pinkse 2008; Pinkse and Kolk 2009). As a result, it remains a challenge to assess whether business is making progress in cutting emissions over and above what would have been achieved under a business-as-usual scenario.

Making sense of adaptation from a business perspective is a complicated endeavour. Although adaptation is interpreted in the climate change literature as the process of adjusting to the physical impact of climate change, this view is not wholly shared by the business community (yet). Of course there are examples of corporate initiatives aimed at adapting to physical impacts such as drought and extreme weather events by those companies active in the insurance, food production and oil and gas industries. Nevertheless, for the vast majority of organisations, adaptation instead means the process of adjusting business processes in response to climate change as an issue of societal concern and/or regulatory constraints. Our research shows that there are two ways in which corporate adaptation is currently unfolding.

The first, and the most economic response, is reflected in business activities to create 'climate-specific' capabilities, which move beyond a focus on green niche markets/customers only, and aim to create competitive advantage and a strategic reorientation (e.g. low-carbon technological trajectories and/or different products/services). Developing and marketing products and services that appeal to a climate-conscious market segment has become increasingly popular over the past few years. Still we can conclude that, as it currently stands, climate-induced capability development may lead to a more radical, competence-destroying reconfiguration of strategic capabilities for a few industries only; most companies stay relatively close to their current activities (Kolk and Pinkse 2008a). A strategic reorientation is most likely to occur in the oil and gas and automotive industries but will not happen in the short run. A reason for this is that companies in these sectors do not agree on the types of technology that will prevail in coming years, and most companies thus first invest in competence-enhancing transition technologies, thereby still relying on existing capability configurations.

Although this form of adaptation has the potential to have business move towards a low-carbon technological trajectory, there are still some significant challenges ahead with regard to further innovation and development of capabilities for climate change. For example, one challenge is whether companies opt to scale up existing technologies, considering the current lack of a 'silver bullet' solution. A related issue is how and in what way climate-friendly solutions can link to and/or build on companies' existing capabilities and further their competitive positions. For a profitable and sustained transition towards more climate-friendly and less carbon-intensive technologies that foster innovation, in the absence of viable markets and concomitant infrastructure, decisive policy steps are needed as well as behavioural changes.

The second way in which companies engage in organisational adaptation relates to this last point on the development of policies. Companies have played an active role in pushing for new ways of environmental governance and different policy modes (Prakash and Kollman 2004), particularly those that are more flexible and accommodative than the 'traditional' forms such as command-and-control regulation or a carbon tax. One of the relatively novel instruments that have become relevant is the widespread use of emissions trading as a way to curb emissions. Companies have played a large role in the development of the carbon market, as they have not just waited for governments to implement trading schemes (Kolk and Hoffmann 2007). That is to say, companies have not only tried to comply with new regulatory constraints, but have also chosen to respond strategically by avoiding them, using their bargaining power to influence actors that enforce new regulations, and acting in voluntary markets to stay ahead or profit from emerging opportunities (Kolk and Pinkse 2008b; Pinkse and Kolk 2007). Both compliance and voluntary carbon markets have generated a surge of corporate activities, even though it is uncertain how things will develop in view of ambiguity about the policy frameworks after 2012. Companies have clearly adapted to climate change by advocating for new, more flexible ways of complying with regulatory constraints on emissions. Although evidence suggests that the carbon price has been integrated in corporate decision-making, its impact on stimulating technological change and emissions reductions appears to be (much) more limited, because the carbon price has still not reached a level high enough to initiate such change.

More generally, we see that climate change has made companies aware of the fact that a more cooperative orientation towards government and civil society eases the process of alleviating public pressure. This cooperative stance is, for example, seen in the many partnership activities of companies with a range of other actors in government and society to address climate change, such as NGOs, national and local governments and other companies. For companies, which face regulatory uncertainty and the complexities of finding an appropriate approach towards a complicated global issue that requires broad involvement, engagement with various stakeholder groups supplements the range of other corporate activities already being undertaken (Hoffman 2005; Kolk and Pinkse 2005). Potentially such partnerships offer ample opportunities to explore options for change in different domains (technology, policy, behaviour, awareness) in a concerted effort. However, while partnerships have received considerable attention in the field of development and social issues more broadly (Kolk, Van Tulder and Kostwinder 2008; Selsky and Parker 2005; Waddock 1991), also considering company involvement, this has not been the case for climate change: there is not much insight into the extent and focus of company engagement in partnerships for climate change. Such information seems to be necessary in order to subsequently assess the pros and cons and the contribution of partnerships as one of the policy modes for addressing climate change.

Therefore, the remainder of this chapter will examine the range of partnerships in which leading companies are involved to shed more light on this topic. The research, in which all three authors were involved, identifies the current state by presenting data from a study on large companies that are part of the *Financial Times* Global 500, a list of the 500 largest companies in the world in terms of market value. In addition, it explores companies' motives and strategic benefits associated with different types of partnership, suggesting areas for further research, also related to limitations of partnerships from a societal perspective. Before moving to the specifics of climate change-oriented partnerships, we will first introduce the concept of partnerships for sustainable development more broadly. We will focus on some peculiarities of this debate that are also relevant for the analysis of partnerships for climate change.

4.1 Partnerships

Partnerships represent what Austin (2000) has called the 'collaboration paradigm of the 21st century' needed to solve 'increasingly complex challenges' that 'exceed the capabilities of any single sector' (see Warner and Sullivan 2004). They have received particular attention with regard to sustainable development, as a result of their inclusion in the Millennium Development Goals, in which a global partnership for development is listed as the eighth goal. At the 2002 World Summit on Sustainable Development (WSSD) partnerships were recognised as a crucial implementation mechanism for sustainable development, in order to make progress on the many ideas launched a decade earlier at the Rio conference that had failed to be translated into concrete measures. Partnerships in a sense aim to address different forms of 'governance' failure in a situation where governments, companies and NGOs are unable to unilaterally achieve desired public objectives, especially when it comes to complex global problems such as protection of the environment (Bäckstrand 2008; Biermann *et al.* 2007; Kolk, Van Tulder and Kostwinder 2008). They can also be seen as sources for new global rule-setting involving non-state actors where 'old' public governance is falling short and regulatory voids need to be filled (Braithwaite and Drahos 2000; Fransen and Kolk 2007).

Partnerships that focus on sustainability have recently been defined as 'collaborative arrangements in which actors from two or more spheres of society (state, market and civil society) are involved in a non-hierarchical process, and through which these actors strive for a sustainability goal' (Van Huijstee *et al.* 2007: 77). The notion is older though; already in the early 1990s, partnerships were, more broadly, conceptualised as 'the voluntary collaborative efforts of actors from organizations in two or more economic sectors in a forum in which they cooperatively attempt to solve a problem or issue of mutual concern that is in some way identified with a public policy agenda item' (Waddock 1991: 481-82). Both definitions highlight the fact that partnerships cut across sectors and involve non-hierarchical processes. Non-hierarchical means that partnerships are based on the idea of shared responsibility (Mazurkiewicz 2005) in which no single actor—for example, the government—regulates the behaviour of other actors. Peculiar to climate change is the complexity of the issue, which seems to require partnerships since one actor cannot solve it alone (Selsky and Parker 2005; Witte *et al.* 2003).

Another notable characteristic of a partnership, particularly as it is formulated by Waddock (1991), is the link to a public-policy agenda item. This raises the question of whether collaborative arrangements between companies concluded to this end could also qualify as a 'partnership'. One might suggest that collaborative firm-only activities could be classified as partnerships and are worthy of examination, though they do not meet the cross-sectoral criteria. For the latter aspect, this would thus mean a deviation from the definitions given above, because it does not involve cooperation with a non-business partner. On the other hand, the concept would then include the so-called 'post-partnerships', a term used by Egels-Zandén and

Wahhlqvist (2007) to refer to the observation that companies tend to prefer cooperating with other business partners after similar efforts in conjunction with NGOs had failed. While this leads to problems of categorisation, cooperation between companies for other than direct market objectives (to this latter the 'strategic alliance' label would apply) seems worthy of inquiry. Particularly in the case of climate change, which has become a rather prominent public-policy issue in many countries, this might be a new development that deserves attention.

In the literature, various types of partnership have been distinguished, looking at the nature of the actors involved (Kolk, Van Tulder and Kostwinder 2008; Selsky and Parker 2005).[2] The partnership form or arena (Selsky and Parker 2005) that has received most attention in the management literature is the one between companies and NGOs: the private–nonprofit partnership or social alliance (Austin 2000; Berger *et al.* 2004; Rondinelli and London 2003). Besides cooperating for broader societal objectives, organisations also have their own motivations to participate in partnerships. By joining forces, organisations may acquire access to 'critical competences' that they do not have individually (Selsky and Parker 2005). Thus a partnership may create advantages, such as greater learning opportunities (e.g. improving employees' interpersonal, technical or reflective skills), increased social capital, access to partners' networks and a better ability to attract, motivate and retain employees (Austin 2000; Kolk, Van Tulder and Kostwinder 2008; Selsky and Parker 2005). For companies, there is also the possibility of enhancing the corporate image or brand reputation and hence boosting sales, preventing potentially negative public confrontations and tapping into new markets (Elkington and Fennell 1998).

There has also been recent interest in collaboration between government (agencies) and companies: the public–private partnership. Particularly in the environmental arena, companies have been cooperating with governments in other ways before, via so-called voluntary agreements. The main distinguishing factor is that in the case of voluntary agreements responsibility for implementation mainly rests with the companies involved under the aegis of the government, while in partnerships this responsibility is shared equally between participants (Mazurkiewicz 2005; OECD 1999). Because of the assumed equality between the actors involved in a partnership, compared to a voluntary agreement, participation will generally be less risky for companies, because the threat of regulation is much smaller. What is more, partnerships can be formed around a relatively narrow topic (Waddock 1991), in the case of climate change, for example, focused on the development of a specific emissions-reducing technology such as biofuels or hydrogen technology. Further, one company may be involved in many different partnerships at the same time. Voluntary agreements, on the other hand, are generally formed around broad

2 We leave public–nonprofit partnerships aside here as the chapter focuses on the ones that involve business. Interestingly, in most of the partnerships included in the WSSD database, business is notably absent, presumably due to the development focus in which public–nonprofit partnerships have long been dominant.

themes such as energy efficiency or greenhouse gas (GHG) emissions reduction for the whole company, and participation typically precludes involvement in other similar initiatives.

The third type of partnership is one in which companies cooperate with actors in both government and society (private–public–nonprofit); such tripartite or multi-stakeholder partnerships are frequently seen as the best way to deal with multifaceted problems in the current epoch. In view of the complexity of climate change it can be seen as requiring cooperation across sectors (and countries) with stakes for all partners as they share a common goal of resolving the issue. However, participants in multi-stakeholder partnerships may also have more strategic motives in mind. Although partnerships are seen as a way for different actors to bundle their knowledge and resources, this is not necessarily merely with the (sole) objective of solving the problem (Waddock 1991). For companies it may well be a means to learn new skills, acquire tacit knowledge that partners possess and share costs. Besides, by working with NGOs or governments, companies typically gain a valuable resource, that is, the reputation that these partners have in the eyes of the public regarding their positive influence on sustainability (Van Huijstee *et al.* 2007). Partnerships can also reduce risks related to climate change, which can be regulatory, reputational, commercial or financial in nature (Innovest 2002; Wellington and Sauer 2005).

On the other hand, partnerships are not always without risk, as, for example, NGOs can draw on business to acquire financial resources and take advantage of corporate skills to create a more general market for sustainable products (Van Huijstee *et al.* 2007). For example, Greenpeace used the near-bankrupt German refrigerator manufacturer Foron Household Appliances to its own advantage to launch ozone-friendly 'Greenfreeze' refrigerators. At first, this led to a successful cooperation for both participating actors. However, after the collaboration Greenpeace basically gave away the ozone-friendly technology to Foron's competitors, thereby destroying Foron's strategic advantage, in the end leading to its bankruptcy (Stafford *et al.* 2000).

To shed some further light on the background, it seems helpful to also consider the focus of partnerships. The partnership literature has distinguished partnerships in terms of the objectives and concomitant relationship: philanthropic, transactional (a specific value transaction) or integrative/strategic (Austin 2000), or, in the case of partnerships for development, of being merely focused on a specific country/activity (micro); at a sector or supply chain (meso), or broad, covering multiple issues and countries/regions (macro) (Kolk, Van Tulder and Kostwinder 2008). These categorisations do not appear to add much for analysing climate change as partnerships target a specific issue, without philanthropy being a main component. The political nature of the climate change debate also makes the setting more complex. In their categorisation of sustainability partnerships involving companies, Steger *et al.* (2009) included some of these peculiarities, distinguishing between quasi-regulation, advocacy, new business and best practices (see also Salzmann *et al.* 2008).

Based on these insights, it might be suggested that companies are engaging in partnerships for climate change with a variety of foci, which could even be seen as forming a continuum of reactive, proactive and innovative ways. A more defensive context may be cases where they work with NGOs on emissions compliance or to avoid litigation. However, partnerships have served more than a reactive purpose, as companies seek to pre-empt regulation with voluntary initiatives, also together with industry groups or NGO partners. Partnerships have also enabled many companies to enter into proactive or exploratory industry-wide or cross-sector dialogues in terms of understanding implications of climate change trends. They have housed innovation in market and technology development.

Hence, companies have sought, through a range of partnership formations, to actively influence policy, reduce emissions, launch new products, undertake research into climate change opportunities and raise public awareness (Hull 2007; Kolk and Pinkse 2004; Pinkse and Kolk 2009). These categories encompass some of the elements also covered by Salzmann *et al.* (2008) in their study of nine climate partnerships, but we focus more on the concrete target (e.g. emissions reduction instead of best practice; policy influence rather than advocacy/quasi-regulation) and also include public education and research partnerships (with research and product launch having a broader reach than their new business category). Obviously, our analysis serves a broader purpose given that we seek to uncover the variety of partnerships in which Global 500 companies are involved.

It is to this range of partnerships that we now turn, presenting exploratory results to obtain more insight into numbers, types and particularities of partnerships in which leading companies are involved. So far, limited attention has been paid to climate change partnerships, let alone offering insight beyond a few individual cases.

4.2 Sample and research method

To obtain insight into partnership involvement by leading companies we analysed the partnership activities of Global 500 companies that have reported their climate change activities to the Carbon Disclosure Project (CDP). We took the fourth CDP survey, the findings of which were released in September 2006, as our starting point. Even though information about participation in partnerships was not explicitly requested by the CDP, an overwhelming number of companies mentioned collaborative efforts of such kind. In addition, we also obtained and verified information about the responding companies from annual and sustainability reports, websites, press coverage and other independent publications in the period July–November 2007. In this way, we were able to identify 183 companies (81 US, 81 European and 21 Asia–Pacific companies, and covering a range of industries) that were involved in a total of 222 different climate change partnerships.

It should be noted that it is near to impossible to come up with a complete list of partnerships, as this is a very dynamic area (Waddock 1991): new partnerships are launched all the time, but many also die a slow death within a few years. Since the objective of this chapter is to obtain insight into the phenomenon with an eye to exploring the strategic dimensions and role of partnerships more broadly, we consider the approach followed as appropriate for the purpose. The data collected was used to identify particularities for each partnership, through a careful content analysis of the available information. Referring to the previous section, this entailed first an examination of the partners involved (business, governmental and NGOs) in order to assess the type (public–private, private–nonprofit, tripartite). Second, we empirically identified the main focus of the partnership, in which we distinguished policy influence, emissions reduction, research, product launch and public education.[3] We give some examples for each of these, and also explore possible motivations for corporate participation that can point to areas for further research.

4.3 Discussion of findings

Tables 4.1 and 4.2 summarise the partnerships for climate change found for the companies that we analysed. The columns list the partnerships according to their main foci, the rows the types of partner. Table 4.1 has a more extensive overview of the partners than Table 4.2, which gives the types of partnership: private–nonprofit, public–private and tripartite. In Table 4.1 we distinguish the collaborations including business, government, NGOs and universities. The last category we added particularly as they turned out to be important participants, mostly in research-oriented partnerships but also in a few others. In Table 4.2, they were included in the nonprofit category, as their main objective is neither profitability (as in the case of companies) nor regulation, the predominant function of government targeted in the partnership literature.

Unlike Table 4.2, Table 4.1 also contains partnerships between companies as we found some that did not resemble the more traditional strategic alliances. Whether or not these activities can be accurately classified as 'partnerships' in the original meaning of the definition is worthy of discussion, as indicated in the previous section. Nevertheless, we thought it worthwhile to list them separately, because partnerships between companies account for a substantial amount of activity and may represent a new trend.[4] When companies work with other companies as partners

3 Some partnerships had multiple goals, but in these cases we classified them according to their primary focus as outlined in the partnership documentation and those of the partners.

4 As product-launch partnerships in which only companies work together seem to be strategic alliances rather than partnerships with a public-policy goal, this category does not exist in Table 4.1.

Table 4.1 **Types of partner and focus of partnerships**

	Focus of partnership					
Type of partner	**Public education**	**Emissions reduction**	**Research**	**Product launch**	**Policy influence**	**Total**
Business		15	5		8	28
Business, government	2	17	27	10	9	65
Business, government, NGO	3	6		2	6	17
Business, government, NGO, university		2	7	1	6	16
Business, government, university		2	11		2	15
Business, NGO	6	14	4	1	10	35
Business, NGO, university			3	2	3	8
Business, university			36	1	1	38
Total	11	56	93	17	45	222

Table 4.2 **Types and focus of partnerships**

	Focus of partnership					
Type of partnership	**Public education**	**Emissions reduction**	**Research**	**Product launch**	**Policy influence**	**Total**
Private–nonprofit	6	14	43	4	14	81
Public–private	2	17	27	10	9	65
Tripartite	3	10	18	3	14	48
Total	11	41	88	17	37	194

in this way, this is often mediated by a business association, such as the World Business Council for Sustainable Development, to prevent potential allegations of collusive behaviour.

In Table 4.2, these activities between companies only have been removed, thus staying in line with the partnership definition included in most of the literature (except for Egels-Zandén and Wahhlqvist 2007). Out of the 194 remaining partnerships, the largest number is between companies and nonprofit organisations, the predominant type studied by management and marketing scholars in the past few years, followed by public–private partnerships. Interestingly, a considerable portion (48 in total) is tripartite in nature, which indicates that partnerships involving private, public and nonprofit partners seem more prevalent in the climate change area than, for example, in the case of partnerships for development (Kolk, Van Tulder and Kostwinder 2008).

If we look at the types of partner that companies chose to work with in the partnerships that we identified, it is interesting to note that government is involved in many of them (113 in total) as one of the partners. This suggests that there is some value in the argument that partnerships are a new governance form that replaces or at least supplements government regulation. There are other factors at play as well, however, as we saw that universities are also frequent partners: in 77 partnerships (around one-third). This is around the same number as the more 'traditional' NGOs, which participate in 74 partnerships (obviously partnerships can have multiple partners). The involvement of research partners suggests that many companies are looking for expertise outside their own organisation. To close this 'knowledge gap', they seem to tap into the climate change-specific knowledge of research institutes and universities to remain ahead of the curve in technological development.

As to focus, tripartite partnerships aim more often at policy influence and less at research than average (this is the other way around for private–nonprofit partnerships), while public–private partnerships pay more attention to emissions reduction and product launch. Overall, it is interesting to see that the largest number of partnerships (over 40%) have research as their main focus, followed at some distance by emissions reduction and policy influence. The relatively limited share (20%) of the policy-influence partnerships may have to do with the fact that the lobbying objective is disguised in the descriptions of goals. However, it appears more likely that the low number of policy-influence partnerships reflects the trend from more antagonistic approaches to more cooperative climate measures. While antagonistic measures prevailed in the 1990s, in the current epoch corporate attempts are more focused on shaping the policy set-up and taking specific steps to find innovative solutions that address processes and products (Kolk and Pinkse 2007; Pinkse and Kolk 2009). If we add up emissions-reduction, product-launch and research partnerships, these represent 75% of the partnerships, with policy influence and public education accounting for the remainder. Below we will discuss the several categories in some more detail with a few examples, and also explore companies' motives.

4.3.1 Policy-influence partnerships

Policy-influence partnerships are generally broad forums in which many companies participate on a multilateral basis. Examples include the Global Roundtable on Climate Change, the Pew Center's Business Environmental Leadership Council, Earthwatch's Corporate Environmental Responsibility Group, The Climate Group and the United States Climate Action Partnership (US-CAP). Companies form broad coalitions to present a united front *vis-à-vis* policy-makers and to increase their political clout (Kolk and Pinkse 2007). There are also collaborative arrangements where only companies participate such as the UK Corporate Leaders Group on Climate Change.

Policy-influence partnerships are the clearest example of 'political partnerships' through which companies try to play a role in domestic politics as well as intergovernmental policy-making on climate change. There are several ways in which they try to do this. One is to use the partnership as a voice to express corporate opinion on climate change. This generally occurs by putting advertisements in newspapers to assure the public that they are taking climate change seriously and are developing measures to tackle the problem and by making public statements to this end. For example, before the G8 meeting in Gleneagles in 2005, the UK Corporate Leaders Group on Climate Change, which includes BP, Scottish Power, Shell, ABN/AMRO and Cisco, wrote a 'public' letter to Britain's then prime minister Tony Blair, arguing that:

> At present, we believe that the private sector and governments are caught in a 'Catch 22' situation with regard to tackling climate change. Governments tend to feel limited in their ability to introduce new policies for reducing emissions because they fear business resistance, while companies are unable to take their investments in low carbon solutions to scale because of lack of long-term policies. In order to help break this impasse, we are proposing to work in partnership with the Government . . .[5]

By making such public statements companies attempt to steer policy-makers in the direction of their most-favoured policy types. Not surprisingly, this typically involves market-based policies. For example, US-CAP has produced a report entitled 'A Blueprint for Legislative Action' that 'lays out a blueprint for a mandatory economy-wide, market-driven approach to climate protection'.[6] Similarly, companies use these partnerships to lobby governments. For example, the US-CAP report served not only as a public statement, but also as a means to lobby the Bush government to implement an emissions trading scheme in the US. Likewise, US utility Public Service Enterprise Group (PSEG) states that it is

> a founding member of a utility-sector coalition known as the 'Clean Energy Group' (CEG) which is actively lobbying the Bush Administration and Congress for a fixed cap on domestic utility-sector GHG emissions to be implemented through an emissions trading program similar to the U.S. program for controlling utility-sector sulphur dioxide emissions.[7]

4.3.2 Emissions-reduction partnerships

Emissions-reduction partnerships have a political function as well, but in addition they seem to be used to seek legitimacy *vis-à-vis* a wider range of stakeholders.

5 www.cpi.cam.ac.uk/leaders_groups/clgcc/uk_clg/2005_letter_to_uk_pm.aspx, accessed 9 March 2010.
6 www.us-cap.org, accessed 27 November 2009.
7 PSEG's response to the Carbon Disclosure Project 2; www.cdproject.net, accessed 7 March 2010.

Besides the government and NGOs, this may include the investment community and the public at large. By voluntarily committing to an emissions-reduction goal, a company can demonstrate living up to a specific norm when it comes to contributing to tackling climate change in a positive way (Selin and VanDeveer 2007). If a company proves to be successful in reducing GHG emissions owing to joining a partnership, in addition to creating reputational benefits, this also places further pressure on competitors not participating in climate change-oriented partnerships. In other words, these successes demonstrate that action on climate change is possible and can be achieved in a cost-effective way (Selin and VanDeveer 2007). Although companies do set reduction targets unilaterally as well, one of the main reasons that many emissions-reduction programmes occur in conjunction with government agencies, NGOs or business consortia is that it gives political meaning to the commitment. Business is generally seen as the main source of GHG emissions and not as part of the solution. Consequently, companies become involved in an emissions-reduction partnership with the aim of changing the adversarial relationship they used to have and thus relieve pressure felt from NGOs and/or governments (Rondinelli and London 2003; Spar and La Mure 2003; Yaziji 2004).

In the case of governments, this means that companies try to pre-empt new regulation that will force business to take account of the issue. For example, under the larger umbrella of the US government's public voluntary programme, Climate Vision, many industry associations have set up initiatives to reduce GHG emissions and often also improve the energy efficiency of their member companies. This includes the American Petroleum Institute Climate Action Challenge (participants include Anadarko Petroleum, ConocoPhillips, Marathon Oil and Occidental Petroleum), the Electric Power Industry Climate Initiative (in which Southern Company is involved), and the Business Roundtable's Climate RESOLVE programme (a partnership with around 160 member companies). In a similar vein, Korean electronics company Samsung has participated in Energy Saving through Partnership, a partnership of the Korean government with the aim of sharing best practices. By the same token, business–NGO partnerships are a way for companies to avoid being victims of more rigorous tactics from NGOs to persuade companies to change their behaviour, such as boycotts, advertisements or sabotage (Yaziji 2004). Examples of NGO-led emissions-reduction programmes in which companies participate are WWF's Climate Savers and Electricité de France's Partnership for Climate Action.

Emissions-reduction partnerships do not exclusively focus on companies' own production activities. There are several initiatives that strive for GHG emissions reductions in the supply chain, transportation and distribution of raw materials and end-products, and the commercial fleet. An example of this is the participation of US pharmaceutical Abbott in the GreenFleet pilot programme, a combined initiative of Environmental Defense and PHH Arval, a company that manages commercial fleets. Notwithstanding the political importance of many emissions-reduction partnerships and their value in seeking legitimacy, a large number also has a market function. That is, next to setting a norm, joint emissions-reduction programmes also act as a way to create more understanding of and develop methodologies for appropriate measurement of GHG emissions. This can lead to considerable cost reductions and enables sharing of best practices between companies that have joined the same initiative. However, how cost-effective and efficient such voluntary partnerships are and whether they indeed create best-practice transfers is a question that has not yet been extensively researched. While the reduction target itself is often unambiguous, to what extent the partnership creates a context in which all participants share knowledge and transfer best practices to facilitate companies in reaching this target is open to discussion.

4.3.3 Public-education partnerships

On the surface, climate change partnerships with a public-education goal are launched to seek legitimacy with specific stakeholder groups, that is, consumers and the public at large. Some of these partnerships, for example those that focus on education projects, show close resemblance to community projects, common in the corporate social responsibility arena. To illustrate, one long-lasting education partnership is the NEED project, which tries to make American students aware of issues around energy, including how using and producing energy has an effect on the environment and society. Several multinationals, including American Electric Power, BP, Chevron, Devon Energy, Duke Energy, Halliburton, PG&E, Royal Dutch/Shell and Schlumberger, take part in this project. An education project like this not only helps companies to gain or maintain legitimacy, but clearly also serves market-oriented goals: in this case developing a young workforce interested in energy issues and raising consumer awareness.

Actually, most public-education partnerships have a role in aiding companies to anticipate a market transition induced by climate change. This can work in different ways. The partnership can be used to strengthen the strategic position in existing markets. For example, Unilever's initiative to set up the Climate Change College with WWF not only educates the public by means of Climate Change Ambassadors trained through the project, but also has a positive influence on the brand image of Unilever's ice cream brand Ben & Jerry's (which organises the Climate Change College and communicates the initiative to the outside world). More common, however, is the case where a public-education partnership is used to create a new market. Although we could identify only 11 partnerships, the vast majority aims to promote sustainable consumption, thereby focusing attention on a particular product group of a company.

For example, in 2005 Home Depot Canada launched, together with the Clean Air Foundation, the Energy Smarts campaign to promote energy efficiency best practices and energy savings upgrades among consumers. It basically attracts consumers' attention to energy-saving products, and thus seems to function as a marketing tool for Home Depot. In 2007, a related and more comprehensive programme—Eco Options—was launched company-wide, but this time not in the form of a business–NGO partnership. Similarly, HBOS cooperates with WWF-UK to support the development of its One Million Sustainable Homes Campaign. The commercial opportunity of this partnership for HBOS is that 'as part of the Sustainable Housing Project, there are plans to develop a range of energy efficiency packages that can be offered to customers in conjunction with their home information package (HIP) and/or mortgage, and provide a series of environmental communication messages to customers'.[8]

4.3.4 Research and product-launch partnerships

Compared to the other foci, partnerships aimed at research and product launch are more directly related to innovation for climate change to contribute to finding solutions to climate change in this way. Both types are generally used to develop and bring to market specific climate-friendly technologies. One technology around which several partnerships have been formed in the past few years is carbon capture and storage (CCS). This technology is attractive for energy-intensive companies as it is one of the only possible 'end-of-pipe technologies' available to reduce carbon emissions, which means that it does not necessitate a change in raw material inputs (i.e. towards non-fossil-fuel-based). However, it is also extremely capital intensive because it requires setting up a new infrastructure to direct carbon dioxide (CO_2) from production sites to underground reservoirs, and cooperation therefore also has the purpose of risk sharing and obtaining (financial) support, usually from government agencies.

A typical example of a research partnership for carbon capture and storage technology is the CO_2 Capture Project. In this partnership several oil companies, including BP, Chevron, Eni, Norsk Hydro, Suncor, Shell and ConocoPhillips, cooperate with the US Department of Energy, the European Commission and the Research Council of Norway 'to develop new, breakthrough technologies to reduce the cost of CO_2 separation, capture, and geologic storage from combustion sources such as turbines, heaters, and boilers'.[9] Besides developing the technology, this partnership has also published a study on public perceptions of CCS, thus trying to improve broader acceptance of this technology. Similar partnerships include the US Department of Energy's Regional Carbon Sequestration Partnerships and the Cooperative Research Centre for Greenhouse Gas Technologies (CO2CRC). In the CO2CRC not only oil companies participate—BP, Shell, Chevron, ConocoPhillips and Schlumberger—but also some coal companies—Rio Tinto, BHP Billiton and Anglo American—as well as universities and international government agencies.

Sustainable mobility and more specifically fuel-cell technology are also a focus of many partnerships, but these do not merely focus on developing hydrogen technologies as many aim to take the first steps to launch hydrogen as a new fuel. To illustrate, the Japanese Ministry of Trade Economy and Industry has set up the Japan Hydrogen Fuel Cell Demonstration Project because they believe that 'devel-

8 HBOS's response to the Carbon Disclosure Project 2; www.cdproject.net, accessed 7 March 2010.

9 www.co2captureproject.org, accessed 7 March 2010.

opments of technologies of fuel cells and hydrogen stations are nearing commercialisation step-by-step'.[10] Another example is the Clean Energy Partnership, an initiative of the German Ministry for Transport which, together with several car, oil and electricity producers such as BMW, DaimlerChrysler, Ford, GM/Opel, Total and Vattenfall, tries to demonstrate that hydrogen can be used for transportation purposes. A first step to launch hydrogen as a fuel was taken by BP and Praxair, which launched the first hydrogen fuelling station at Los Angeles Airport together with the local government. Other technologies around which partnerships have been formed for their development and marketing are coal gasification (for example, the US Department of Energy Integrated Gasification Combined Cycle with Praxair and Southern Company), the installation of solar panels (for example, Exelon with City of Chicago), and methane and F-gases (for example, PG&E initiative for methane recapture with the City of São Paulo).

However, not all partnerships focus on one specific technology as some aim to develop a wider range of climate-friendly technologies simultaneously. An example is the Carbon Mitigation Initiative in which Ford and BP cooperate with Princeton Environmental Institute. In addition, many other partnerships are unclear about which specific technologies they try to pursue, instead stating a general orientation on reduction of GHG emissions or enhancement of energy efficiency. Finally, some research or product-launch partnerships are not in the technological domain at all, but have been set up to develop particular financial products linked to climate change. Prominent examples include the launch of several (carbon) funds that have been set up to invest in GHG reduction projects often combined with the acquisition of credits under the Clean Development Mechanism and Joint Implementation. Examples include the California Clean Energy Fund and Start Green.

While most examples given above involve partnerships in which multiple companies cooperate with one or more other nonprofit or government partners, more often than not, companies become involved on a unilateral basis. This has to do with the focus of these types of partnership, which predominantly aim at developing new technologies or products that directly or indirectly contribute to a reduction of GHG emissions. As it frequently involves developing strategically valuable knowledge on specific technologies, a unilateral approach is understandable because this makes it much easier to create a first-mover advantage based on deployment of a new and climate-friendly technology.

4.4 Conclusions

While there continues to be considerable uncertainty about the future of post-Kyoto climate policy, our study suggests that partnerships have become an impor-

10 www.jhfc.jp/e, accessed 7 March 2010.

tant part of companies' climate change approaches and thus also one of the policy modes that can be distinguished. The findings show that companies are active in collaborative efforts with a range of other actors in government and society to address climate change, often engaging in multiple partnerships that can have different foci. The popularity of partnerships may have to do with the strategic benefits they offer to companies in particular. Companies can use partnerships to influence the direction and shape of the climate change debate. Unlike the political activities in the 1990s when antagonistic approaches prevailed, partnerships serve to show that companies are willing and able to work cooperatively with other actors on this issue. Besides, companies are becoming aware that not only regulatory but also societal attention to climate change is increasing. As a consequence, partnerships can play a role in anticipating corporate loss of legitimacy in the eyes of the public by demonstrating concrete action rather than only taking a position in the policy debate. Climate change has at last started to affect markets in which companies operate. It induces a market transition that works against carbon-intensive products and production processes, thus stimulating the development of more climate-friendly technologies. While corporate involvement in partnerships can be linked to political, legitimacy-seeking and market-oriented intentions, what driver(s) will be most important, and in which cases, deserves further study. This also applies to the role of partnerships in companies' overall climate change strategies.

In more practical terms it can be said that partnerships offer a considerable degree of flexibility in scope, resource commitment, intensity of involvement and firm-specificity. They allow companies to increase involvement in climate change via cooperative measures from low to high commitment of resources and expertise, flexibility to join existing societal-level dialogues on climate change and policy channels as well as to instigate highly individualised product development schemes intimately aligned with core capabilities. Partnerships also provide opportunities to share risk in researching consequences and innovations in market and technology development related to climate change. These advantages explain the popularity of climate change partnerships. It is unclear, however, to what extent companies are able to reap (assumed) benefits. It might be a worthwhile topic for follow-up research, also considering what role the specificities of the partnerships, the partners and the companies play in this regard. It is also not yet clear whether (and which) companies (deliberately) develop a portfolio of climate change partnerships, for example in relation to their individual (core) activities (Kolk, Van Tulder and Kostwinder 2008). This would require a more in-depth investigation at the firm-level of decision-making processes surrounding engagement in particular existing initiatives and/or helping to set up others.

From a policy and societal perspective, and thus also as input for the (possible future) partners involved in these collaborative agreements, the activities and outcomes rather than intentions deserve further attention. As reputational and issue management considerations seem to play a clear role for companies, it is a critical question whether and to what extent partnership involvement has real substance (and goes beyond what some might label as 'greenwashing'). However, as the partnership phenomenon is rather novel, it is difficult to speak about effectiveness. Compared to voluntary agreements, where assessing effectiveness has not been an easy issue either and research is still under way (Khanna and Ramirez 2004; Morgenstern and Pizer 2007; Welch *et al.* 2000), it is even more complicated to assess this for climate change partnerships. The reason is that partnerships, unlike voluntary agreements, usually do not have a clear, defined target. Many partnerships are not directly aimed at emissions reduction, but have a range of 'softer' targets such as the development of a specific technology or persuading policy-makers to move in a certain direction. It is therefore rather difficult to establish the success of a partnership. An indicator that might be used to analyse effectiveness is to look at the lifetime of a partnership and the satisfaction of participants with the way it has been functioning, also in comparison to the original aims (Kolk, Van Tulder and Kostwinder 2008). At this moment, a straightforward assessment is complicated by the wide range of objectives and partners, the multitude of corporate, sector and country settings, as well as issues related to definitions and boundaries. While still a mostly unexplored field of research, the prominence that partnerships have gained over the last few years in the climate change arena begs for more attention in the years to come.

References

Austin, J.E. (2000) 'Principles for Partnership', *Leader to Leader* 18: 44-50.

Bäckstrand, K. (2008) 'Accountability of Networked Climate Governance: The Rise of Transnational Climate Partnerships', *Global Environmental Politics* 8.3: 74-102.

Berger, I.E., P.H. Cunningham and M.E. Drumwright (2004) 'Social Alliances: Company/Nonprofit Collaboration', *California Management Review* 47.1: 58-90.

Biermann, F., M.S. Chan, A. Mert and P. Pattberg (2007) 'Multi-stakeholder Partnerships for Sustainable Development: Does the Promise Hold?', paper presented at the *Conference on the Human Dimensions of Global Environmental Change*, Amsterdam, 24–26 May 2007.

Braithwaite, J., and P. Drahos (2000) *Global Business Regulation* (Cambridge, UK: Cambridge University Press).

Egels-Zandén, N., and E. Wahhlqvist (2007) 'Post-partnership Strategies for Defining Corporate Responsibility: The Business Social Compliance Initiative', *Journal of Business Ethics* 70.2: 175-89.

Elkington, J., and S. Fennell (1998) 'Partners for Sustainability', *Greener Management International* 24: 48-60.

Fransen, L., and A. Kolk (2007) 'Global Rule-setting for Business: A Critical Analysis of Multi-stakeholder Standards', *Organization* 14.5: 667-84.

Hoffman, A.J. (2005) 'Climate Change Strategy: The Business Logic behind Voluntary Greenhouse Gas Reductions', *California Management Review* 47.3: 21-46.

Hull, L. (2007) 'Multinational's Response to Climate Change; Focus on Multi-stakeholder Partnerships' (MBA thesis, University of Amsterdam Business School).

Innovest (2002) *Value at Risk: Climate Change and the Future of Governance* (Boston, MA: CERES).

Khanna, M., and D.T. Ramirez (2004) 'Effectiveness of Voluntary Approaches: Implications for Climate Change Mitigation', in A. Baranzini and P. Thalmann (eds.), *Voluntary Approaches in Climate Policy* (Cheltenham, UK: Edward Elgar): 31-66.

Kolk, A. (2008) 'Developments in Corporate Responses to Climate Change within the Past Decade', in B. Hansjurgens and R. Antes (eds.), *Economics and Management of Climate Change: Risks, Mitigation and Adaptation* (New York: Springer): 221-30.

—— and V. Hoffmann (2007) 'Business, Climate Change and Emissions Trading: Taking Stock and Looking Ahead', *European Management Journal* 25.6: 411-14.

—— and D. Levy (2004) 'Multinationals and Global Climate Change: Issues for the Automotive and Oil Industries', in S. Lundan (ed.), *Multinationals, Environment and Global Competition* (Oxford, UK: Elsevier): 171-93.

—— and J. Pinkse (2004) 'Market Strategies for Climate Change', *European Management Journal* 22.3: 304-14.

—— and J. Pinkse (2005) 'Business Responses to Climate Change: Identifying Emergent Strategies', *California Management Review* 47.3: 6-20.

—— and J. Pinkse (2007) 'Multinationals' Political Activities on Climate Change', *Business & Society* 46.2: 201-28.

—— and J. Pinkse (2008a) 'A Perspective on Multinational Enterprises and Climate Change: Learning from an "Inconvenient Truth"?', *Journal of International Business Studies* 39.8: 1,359-78.

—— and J. Pinkse (2008b) 'The Influence of Climate Change Regulation on Corporate Responses: The Case of Emissions Trading', in R. O'Sullivan (ed.), *Corporate Responses to Climate Change: Achieving Emissions Reductions through Regulation, Self-regulation and Economic Incentives* (Sheffield, UK: Greenleaf Publishing): 53-68.

——, D. Levy and J. Pinkse (2008) 'Corporate Responses in an Emerging Climate Regime: The Institutionalization and Commensuration of Carbon Disclosure', *European Accounting Review* 17.4: 719-45.

——, R. Van Tulder and E. Kostwinder (2008) 'Partnerships for Development', *European Management Journal* 26.4: 262-73.

Mazurkiewicz, P. (2005) 'Corporate Self-regulation and Multi-stakeholder Dialogue', in E. Croci (ed.), *The Handbook of Environmental Voluntary Agreements* (Dordrecht, Netherlands: Springer): 31-45.

Morgenstern, R.D., and W.A. Pizer (eds.) (2007) *Reality Check: The Nature and Performance of Voluntary Environmental Programs in the United States, Europe, and Japan* (Washington, DC: RFF Press).

OECD (Organisation for Economic Cooperation and Development) (1999) *Voluntary Approaches for Environmental Policy: An Assessment* (Paris: OECD).

Pinkse, J., and A. Kolk (2007) 'Multinationals Enterprises and Emissions Trading: Strategic Responses to New Institutional Constraints', *European Management Journal* 25.6: 441-52.

—— and A. Kolk (2009) *International Business and Global Climate Change* (London: Routledge).

Pleon (2007) *Pleon Climate Change Stakeholder Report: Multi-stakeholder Partnerships in Climate Change* (Amsterdam: Pleon).

Prakash, A., and K. Kollman (2004) 'Policy Modes, Firms, and the Natural Environment', *Business Strategy and the Environment* 13: 107-28.

Rondinelli, D.A., and T. London (2003) 'How Corporations and Environmental Groups Cooperate: Assessing Cross-sector Alliances and Collaborations', *Academy of Management Executive* 17.1: 61-76.

Salzmann, O., U. Steger and A. Ionescu-Somers (2008) 'Climate Protection Partnerships: Activities and Achievements', in R. O'Sullivan (ed.), *Corporate Responses to Climate Change: Achieving Emissions Reductions through Regulation, Self-regulation and Economic Incentives* (Sheffield, UK: Greenleaf Publishing): 151-67.

Selin, H., and S.D. VanDeveer (2007) 'Political Science and Prediction: What's Next for U.S. Climate Change Policy?', *Review of Policy Research* 24.1: 1-27.

Selsky, J.W., and B. Parker (2005) 'Cross-sector Partnerships to Address Social Issues: Challenges to Theory and Practice', *Journal of Management* 31.6: 849-73.

Spar, D.L., and L.T. La Mure (2003) 'The Power of Activism: Assessing the Impact of NGOs on Global Business', *California Management Review* 45.3: 78-101.

Stafford, E.R., M.J. Polonsky and C.L. Hartman (2000) 'Environmental NGO–Business Collaboration and Strategic Bridging: A Case Analysis of the Greenpeace–Foron Alliance', *Business Strategy and the Environment* 9: 122-35.

Steger, U., A. Ionescu-Somers, O. Salzmann and S. Mansiourian (2009) *Sustainability Partnerships: The Manager's Handbook* (Basingstoke, UK: Palgrave Macmillan).

Van Huijstee, M.M., M. Francken and P. Leroy (2007) 'Partnerships for Sustainable Development: A Review of Current Literature', *Environmental Sciences* 4.2: 75-89.

Waddock, S.A. (1991) 'A Typology of Social Partnership Organizations', *Administration and Society* 22.4: 480-515.

Warner, M., and R. Sullivan (eds.) (2004) *Putting Partnerships to Work: Strategic Alliances for Development between Government, the Private Sector and Civil Society* (Sheffield, UK: Greenleaf Publishing).

Welch, E.W., A. Maz and S. Bretschneider (2000) 'Voluntary Behavior by Electric Utilities: Levels of Adoption and Contribution of the Climate Challenge Program to the Reduction of Carbon Dioxide', *Journal of Policy Analysis and Management* 19.3: 407-25.

Wellington, F., and A. Sauer (2005) *Framing Climate Risk in Portfolio Management* (Washington, DC: CERES/World Resources Institute).

Witte, J.M., C. Streck and T. Benner (2003) 'The Road from Johannesburg: What Future for Partnerships in Global Environmental Governance?', in T. Benner, C. Streck and J.M. Witte (eds.), *Progress or Peril? Networks and Partnerships in Global Environmental Governance: The Post-Johannesburg Agenda* (Berlin: Global Public Policy Institute): 59-84.

Yaziji, M. (2004) 'Turning Gadflies into Allies', *Harvard Business Review*, February 2004: 110-15.

5

Energy conservation in Dutch housing renovation projects

Thomas Hoppe, Hans Bressers and Kris Lulofs
Centre for Studies in Technology and Sustainable Development (CSTM), The Netherlands

5.1 Background and problem definition

In the synthesis volume of its Fourth Assessment Report, the Intergovernmental Panel on Climate Change (IPCC) made it clear that the effects of climate change could become catastrophic if no action is taken to reduce greenhouse gas emissions as soon as possible. The report also boldly concluded that global warming is to a large extent caused by human activities (IPCC 2007). In order to achieve substantial cuts in greenhouse gas emissions, governments are attempting to mitigate energy consumption in several major sectors of the economy.

The built environment is one such man-made sector, and one that theoretically provides ample opportunity for significant energy conservation. The application of such technical measures as insulation and innovative, high-yield heating systems means that the energy efficiency levels of dwellings can be dramatically improved. In the Netherlands the built environment is responsible for 34% of total greenhouse gas emission. Of this emission, 56% comes from the residential sector (VROM 2005). Greenhouse gas emissions in the residential sector are primarily caused by decentralised combustion of gas (in houses) and combustion of coal in power plants. In Dutch houses gas is used for space heating, water heating and cooking. Electricity is used for lighting and domestic appliances. The housing stock in the Netherlands is generally old. The energy quality of these old houses is dramatically poorer than those that have been built more recently. To a large extent this is because legislation on energy efficiency was only implemented after 1975. Before that time, there were no standards that prescribed insulation and the installation of high-yield condensing boilers (De Jong *et al.* 2005). Since 1975, regulation of the energy quality of new houses has gradually become more ambitious, even though it only impacts houses in their construction phase. Resolving the problem by replacing old houses with new ones is seriously hampered by the fact that the annual housing turnover ratio is less than 1% (CBS 2008). If a substantial reduction of greenhouse gas emissions in the residential sector is to be achieved, effective measures should be targeted on the existing housing stock.

This is not exactly hot news; as early as the year 2000 the Dutch Ministry of Housing, Spatial Planning and the Environment (Ministerie van Volkshuisvesting, Ruimtelijke Ordening en Milieubeheer—VROM) proclaimed that a reduction in greenhouse gas of approximately two-thirds should be achieved in the existing housing stock (VROM 2000). Technically speaking, adequate solutions are available to solve the problem. Domestic energy conservation up to a level of 90% is currently feasible (Trecodome 2008). Unfortunately, it is the owners and occupiers who decide whether the application of such technical measures is desirable. When owners or occupiers consider renovating their homes they rarely prioritise energy efficiency, especially when energy costs are but a small part of the total cost of living (Lulofs and Lettinga 2003; SenterNovem 2005; Sunnika 2001). Moreover, the owners and occupiers have other needs, such as comfort, health and return on investment. Thus, owners and occupiers need to be pushed towards those alternatives that benefit the climate. However, in local settings it is far from easy in practice to implement policies on energy efficiency and the climate. This implies that efforts need to be aimed at local stakeholders to negotiate trade-offs. These efforts can influence the decision-making of house owners and occupiers.

In this chapter, the central question we seek to answer is the variation in the energy conservation achieved in renovation projects on existing housing sites. The chapter's particular focus is housing locations in post-war neighbourhoods, which are characterised by relatively low-value houses, predominantly owned by semi-public housing associations. We seek the explanation in six factors: the influence exercised by policy instruments, the influence exercised by local governments, the influence exercised by housing associations, the influence of collaborative efforts between actors, cognitive cohesion between actors, and the influence of contextual factors. The central research question is analysed by applying both qualitative and quantitative research methods in a comparative design. Our analysis is relevant in the context of the urgent policy challenge of meeting the 2020 Dutch climate mitigation policy goal of a 30% reduction in energy consumption.

Our analysis is reported as follows: Section 5.2 presents a literature review of instruments and programmes that have been applied in the past to influence energy conservation in the existing housing stock. The review includes national and local policy programmes and instruments, and international experience is also included. Next, Section 5.3 describes the arena, presenting a list of the main actors, their interests and resources, and the ways they interact with each other. Section 5.4 presents the theoretical framework. The research design and methodology are summarised in Section 5.5, and Section 5.6 reports the results of the analysis. Section 5.7 reports the conclusions of the empirical study, and looks at the position of this research both in relation to the Vulnerability, Adaptation and Mitigation (VAM) research programme and in the context of Dutch and European Union policy.

Because of this present chapter's focus on policy implementation at the local level and the role played by local authorities, we would like to draw attention to Chapter 8 in this book.

5.2 Literature review of energy conservation in the existing housing stock

Since the first oil crisis in 1973, many programmes have been implemented to attempt to conserve energy in the residential sector. This has permitted comprehensive studies on the effects of programmes and several autonomous instruments, often in quasi-experimental designs. Here we present what is known about the effectiveness of policy mixes as applied in the Netherlands and internationally, followed by our analytical observations.

5.2.1 Policy programmes

National energy conservation policy programmes often contain a 'policy mix' that comprises different instruments in which economic incentives are facilitated by communicative instruments, such as information campaigns, workshops, education and home audits (NIP 1988; Clinch and Healy 2000; Balthasar 2000; Lulofs and Arentsen 2001). Programmes have often been characterised by decentralised sub-programmes and the incorporation of economic policy objectives to stimulate employment. In general, policy objectives were formulated ambitiously, but were realised only in part. Moreover, many problems were observed in policy programmes aimed at energy conservation in the residential sector. In the Netherlands, unfavourable macroeconomic developments and government change were the main causes of discontinuity in a large-scale energy conservation programme that lasted from 1978 to 1987 (NIP 1988). Deteriorating economic prospects led to a decline both in programme budgets and in the purchasing power of house owners and occupiers, and hence a decrease in the uptake of domestic appliances that would further energy conservation. Another problem concerned the sudden increase in market demand, which both suppliers and contractors experienced because they had to adapt rapidly. This proved problematic in Ireland (Clinch and Healy 2000) and comparable problems were also experienced in the Netherlands (NIP 1988).

Little evidence exists about the actual effectiveness of national and state policy programmes. In general, little data has been collected that precisely measures programme effectiveness. For instance, the impact of the federal Residential Conservation Service programme on national energy consumption in the United States was very limited and varied across states. While estimates of programme savings showed some consistency across states and utilities, the variation in estimates of programme cost-effectiveness was very great. Moreover, many of the inputs to a cost-effectiveness analysis, such as fuel prices and programme costs, turned out to be highly site-specific (Hirst 1984). An evaluation of government programmes to encourage home energy conservation in the 1970s—through such means as tax deductions, regulation, energy audits and weatherisation programmes[1]—showed that the effects were marginal in size and minimal in impact. Most of the decline in consumption has been achieved through a combination of reduced living standards and coincidental changes in lifestyles that lowered the demand for energy. These changes included a shift of the population from cold to warmer regions of the country, a trend towards smaller households, and a decline in real household income (Frieden and Baker 1983). It is also useful to mention that when follow-up data collections were included (after the initial programme had been terminated), it appeared that the positive effects of the intervention were not maintained (Abrahamse *et al.* 2005: 282). Furthermore, interventions might actually lead to adverse effects in the energy consumption by households. For instance, the 'rebound effect' holds that households might increase their energy consumption in reaction to expected lower energy costs (Goetschel *et al.* 1995; Sorrell 2007).

There is also evidence that policy programmes can have dangerous side effects. Some programmes that relied on the increase of energy prices (by taxation) to stimulate energy conservation have been held responsible for serious hazards to health and safety. For instance, 96 deaths in the states of Maine and Vermont in 1979 were attributed to wood-stove fires. The wood stoves were used by households as a cost-efficient alternative to safer oil heating systems. Another programme-related threat to health involved the large-scale use of urea formaldehyde foam, a popular insulation material. Only after it had been applied on many sites was it identified as a cause of cancer (Frieden and Baker 1983).

The households' approach turned out to form a considerable difficulty in the programmes. This was due to information gaps, the unwillingness of household members to discount long-term benefits, and a lack of acceptance of the high transaction costs involved in investments in energy conservation in one's home (Clinch and Healy 2000). It is notable that it was especially difficult to approach low-income households (Clinch and Healy 2000; Van der Waals *et al.* 2003). In this regard it is worth mentioning that traditional evaluation methods were found to be subject to considerable bias in the selection of the experiment group members

1 Weatherisation refers to modifying a dwelling to reduce energy consumption and optimise energy efficiency. In the US weatherisation is part of an energy consumption reduction policy. Local weatherisation service providers install energy efficiency measures in the homes of qualifying homeowners free of charge. The target group is low-income households (DOE 2009).

because they clearly favoured 'upscale' households (e.g. high-income, high-education, owner-occupied single house). An analysis that took this bias into account led to the result that the amount of energy saved that could be attributed to the policy programmes was found to be considerably less than had traditionally been believed (Hirst 1984; Hartman 1988). In a price-driven economy, there is evidence that the poor are especially responsive to home energy conservation programmes. Strikingly, one programme evaluation showed that even when measures involved spending money, as in the case of conservation investments, the poor still outranked all other income groups. By contrast, programmes aimed at higher-income groups showed rather the opposite: the greater awareness of the better-educated, higher-income group was not linked to greater efforts to reduce their own energy use. The lower- and middle-income group, less knowledgeable about energy matters, felt harder hit by price increases in the 1970s and outdid the high-income group in their attempts to reduce energy consumption (Frieden and Baker 1983).

5.2.2 Analytical elaboration on factors explaining the degree of energy conservation in the existing housing stock

On the basis of a national programme evaluation in the United States, Frieden and Baker (1983) mentioned four necessary conditions for a house owner to be persuaded to invest rationally as a result of market stimuli. First, the house owner has to have some basis on which to estimate the return. Second, the house owner must find energy costs sufficiently onerous to warrant taking some action. Third, the house owner must have adequate financial assets to be able to invest in home improvements. Fourth, the house owner will need a credible source of advice. These factors present some formidable barriers to the operation of an efficient market in conservation investments, such as lack of financial resources, uncertainty about returns on investment and uncertainty about the nature of the technological fix needed. Furthermore, some lessons can be learned from the policy programmes about how to approach households. A household's assessment of investment in measures to reduce energy use depends on: the ability to carry out home repairs, past experiences of shortages, and awareness of a social norm or conservation ethic. In addition, neighbourhood networks could prove to be effective channels for the diffusion of energy conservation practices (Leonard-Barton and Rogers 1979).

Information alone is not considered an effective strategy. It only becomes effective when other instruments become involved, such as subsidies, loans and feedback. Stern *et al.* (1986) argue that in order to reach low-income households one needs a combination of strong financial incentives and information campaigns, as well as careful design and tailoring of the actual implementation. The argument that a combination of instruments is necessary does not only apply to the policy field of energy conservation in the existing residential sector. Other environmental policy domains also provide evidence that policy instruments are seldom implemented independently. Rather, they are implemented together with other (types of) instruments and even seem to be effective only when they form part of a policy mix (Bressers and O'Toole 2005).

Finally, it should be mentioned that the studies on the instruments and programmes that were designed for energy conservation suffer from several methodological shortcomings. For instance, studies that used a survey design addressed small numbers of households. The poor statistical power gave little reason for valid statistical generalisation. Another pitfall is that only the effects of autonomous instruments were studied, rather than combinations of instruments (Abrahamse *et al.* 2005). Although many studies have shown that a combination of strategies is generally more effective than applying one single strategy, few have been conducted to see which strategies actually contributed to the overall effect. More systematic research on the effectiveness of interventions under various circumstances would be advisable in this respect (Abrahamse *et al.* 2005). Structural measurement of the key dependent variable (energy consumption) also proved difficult.

Furthermore, the majority of studies were limited to households as units of observation. Housing estates or even entire cities have only rarely been mentioned in journal articles. In other words, a valid glance at the 'real-life' process has been neglected. This becomes clear when one considers that only a few studies include case study designs (e.g. Hirst 1987; Van der Waals 2001; Van der Waals *et al.* 2003). The case studies that have been conducted have been overwhelmingly descriptive and not really useful for systematic analysis, especially since no data on the dependent variable was reported. Nor have any international comparative studies been conducted thus far. In general, the field lacks an analytical body that can be used for theoretical elaboration of factors that influence levels of energy conservation in natural local contexts.

5.3 The institutional context in the Netherlands

In order to grasp the environment in which the energy efficiency of current houses can be improved it is necessary to gain some insight into the roles of the local actors involved, their interests, the resources they possess and exchange, and the ways in which they interact. Opportunities for large-scale energy conservation in the current housing stock lie in large-scale renovation projects in relatively old, post-war neighbourhoods. The houses and their environments are often characterised by poor-quality, obsolete physical construction. An additional characteristic is that the poor-quality buildings are accompanied by a poor-quality social structure. Neighbourhoods are characterised by a high degree of unemployment, above-average crime rate and a high proportion of ageing residents. The population on average also has a relatively low socioeconomic status. Renovation projects are primarily meant to improve both social and physical structures in neighbourhoods. Energy conservation is considered no more than a secondary objective in that endeavour. The houses in the neighbourhood are for the greater part owned by one or more former public or semi-public housing associations. The housing associations manage the houses with the public objective of delivering quality housing for consumers who do not have the means to buy houses themselves. Until 1995 housing associations in the Netherlands were public or semi-public institutions, largely financed by central government. In 1995 they were liberalised, receiving financial decision-making autonomy. However, they have maintained their key public task of providing quality housing to those who cannot afford to buy their own homes (Koffijberg 2005).

Considerable decision-making is involved when a large-scale neighbourhood or building-block renovation plan is being scheduled. Covenants are often used to cover agreements of intent between local governments and housing associations. Local governments are able to exercise influence and encourage the uptake of energy conservation appliances by making trade-offs with housing associations, with a strategic use of urban renewal subsidies and legal permits. However, the local authorities remain firmly dependent on the willingness of housing associations to cooperate. Housing associations have the most significant resources because they own the housing stock and have the financial reserves to make the investments required. Moreover, in renovation projects, legal consent is required from the tenants who live in the houses. The legal standard holds that at least 70% of the tenants must approve the renovation project plans. The legal approval rate gives the tenants some room to negotiate with their housing association. It is not surprising, therefore, that housing associations take great pains to persuade their tenants to approve their (and the local authority's) plans. However, local governments and tenants have few means to negotiate with housing associations in order to encourage them to install technical appliances that significantly improve energy efficiency in the houses. The power imbalance is key to the housing association's advantage. In the end the housing associations decide whether or not and how much to invest in energy efficiency (Hoppe and Lulofs 2008).

Parts of the post-war neighbourhoods also contain private house owners. The owner-occupiers are often former tenants of housing associations that sold them their houses in the years prior to the renovation project. When renovation projects are scheduled where many owner-occupiers reside in the neighbourhood, the housing association(s) and municipality are often inclined to have them participate in the project. Compared to the public-housing occupants, the owner-occupiers can only participate if they decide to invest their own funds (housing associations make the investments for their tenants and are often only compensated by a small monthly rent increase, if they are compensated at all). Loans and mortgages are often so high that low-income house owners have problems acquiring them. This means that access to loans and mortgages represents a serious barrier to persuading house owners to invest and participate in the neighbourhood renovation project (Clinch and Healy 2000). Even when national government offers additional means to further encourage this group, the actual effect is marginal. In short, several institutional barriers exist that prevent the large-scale adoption of technical appliances to stimulate energy efficiency in existing housing (Hoppe and Lulofs 2008).

5.4 Theoretical framework

Several theoretical insights are useful to show us how to perceive and explain the phenomenon of energy conservation in the existing housing stock. These theoretical insights originate from different disciplines, such as environmental economics and environmental psychology, studies in diffusion of innovation and policy studies. The last two fields are especially useful thanks to their emphases on the application of innovative measures and their diffusion in local settings.

Insights from diffusion of innovation studies allow us to look into the processes that underlie the dissemination and acceptance of innovative concepts in social communities (e.g. Burt 1987; Granovetter 1973, 1978; Rogers 1962). The acceptance and adoption of innovative measures is necessary to approach a sustainable society, which also involves the replacement of fossil fuels by sustainable alternatives. This turns out to be difficult because conventional technologies, such as those surrounding fossil fuels, are 'locked in' by means of a cluster of socially accepted system factors that represent barriers to innovative alternatives, such as sustainable energy carriers (Unruh 2000). The focus in diffusion of innovation studies—and to a lesser degree sociotechnical studies—traditionally lies on the supply side of the market and *getting* processes of diffusion and change going, as compared to the demand side of the market and *keeping* a process of diffusion and change going. The process of diffusion is more complicated because early-market customers have already adopted the concept, whereas mainstream-market customers still need to be convinced to adopt the concept. As Bressers (1989) has stated, it is very difficult to convince the majority of potential adopters. The exemplary minority has already been convinced. Conventional behaviour and the existence of institutional barriers (such as sectoral policies) limit further adoption. A facilitating institutional setting is considered a precondition for continuing the process of acceptance. Several strategies exist that encourage the process of acceptance, some of which have become part of policy strategies and instruments. Such incentives are implemented broadly in settings where they have to deal with serious setbacks when competing with several constraints that have their backgrounds in traditional policy domains, such as housing and spatial planning. This means that successful implementation of policy instruments aimed at the diffusion of innovative or sustainable energy appliances is seldom self-evident.

Policy implementation studies look at factors that explain the effectiveness of policy implementation and its products. Implementation studies originate from the 1970s and are characterised by a broad range of theoretical developments (O'Toole 2000). During the 1990s attention became centred around a few topics, such as 'policy networks' (Marsh and Rhodes 1992; Bressers 1993; Dowding 1995; Smith 1997; Klijn 1996; Börzel 1998; Bressers and O'Toole 1998), 'network management' (De Bruijn and Ten Heuvelhof 1995; Kickert *et al.* 1997), and the prospect that the horizontal 'governance' model was to replace the hierarchic-traditional 'government' model (Bressers and Kuks 2003). In order to encompass the broad continuum of theoretical developments in environmental implementation studies,

Bressers (2004, 2009) developed the contextual interaction theory, which assumes that the choice and implementation of policy instruments depend on the cognition, motivation and resources of local actors, the distribution of power between them, and the way they interact with each other in a local policy arena. Furthermore, the theory lays a strong emphasis on contextual factors. It also holds that environmental policy is seldom prioritised in the list of preference held by local actors in the local context.

The study presented here uses many elements of the contextual interaction theory. The relevance of the theory for the domain of energy conservation in existing housing sites is that it involves the implementation of a type of environmental policy, in this case as an incentive to stimulate energy conservation. The contextual interaction theory facilitates a systematic analysis of environmental policy implementation processes.

The insights presented in the literature review of previous energy conservation policies led us to choose an approach that applies multiple theoretical points of view. We prefer a multi-theoretical approach to a mono-theoretical one because we assume that a multi-theoretical approach leads to a larger explained variation. It is therefore useful to name several clusters of independent variables in order to test them at a later stage. We aim to discover which cluster of independent variables deliver the most powerful explanations. We present a graphical view of our research model in Figure 5.1.

Figure 5.1 **Research model**

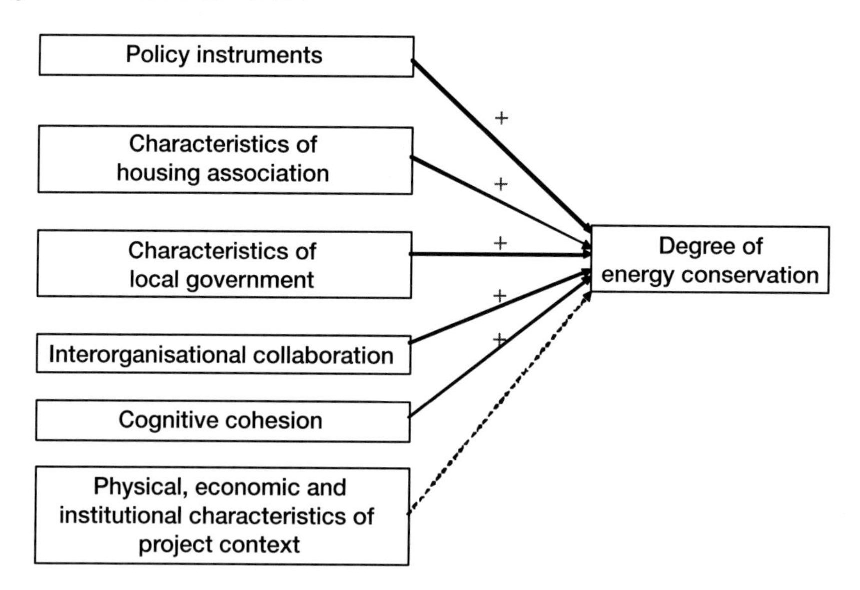

All variables clusters are subdivided by a number of different items. These items are used as indicators for the specific explanatory model of the particular independent variable. The six clusters of independent variables concern: (1) the use of policy instruments in the domain of energy policy, (2) characteristics of local governments, (3) characteristics of housing associations, (4) interorganisational collaboration between actors, (5) cognitive cohesion, and (6) physical, economic and institutional characteristics of the project context. This last cluster was added to the research model as a contextual component next to the other variables, which are more theoretically oriented. Without specific knowledge of the project context it is useless to analyse the outcome of policy implementation processes. Below we present an overview of the main hypotheses in the research model. The hypotheses concern the main propositions in the analytical framework. Since the main independent variables might be operationalised as scales constructed of a number of indicators, the sub-set items are mentioned, too.

- The greater the number of policy instruments in the climate policy domain that are being implemented in the local project arena, the better the degree of energy conservation in the renovated housing stock is likely to be. The variable comprises the following items: the presence of local or regional energy conservation covenants; the use of subsidy schemes; and the use of communicative policy instruments
- The more the organisational characteristics of local governments favour energy conservation, the better the degree of energy conservation in the renovated housing stock is likely to be. The variable comprises the following items: orientation towards environmental problems; the presence of formal energy conservation policy; personal capacity; the presence of advocates of energy conservation in housing; the degree of organisational tuning; the political orientation of the local officials; the size of budget for the subject as provided by national government; and the size of the municipality
- The more the organisational characteristics of housing associations favour energy conservation, the better the degree of energy conservation in the renovated housing stock is likely to be. The variable comprises the following items: orientation towards environmental problems; the presence of formal energy conservation policy; personal capacity; the presence of advocates of energy conservation in housing; the degree of organisational tuning; the financial position (company capital); and the size of the housing stock owned
- The more interorganisational collaboration efforts that are undertaken, the better the degree of energy conservation in the renovated housing stock is likely to be. The variable comprises the following items: the presence of opinion leaders; the frequency of visits to professional meetings on the subject; and the size of the project configuration over time
- The more cognitive cohesion that exists between organisations, the better the degree of energy conservation in the renovated housing stock is likely to be. The variable comprises the following items: cohesion towards environment

and sustainable development; cohesion towards technological innovation adoption; and cohesion towards the national climate policy strategy

- Contextual factors are used as control variables; therefore, the dotted line is used in Figure 5.1. The 'project context' cluster comprises the following control variables: division of ownership rights in houses on site; total investment per house; lengthening of the exploitation term per house; type of heating system; distance to city heating facility; equilibrium in supply and demand in the market for public housing; initial energy quality of houses on location; type of house; number of houses to be renovated on site; address density; degree of public participation in the project; and degree to which energy conservation policy is institutionalised in the project's management structure

5.5 Research design and methodology

In this research the dependent variable is the degree of energy conservation realised at the end of renovation projects at existing housing sites. The cases relate to substantial renovation projects at existing housing locations and the research domain is the Netherlands. The sample is not randomly selected. The number of cases (11) is too small for formal statistical generalisation; however, due to the case selection method—cases most resembling the population of interest with regard to background variables—it is useful to explore whether it is possible to generalise to a sample of 33 housing locations, present in a national monitored dataset, the EPL monitor developed by SenterNovem, the national energy agency (SenterNovem 2007).[2] EPL stands for energy performance on location. SenterNovem aimed to select representative sites. For the criterion 'geographical spread over the country' the agency did indeed succeed. We raised our ambition and checked for significant differences in 14 background variables. The result was that no significant dissimilarities were found between the sample of 11 cases and the sample of sites monitored by national government.[3]

2 SenterNovem is mandated by VROM to implement climate policy measures.

3 The following variables were checked for significant differences in means between sample and sites in national monitor dataset: size of the municipality ($F = .841$; d.f. = 32; $p = .366$); local authority's financial support provided by national government ($F = .022$; d.f. = 32; $p = .882$); size local authority's urban renewal budget ($F = 3.863$; d.f. = 29; $p = .059$); number of houses on site ($F = .172$; d.f. = 32; $p = .682$); ambition formulated for energy performance of site ($F = .436$; d.f. = 28; $p = .515$); share of houses on site to be newly built ($F = .464$; d.f. = 32; $p = .501$); number of houses on site to be newly built ($F = 2.402$; d.f. = 32; $p = .131$); type of energy provision ($F = .000$; d.f. = 32; $p = 1.000$); local authority's score on national sustainable development index ($F = .059$; d.f. = 29; $p = .810$); local authority's collaboration efforts towards local actors ($F = 1.289$; d.f. = 29; $p = .266$); political orientation

5.5.1 Data collection

The data collection encompassed the collection of different kinds of data. When the study started, only quantitative data was available from the previous study on ambition-setting and energy conservation on existing housing sites (Hoppe *et al.* forthcoming). Moreover, data on the dependent variable—achieved energy conservation—was not yet available. After the case selection stage and the pilot study stage were finalised, contact was made with people involved in the housing sites of interest. We partially applied the 'snowballing' method to contact other key people in the cases. Subsequently, dates were set for in-depth interviews, and 40 on-site interviews and 30 telephone interviews were conducted. Additional documentation on the cases was partially traced before the interviews were conducted, but also after access was provided by the interviewees. Project documentation comprised formal policy documents, advisory reports, annual reports, specific information papers, websites, feasibility studies and geographical maps of project locations.

The group of interviewees predominantly featured people from the following professions: project manager at the housing association, project leader in the local authority (urban renewal, property development), civil servant dealing with environmental or energy/climate affairs in the local authority, or energy associate at the housing association. The high incidence of these professions was beneficial to the researcher for three reasons: (1) most interviewees were involved in decision-making in the projects of interest on the subject of energy conservation, (2) they were often involved in the project for relatively long periods, which made them very knowledgeable and experienced, and (3) they possessed good networks with many contacts that were of interest to the researcher. Finally, it is noteworthy that most interviewees were males in the age category 40–50, with higher education (most frequently in civil engineering).

5.5.2 Data treatment

The quest for comparison of 11 cases meant that analysis by qualitative means alone was out of the question. The number of cases required a predominantly quantitative analysis. Data treatment was of great importance to the comparative analysis.[4]

First of all, the interview recordings were written down in transcription reports. The decision was made to do this in near-literal transcription reports in order to make full use of the richness of the data collected. After data collection, transcription reporting, supplementing ambiguities in data sources and story lines,

---

of local officials ($F = 3.242$; d.f. = 32; $p = .081$); mean financial value of dwellings on site ($F = .242$; d.f. = 32; $p = .626$); address density on site ($F = 2.555$; d.f. = 32; $p = .120$); participation of site in national urban renewal programme '56 *wijken*' ($F = 1.292$; d.f. = 32; $p = .264$). The confidence level used was 95%.

4 This does not imply that we regard data treatment for qualitative analysis as unimportant.

case histories were reconstructed. After finishing the case histories the phase of quantification of qualitative data began (however, much quantitative data was already present from the cases). Ten-point scales were constructed and scores were assigned per case. In this way a data matrix was created, which meant that careful attention had to be paid to case histories and case-specific data in order to fill in the data reliably. A code document was designed for construct-validity reasons and the reliable assignment of scores. In order to carry out this process in a trustworthy manner, all score assignments were accommodated with textual argumentation. Subsequently, the score assignment was replicated to check reliability and consistency.

Because the scores on the dependent variables were not clear per case, data needed to be collected, which required the collection of building-related technical data on specific construction elements in renovation activities that influence energy performance in houses. Following that, specific software was required to calculate energy performance. The software program used (OEI 2.1) had also been used in the calculation of feasibility reports during the planning stages. It had been developed in 2003 by SenterNovem to help local actors conduct energy audits of their specific locations. The software assumes average energy consumption per household based on monitoring data with a stratified sample. Variation in energy consumption between residential sites is therefore based on the technical measures that have been applied in the houses at the sites. Therefore, behavioural phenomena—such as the 'rebound effect'—are outside the scope of this study.

The software program uses the indicator EPL to measure energy performance of residential sites. The EPL indicator applies a 10-point scale, where '10' expresses high energy efficiency and '0' low energy efficiency, although scores can even become negative when very low energy efficiency is encountered (Hoiting *et al.* 2004). The reliability of calculations for the 11 sites of interest was assessed by discussing extreme outcomes with persons involved with the local projects.

5.5.3 Data analysis

Data analysis in the comparative research design was characterised by phasing, the use of different types of research method and the use of different types of data. The analysis featured both qualitative and quantitative methods in order to compare the cases. Qualitative and quantitative methods were used to compensate, meaning that 'mixed methods' were applied, a methodology that derives from an epistemologically pragmatic stance (Johnson and Onwuegbuzie 2004). The objective of applying both qualitative and quantitative methods in comparative research is, among others, to aim to confirm analytical results (triangulation), improve the researcher's interpretation and optimise the sample.

Because a comparative analysis of 11 cases with qualitative data only was impossible, the decision was made to conduct a quantitative analysis. This required the quantification of qualitative data and the creation of a data matrix by adding already quantitative data. This treatment made multiple regression analysis possible, although the small number of cases corresponds to poor statistical power and therefore little opportunity for statistical generalisation. The small number of cases also meant that a confidence interval of 90% was used for correlation and multiple regression analysis, and that a very limited number of variables (three) were allowed in the regression model. The regression analysis 'forward' method was applied to take the small number of cases into account. The multivariate analysis made it possible to provide insight into the relative contribution of each independent variable. However, before the regression analysis could be conducted, careful attention needed to be paid to the independent variables that were to be used in the regression model. Therefore, information on correlations is provided in the sections prior to the presentation of the results from the multiple regression analysis (sub-sections 5.6.2 and 5.6.3).

Scaling was applied to calculate the scores per independent variable (cluster) and also to allow for data reduction.[5] Before scaling, bivariate correlations were checked. The tests we applied for bivariate correlations were one-tailed, because we assumed we know the direction of the correlations from the hypotheses mentioned earlier in Section 5.4. To take care of construct validity and statistical validity, scales were designed that had to meet the criteria of Cronbach's alpha (Cronbach 1951): theoretical intersubjectivity between items and a reliability score larger than 0.5. In essence, this particular test has to be carried out every time a scale consists of multiple items (Carmines and Zeller 1979). Furthermore, scales were allowed to be constructed only when at least some of the items correlated significantly and were predicted in the right direction following the previous analysis with bivariate correlations.

5.6 Analysis and results

This section presents the results of the comparative analysis. We have chosen to present the results in stages. In the first place, an overview is created by presenting descriptive statistics, involving means, extremes, range, standard deviations and skewness of distribution. Furthermore, important data per case are presented in an inter-case matrix. Subsequently, the results of the multiple regression analysis are presented. To start the overview of the results, a geographic map of the Netherlands is presented in Figure 5.2, which contains the locations of the sites studied.

5 Data reduction here refers to limiting the number of independent variables. This was necessary in order to run a multivariate regression analysis, given the limited number of cases.

Figure 5.2 **Map of the Netherlands with the locations of the sites studied**

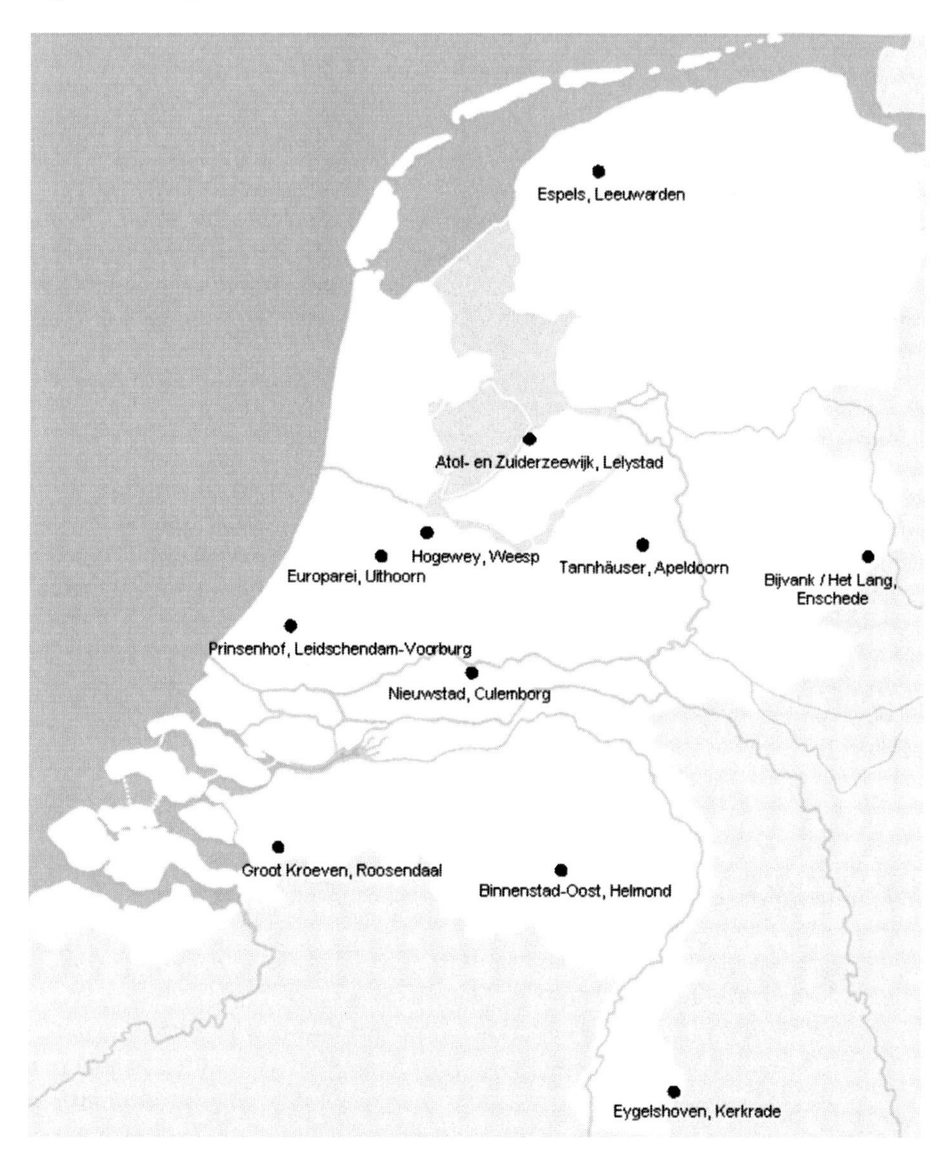

5.6.1 Descriptive statistics

An overview of important numerical data per case is presented in Table 5.1. The data includes the number of houses renovated, the type of house, the ambition set for energy conservation (in EPL-scale points), the scale of the energy performance achieved (in EPL-scale points), the actual energy conservation achieved (in EPL-scale points) and the relative energy conservation achieved (as a percentage). Table

5.1 is structured in descending sequence according to the category 'relative energy conservation achieved'.

Table 5.2 gives an overview of descriptive data according to nine key categories of the 11 sites of interest. Two types of descriptive statistics are of immediate interest: the average energy conservation achieved, and the difference between energy conservation ambitions and actual achievements. The average energy conservation is 39.7%. The site with the smallest amount of energy conservation achieved showed a value of 26.5% (Atol- en Zuiderzeewijk). The site with the greatest amount of energy conservation achieved a value of 69.8% (Groot Kroeven). In the latter case the innovative concept of passive renovation (the renovation variant to passive housing) had been applied, a technology that features extreme insulation standards and the use of passive solar energy. The technique has only rarely been applied in the Netherlands, most often in renovation locations. Furthermore, a difference was found between the energy performance enhancement aimed for (2.39 in EPL-scale points) and actually achieved (1.74). It is worth mentioning that the level of energy conservation actually achieved was lower than the targets previously set. This could indicate that ambitions do not predict project outcomes adequately, and targets can be considered overly optimistic.

On average the sites featured 440 houses under renovation. The location with the smallest number of houses under renovation featured 100 houses (Tannhäuser), whereas the largest site featured 1,628 houses under renovation. The standard deviation is rather large (459), due to the fact that the distribution of the category is very right-asymmetric (2.071). Most cases are close to the mean, whereas one site (Prinsenhof) has many houses. The differences in types of house are also worth mentioning. On seven sites the houses were family houses built between the 1930s and the 1970s. On four sites the dwellings were apartment buildings, built in the 1960s. When selecting the cases it was not possible to pay attention to variance in the type of housing. Nonetheless, the variance analysis did not show significant difference from analysis of a larger sample of housing sites according to the type of house. The average total investment per house was quite high at €62,383 (given that €100,000 is sufficient to build a new family house).[6] The investments ranged between extremes of €25,000 and €105,078. This too is quite a variation. The last category in Table 5.2 refers to exploitation term lengthening.[7] This category is part of the table for the following reason: lengthening the exploitation term is often used by housing associations as a means to compensate for less profitable investments (equipment to encourage energy efficiency is often considered to lie in this category). Exploitation term lengthening was 34 years on average, which is more than half a lifetime of a house in the Netherlands.

6 Investments here refer to housing renovation in general, and not specifically to energy conservation.

7 Lengthening the exploitation term of houses enables high investment to be recovered within the term. This method is sometimes applied as a means to stimulate investments in measures that increase the level of energy efficiency in houses.

Table 5.1 **Key numerical data per site**

#	Name of site	Name of town	Number of dwellings	Dwelling type	Target energy performance enhancement (in EPL-scale points)	Achieved energy performance enhancement (in EPL-scale points)	Degree of energy conservation	Height of achieved energy performance (in EPL-scale points)
1	Groot Kroeven	Roosendaal	246	Family house 1960s	0.13	2.44	69.8%	7.06
2	Eygelshoven	Kerkrade	300	Family house 1950s	1.50	2.40	51.1%	5.85
3	Europarei	Uithoorn	635	Apartment 1960s	4.30	2.20	50.2%	4.99
4	Prinsenhof	Leidschendam-Voorburg	1,628	Apartment 1960s	3.50	2.14	43.8%	4.60
5	Hogewey	Weesp	258	Apartment 1960s	1.50	1.40	35.0%	5.00
6	Espels	Leeuwarden	117	Family house pre-World War II	–	1.55	34.0%	5.48
7	Binnenstad-Oost	Helmond	121	Family house pre-World War II	2.20	1.73	32.9%	5.18
8	Tannhäuser	Apeldoorn	100	Apartment 1960s	2.70	1.39	32.9%	4.77
9	Bijvank het Lang	Enschede	854	Family house 1970s	–	1.70	30.5%	6.40
10	Nieuwstad	Culemborg	200	Family house 1970s	3.30	1.26	30.1%	5.13
11	Atol- en Zuiderzeewijk	Lelystad	380	Family house 1960s	–	0.92	26.5%	5.54

Table 5.2 **Descriptive statistics**

	N	St. mean	Minimum	Maximum	Range	Standard deviation	Skewness
Energy conservation target (in EPL-scale points)	8	2.3913	0.13	4.30	4.17	1.33803	-0.309
Achieved energy conservation (in EPL-scale points)	11	1,7391	0.92	2.44	1.52	0.49758	0.047
Achieved relative energy conservation (in %)	11	39.72	26.50	69.80	43.30	12.88	1.430
Height of achieved energy performance (in EPL-scale points)	11	5.4545	4.60	7.06	2.46	0.73741	1.174
Number of renovation houses	11	440.09	100	1,632	1,532	458,72	2.071
Investment per renovation house	11	62,383.73	25,000	105,078	80,078	31,121.29	0.407
Exploitation term lengthening	11	33.86	15	55	40	11,80	0.390

Except for two categories, few skew distributions were found. In particular, the category 'energy conservation achieved' (in EPL-scale points) is hardly skewed at all. For that reason this category is used as the dependent variable further on in the analysis.

5.6.2 Bivariate correlations

The bivariate correlations are presented in Table 5.3. The table is organised according to the clusters of variables in the analytical framework (see Fig. 5.1).

Due to the small number of cases we paid careful attention to further investigating the significant correlations that resulted from the analysis. On some occasions this led to the conclusion that correlations were no longer suitable for further inquiry in the multiple regression analysis to be conducted in the next research phase. The following correlations withstood close investigation:

- energy conservation * frequency of visits to discussion meetings
- energy conservation * size of the project configuration over time
- energy conservation * scale of interorganisational collaboration
- energy conservation * financial position of the housing association

Table 5.3 **Presentation of bivariate correlations**

Cluster 'policy instruments from the domain of climate policy'

Indicator name	r	p
Scale 'policy instruments'[a]	.722	.006**
Use of communicative policy instruments	.689	.009**
Presence of local or regional covenants	.551	.039*
Use of subsidies	.307	.179

Cluster 'characteristics of the local authority'

Indicator name	r	p
Financial support by national government	-.580	.031*
Political orientation of the officials	.371	.131
Orientation towards the environment	-.188	.290
Membership of Climate Treaty	-.147	.333
Size of the municipality	-.117	.366
Organisational fine-tuning	.060	.431
Personnel capacity	-.058	.432
Formal climate policy	-.045	.447
Presence of energy efficiency advocates	.017	.480

Cluster 'characteristics of the housing association'

Indicator name	r	p
Financial position	.750	.004**
Scale 'characteristics of the housing association'[b]	.734	.005**
Organisational fine-tuning	.653	.015*
Number of houses in property (size of stock)	.487	.064
Formal climate policy	-.372	.130
Presence of energy efficiency advocates	-.028	.468
Orientation towards the environment	-.011	.487
Personnel capacity	.005	.494

Cluster 'interorganisational collaboration'

Indicator name	r	p
Scale 'interorganisational collaboration'[c]	.764	.003**
Frequency of visits to thematic meetings	.760	.003**
Size of the project configuration over time	.668	.012*
Opinion leadership	.301	.184

Cluster 'cognitive cohesion'

Indicator name	**r**	**p**
Cohesion towards the national climate policy strategy	.399	.112
Cohesion towards environment and sustainable development	.176	.302
Cohesion towards technological innovation adoption	.111	.373

Cluster 'project context'

Indicator name	**r**	**p**
Exploitation term lengthening of renovated houses	.504	.057
Distance to district heating facility	-.489	.063
Support by urban renewal policies	-.447	.084
Number of houses	.321	.168
Equilibrium in public housing market	.291	.193
Institutionalisation of energy efficiency in decision-making process	.213	.265
Initial energy quality of houses	-.209	.268
Degree of public participation	-.194	.284
Type of heating system	.182	.296
Distribution of property ownership	.179	.299
Address density	.056	.435
Share of newly built houses	-.053	.439
Investment per house	-.009	.489

a Cronbach's alpha = .552. Items: use of subsidies, presence of local and regional covenants and use of communicative policy instruments.

b Cronbach's alpha = .747. Items: organisational fine-tuning, number of houses in property and financial position.

c Cronbach's alpha = .615. Items: opinion leadership, frequency of visits to thematic meetings, size of the project configuration over time.

* significant at .05 level ** significant at .01 level.

- energy conservation * organisational fine tuning within housing associations
- energy conservation * presence of local or regional covenants
- energy conservation * use of communicative policy instruments
- energy conservation * exploitation term lengthening of houses on site

5.6.3 Expected correlations that were not found

A number of correlations that for theoretical reasons were expected, were not found. This is explained below.

In the variable cluster on instruments from climate policy it was striking that subsidies did not correlate significantly with the degree of energy conservation achieved. The explanation relates to the argument that subsidies are not given to projects in which the most far-reaching, innovative energy efficiency techniques are applied. On the one hand, evidence from the case studies showed that high energy conservation figures were achieved on sites where substantial subsidies were absent. On the other hand, rather low degrees of energy conservation were achieved on sites to which substantial subsidies had been provided. Another argument for this finding is that government agencies would rather give subsidies to sites on the basis of expected absolute energy conservation numbers and CO_2 emissions avoided, than the relative decline in fossil energy use or CO_2 emissions avoided.

In the variable cluster on characteristics of housing associations several items were found that did not correlate significantly with the dependent variable, even though theoretical expectations predicted they would. Orientation towards the environment and formal policies did not correlate. Case evidence teaches us that housing associations do care about environmental issues such as energy efficiency, but they prefer to focus on the application of innovative sustainable energy systems in new construction and project development rather than energy conservation in their existing housing stock. The argument here might be that housing associations are more concerned with energy efficiency in new houses rather than the old stock. Investment in new houses is considered profitable whereas investment in old houses is considered a waste of capital. In their annual reports, housing associations do mention the existing stock in regard to maintenance activities and the replacement of conventional heating boilers by high-yield condensing boilers. However, this does not find a reflection in energy conservation in renovation projects. Maintenance and renovation are also considered as two different items in accounting terms, which is the way housing associations manage their business. Nor were significant correlations found with regard to specialised personnel capacity and the presence of energy advocates.

Several items in the variable cluster characteristics of local governments were also identified that did not correlate significantly with energy conservation. Many items correlated negatively, but were not significant. It could be argued that the cluster variable did not correlate owing to the local government's relative lack of involvement in the project's implementation stage. The significant (negative) correlation—which did not withstand the scatter plot inspection test—of the budget provided by central government to stimulate local climate policy ('bestuursakkoord nieuwe stijl'[BANS])[8] was a surprise. A qualitative argument for the direction of the correlation lies with the emphasis local governments seem to place on the realisation of energy efficiency goals in new construction sites. The evidence from the case supports this finding. The sites with poor levels of energy conservation realised in the existing stock were also those where local governments were particularly active in promoting district heating and sustainable energy for new construction sites, and where large sums from central government funding were available (such as the municipalities of Lelystad and Apeldoorn). By contrast, high energy conservation rates were achieved in municipalities where the local governments were relatively poorly supported by national government budgeting on local climate policy (such as the municipalities of Roosendaal, Uithoorn and Kerkrade). Another finding related to this phenomenon is that revitalisation of housing sites often involved both the renovation of old houses and the construction of new ones. When formulating energy efficiency objectives this had the side-effect that lofty ambitions were set especially for those areas where new construction was planned, rather than for the areas containing houses that were scheduled for renovation. The energy efficiency ambitions set for houses that were to be renovated were moderate or even low. This effect was displayed in five of the sites studied (Tannhäuser, Binnenstad-Oost, Nieuwstad, Espels and Prinsenhof). In these cases the energy conservation in the existing houses was moderate, not high (close to the statistical mean of 39.7%), while new construction nearby or in the specific locations featured housing blocks that had heat pumps installed or were connected to collective district heating facilities.

The variable cluster in the project context also features several items that can be considered as a surprise. The first item to mention that did not correlate was the initial energy performance of houses. A previous study found evidence that this item correlated closely and negatively with ambition-setting for energy conservation. The explanation for this correlation was found in the calculating behaviour of local governments and the way they approached and interacted with local actors (Hoppe *et al.* forthcoming). Nonetheless, the item does not seem to correlate in any way with the degree of energy conservation achieved. For example, the greatest degree of energy conservation achieved occurred on a site (Groot Kroeven) where the initial energy performance of the houses was already high (contrary to the prediction on which ambition-setting was based). However, the site is a perfect example of a case where a project's implementation contrasted strongly with the original planning and early project objectives.

8 This means 'new style policy arrangement' in English. It refers to a covenant between national government and provincial and local authorities.

Another notable finding was that investment per house did not correlate with the degree of energy conservation achieved. Nor did public participation (of tenants) correlate significantly. From a theoretical stance it was expected that tenants would be interested in significant energy efficiency improvement of the houses in which they lived. However, this would require them to discount the long-term benefits of the application of technical measures that would reduce their annual energy costs. Because housing associations usually calculate a small monthly rent increase for their tenants to cover the cost of improvements, tenants are negatively affected, since they emphasise short-term costs over long-term benefits (even when the net effect clearly favours the long term benefits). Tenants often only perceive a rise in living costs and thus a decline in personal purchasing power. The cases also show that tenants are actually inclined to stick with conventional systems, even when information about direct benefits are offered on innovative and sustainable energy systems.

It is useful to mention that the sites' distance from potential district heating facilities did not correlate significantly. It was expected that the distance would be of interest to local actors in the planning stage and that connection to a district heating facility would be a serious alternative to consider. However, district heating was only implemented in one case (Bijvank en het Lang). In this case the houses were already connected to the district heating system. In six other cases, connection to district heating facilities was considered, but abandoned when the project plans evolved. So they were never considered as a serious option. However, connection was considered in one case. This did not materialise as the district heating facility could not be built. The legal permit for the facility was not granted, which terminated plans for applying district heating on the residential site (Tannhäuser case).

Finally, it needs to be mentioned that although targets for improvement in energy conservation had been set for eight sites, no correlation was found between target-setting and the actual energy conservation achieved.

To summarise: no contextual factor was found that structurally influenced the achieved levels of energy conservation on sites. The same holds for intra-organisational characteristics of local governments.

5.6.4 Results of the multiple regression analysis

This sub-section reports the results of the multiple regression analysis. Following the construction of scales, a regression analysis could be conducted. The previous analysis of bivariate correlations allowed the construction of three scales, concerning the variables 'policy instruments from the domain of climate policy', 'characteristics of the housing association' and 'interorganisational collaboration'. Following the regression analysis, three significant variables were identified: 'interorganisational collaboration' ($\beta = .453$; $p = .014$; contribution to $R^2 = 58.4\%$), 'characteristics of the housing association' ($\beta = .466$; $p = .009$; contribution to $R^2 = 26\%$), and 'the use of policy instruments from the domain of climate policy' ($\beta = .299$; $p = .075$; contribution to $R^2 = 6\%$). The model explains 90.4% of the variance (adjusted $R^2 =$

86.3%) and the F-value is 22.000. The result of the analysis shows the importance of interorganisational collaboration between actors in local contexts. The result also provides the insight that housing associations exercise an important role in the achievement of energy conservation levels on residential sites with public housing. Finally, instruments from the domain of climate policy can still be considered a significant factor, albeit limited in scope. It is fair to state that much depends on close collaboration between local-level actors, especially with regard to the active involvement of housing associations.

5.7 Conclusion

Despite the technical opportunities to achieve significant energy conservation levels (up to 90%) in the existing housing stock, little has been achieved recently. Conventional appliances and energy systems are often used for a number of reasons, having to do with different barriers that block the adoption of more innovative and sustainable alternatives to improve energy efficiency. In this chapter we have tried to answer the following research question: which factors explain the variation in the energy conservation achieved in renovation projects on existing housing sites? We sought six theoretical explanations: the influence exercised by policy instruments, the influence exercised by housing associations, the influence exercised by local governments, collaborative efforts between actors, cognitive cohesion between actors and contextual factors. The central research question was analysed using a comparative research design. Both qualitative and quantitative research methods were applied.

The study provides insights into the factors that significantly explain energy conservation levels achieved in existing housing sites. First of all, it should be mentioned that, although reduction potentials of 90% are technically feasible, the projects analysed averaged only a 39.7% decrease in energy consumption, varying between 26.5% and 69.8%. Thus, it can be stated that the technical potential is far from being adequately harvested. Second, it was striking that goal-setting in projects did not correlate with actual project outcome. Goals were set higher than the energy conservation levels that were eventually achieved in the projects. The question of why no evident relationship was found between ambition and outcome needs further investigation. However, it is likely that it might have to do with commitment by local actors to goals formulated in agreements with local authorities, since it was clear that ambitions were downscaled during the process. By conducting further analysis it was found that three factors significantly influenced the levels of energy conservation achieved: interorganisational collaboration, the characteristics of the housing association and the use of instruments from the domain of climate policy. Whereas close collaboration between local actors and the influence of housing associations' characteristics were found to be of great importance, the influence of policy instruments was limited.

Since the role of local partners (housing associations) was found to be of great significance, further examination should be devoted to them. The analysis showed that 'rich' housing associations were eager to invest in energy-efficient homes, while the 'poor' ones did so to a lesser extent. It was also found that housing associations were primarily involved with social issues (their prime business), and when they were involved with energy efficiency they preferred innovative, state-of-the-art techniques in new houses under construction to the relatively cheaper measures that could significantly reduce energy demand in the existing stock. This also explains why the presence of corporate energy/climate policies did not correlate with energy conservation in projects: energy conservation in existing houses remains a low priority on the housing associations' agenda. Although the financially better-off housing associations had better outcomes in energy conservation, it should be mentioned that it was not easy for them to find the necessary budget: they remained dependent on additional external funding. Furthermore, it should be noted that the role of local governments was limited, as they only influenced target setting, not goal achievement.

Tenants were found to have little concern with the energy efficiency of their homes. In some cases they actively preferred conventional suboptimal systems to more energy-efficient alternatives, even though additional information was provided on direct financial benefits. Although the US literature claims that low-income households are responsive to programmes, this is hardly the case in our study. Furthermore, the study showed that it is difficult to persuade low-income house owners, even when substantial policy efforts were devoted to them, with careful implementation (local programmes involved loans, subsidies, mortgages, green funds, campaigns, personal assistance). It remains very hard for policy agencies to approach this group. Policy-makers should pay careful attention to how to approach this group in the near future.

According to the study's results, the policy plans by the Dutch national government for the period 2009–2020 are too optimistic and need revision. In particular, the link between the policy budget and the 2020 goals falls short. Furthermore, the goal set for the public housing sector is considered too ambitious. The policy goal to reduce energy consumption by 46–50% (WWI *et al.* 2008) can be considered too ambitious, as the study has shown that the actual reduction rate is less than 40%. In some cases it has been shown that higher energy conservation levels could be achieved. Nonetheless, they remain 'best practices', which cannot easily be generalised. A 'bottom up' approach that takes contextual factors into account—like this study—could help validate future policy-making ('backward mapping'; Elmore 1980).

The study has shown that implementing climate mitigation measures in the existing housing stock remains a very complex and difficult task. Harvesting the technical reduction potential is far from being an accomplished goal; even the mid-term policy goals set by the Dutch government seem to be overly ambitious. Future research and policy-making should devote careful attention to the way local actors should be addressed in order to gain commitment (from both households and housing associations), and how local authorities may actively facilitate the project of energy conservation in housing renovation sites, and not just by setting ambitious goals. Furthermore, systematic, in-depth comparison of local-level projects, as well as international comparative analyses, are necessary to assist the European Union in developing policy instruments and strategies that stimulate actual, local housing renovation projects.

References

Abrahamse, W., L. Steg, C. Vlek, and T. Rothengatter (2005) 'A Review of Intervention Studies Aimed at Household Energy Conservation', *Journal of Environmental Psychology* 25: 273-91.

Balthasar, A. (2000) *Energie 2000: Programmwirkungen und Folgerungen aus der Evaluation* (Zurich: Verlag Ruegger).

Börzel, T.A. (1998) 'Organizing Babylon: On the Different Conceptions of Policy Networks', *Public Administration* 75 (Summer 1998): 253-73.

Bressers, H. (2009) 'From Public Administration to Policy Networks: Contextual Interaction Analysis', in S. Nahrath and F. Varone (eds.), *Rediscovering Public Law and Public Administration in Comparative Policy Analysis: Tribute to Peter Knoepfel* (Bern, Switzerland: Haupt Verlag).

Bressers, H.Th.A., and S.M.M. Kuks (2003) 'What Does "Governance" Mean? From Conception to Elaboration', in H. Bressers and W. Rosenbaum (eds.), *Achieving Sustainable Development: The Challenge of Governance across Social Scales* (London: Praeger): 65-89.

—— and L.J. O'Toole, Jr (2005) 'Instrument Selection and Implementation in a Networked Context', in P. Eliadis, M. Hill and Michael Howlett (eds.), *Designing Government: From Instruments to Governance* (Montreal/Kingston, Canada: McGill-Queen's University Press): 132-53.

Bressers, J.Th.A. (1989) *Naar een cybernetica in de beleidswetenschap: Leren met het oog op de milieucrisis* (Enschede, Netherlands: Universiteit Twente).

—— (1993) 'Beleidsevaluatie en beleidseffecten', in A. Hoogerwerf (ed.), *Overheidsbeleid; Een inleiding in de beleidswetenschap* (Alphen aan de Rijn, Netherlands: Samsom Tjeenk Willink): 161-79.

—— (2004) 'Implementing Sustainable Development: How to Know What Works, Where, When and How', in W. M. Lafferty (ed.), *Governance for Sustainable Development: The Challenge of Adapting Form to Function* (Cheltenham: Edward Elgar Publishing): 284-318.

—— and L.J. O'Toole (1998) 'The Selection of Policy Instruments: A Network-Based Perspective', *Journal of Public Policy* 18.3: 213-39.

Burt, R.S. (1987) 'Social Contagion and Innovation: Cohesion versus Structural Equivalence', *American Journal of Sociology* 92.6: 1,287-335.

Carmines, E.G., and R.A. Zeller (1979) *Reliability and Validity Assessment* (Beverly Hills, CA./London: SAGE Publications).

CBS (Centraal Bureau voor de Statistiek) (2008) *Veranderingen in de woningvoorraad* (Voorburg, Netherlands: CBS).

Clinch, P.J., and J.D. Healy. (2000) 'Domestic Energy Efficiency in Ireland: Correcting Market Failure', *Energy Policy* 28: 1-8.

Cronbach, L.J. (1951) 'Coefficient Alpha and the Internal Structure of Tests', *Psychometrika* 16: 297-334.

De Bruijn, J.A., and E.F. ten Heuvelhof (1995) Netwerkmanagement; Strategieën, instrumenten en normen (Utrecht, Netherlands: Lemma Uitgeverij BV).

De Jong, J.J., E.O. Weeda, Th. Westerwoudt and A.F. Correlje (2005) *Dertig jaar Nederlands energiebeleid* (The Hague: Clingendael).

DOE (US Department of Energy) (2009) *About the Weatherization Assistance Program* (Washington, DC: US Department of Energy; apps1.eere.energy.gov/weatherization/about.cfm, accessed 12 March 2010).

Dowding, K. (1995) Model or Metaphor? A Critical Review of the Policy Network Approach', *Political Studies* 43: 136-58.

Elmore, R.F. (1980) 'Backward Mapping: Implementation Research and Policy Decisions', *Political Science Quarterly* 94: 601-16.

Frieden, B.J., and K. Baker (1983) 'The Market Needs Help: The Disappointing Record of Home Energy Conservation', *Journal of Policy Analysis and Management* 2.3: 432-48.

Goetschel, U., *et al.* (1995) 'Lead-cost-planning Case Study: Optimization of Supply and Demand in Flanitzhutte', paper presented at the *Fourth International Energy Efficiency and DSM Conference*, Berlin, 10–12 October 1995.

Granovetter, M.S. (1973) 'The Strength of Weak Ties', *American Journal of Sociology* 78.6: 1,359-80.

—— (1978) 'Threshold Models of Collective Behavior', *American Journal of Sociology* 83.6: 1,420-43.

Hartman, R.S. (1988) 'Self-selection Bias in the Evolution of Voluntary Energy Conservation Programs', *Review of Economics and Statistics* 70.3: 448-58.

Hirst, E. (1984) 'Household Energy Conservation: A Review of the Federal Residential Conservation Service', *Public Administration Review* 44.5: 421-30.

—— (1987) 'Cooperation and Community Conservation; Final Report, Hood River Conservation Project' (DOE/BP-11287-18; Portland, OR: Pacific Power and Light and Bonneville Power Administration).

Hoiting, H., G.J. Donze and P.W.G. Nuiten (2004) 'Energieprestatiemethoden: samen sterk?', *Bouwfysica* 17.1: 24-28.

Hoppe, T., and K. Lulofs (2008) 'The Impact of Multi-level Governance on Energy Performance in the Current Dutch Housing Stock', *Energy & Environment* 19.6: 819-30.

——, H. Bressers and K. Lulofs (forthcoming) 'Local Government Influence on Energy Conservation Ambitions in Existing Housing Sites: Plucking the Low-Hanging Fruit?'.

IPCC (Intergovernmental Panel on Climate Change) (2007) *Climate Change 2007: Synthesis Report* (Geneva: IPCC).

Johnson, R.B., and T. Onwuegbuzie (2004) 'Mixed Methods Research: A Research Paradigm Whose Time has Come', *Educational Researcher* 33.7: 12-26.

Kickert, W.J.M., E.-H. Klijn and J. F.M. Koppenjan (1997) *Managing Complex Networks; Strategies for the Public Sector* (London: SAGE Publications).

Klijn, E.-H. (1996) 'Regels en sturing in netwerken; de invloed van netwerkregels op de herstructurering van naoorlogse wijken' (PhD thesis, Rotterdam: Erasmus Universiteit).

Koffijberg, J.J. (2005) 'Getijden van beleid: omslagpunten in de volkshuisvesting; Over de rol van hiërarchie en netwerken bij grote veranderingen' (PhD thesis, Technische Universiteit Delft).

Leonard-Barton, D., and E.M. Rogers (1979) 'Adoption of Energy Conservation among California Homeowners', paper presented at the *Annual Meeting of the International Communication Association*, Philadelphia, 1–5 May 1979.

Lulofs, K.R.D., and M.J. Arentsen (2001) *Improving Quality and Learning Performance of 'Energie 2000' Evaluation Research* (Enschede, Netherlands: Universiteit Twente, CSTM).
Lulofs, K., and B. Lettinga (2003) *Instrumenten 'Mainstream Market': CO_2-reductie in de gebouwde omgeving* (Enschede, Netherlands: Universiteit Twente, CSTM).
Marsh, D., and R.A.W. Rhodes (1992) 'Policy Communities and Issue Networks: Beyond Typology', in D. Marsh and R.A.W. Rhodes (eds.), *Policy Networks in the British Government* (Wotton-under-Edge, UK: Clarendon Press).
NIP (Nationaal Isolatie Programma) Stuurgroep (1988) 'Evaluatie 10 jaar Nationaal Isolatie Programma' (The Hague: Ministerie van VROM/Ministerie van Economische Zaken).
O'Toole, L.J., Jr. (2000) 'Research on Policy Implementation: Assessment and Prospects', *Journal of Public Administration Research and Theory* 10.2: 263-88.
Rogers, E.M. (1962) *Diffusion of Innovations* (New York: Free Press).
SenterNovem (2005) 'EPL Monitor 2004: Herstructureringlocaties' (Utrecht, Netherlands: SenterNovem).
—— (2007) 'EPL Monitor 2006: Herstructureringlocaties' (Utrecht, Netherlands: SenterNovem).
Smith, M.J. (1997) 'Policy Networks', in M. Hill (ed.), *The Policy Process* (London: Prentice Hall): 76-86.
Sorrell, S. (2007) 'The Rebound Effect: An Assessment of the Evidence for Economy-wide Energy Savings from Improved Energy Efficiency' (Report of the Sussex Energy Group for the Technology and Policy Assessment function of the UK Energy Research Centre; London: UKERC).
Stern, P.C., E. Aronson, J.M. Darley, E. Hirst, W. Kempton and T.J. Wilbanks (1986) 'The Effectiveness of Incentives for Residential Energy Conservation', *Evaluation Review* 10.2: 147-76.
Sunnika, M. (2001) 'Policies and Regulations for Sustainable Building: A Comparative Study of Five European Countries' (Housing and Urban Policy Studies, No. 19; Delft, Netherlands: OTB).
Trecodome (2008) 'Passive Renovation in the Netherlands; Innovation in Construction', presentation at the *Corpovenista Conference*, The Hague, 14 May 2008.
Unruh, G.C. (2000) 'Understanding Carbon Lock-in', *Energy Policy* 28.12: 817-30.
Van der Waals, J.F.M. (2001) 'CO_2-reduction in Housing; Experiences in Building and Urban Renewal Projects in the Netherlands' (PhD thesis, Universiteit Utrecht).
——, W. J. V. Vermeulen and P. Glasbergen (2003) 'Carbon Dioxide Reduction in Housing: Experiences in Urban Renewal Projects in the Netherlands', *Environment and Planning C: Government and Policy* 21: 411-27.
VROM (Ministerie van Volkshuisvesting, Ruimtelijke Ordening en Milieubeheer) (2000) 'Nota Wonen: Mensen, wensen, wonen: Wonen in de 21ste eeuw' (The Hague: Ministerie van VROM).
—— (2005) 'Evaluatienota Klimaatbeleid 2005; Onderweg naar Kyoto; Een evaluatie van het Nederlandse klimaatbeleid gericht op realisering van de verplichtingen in het Protocol van Kyoto' (The Hague: Ministerie van VROM).
WWI (Ministerie van Wonen, Wijken en Integratie), Ministerie van Volkshuisvesting, Ruimtelijke Ordening en Milieubeheer (VROM), Aedes and Vereniging Nederlandse Woonbond (2008) 'Convenant Energiebesparing corporatiesector' (Ede, Netherlands: WWI).

6

Natural hazards, poverty traps and adaptive livelihoods in Nicaragua

Marrit van den Berg and Kees Burger
Development Economics, Wageningen University, The Netherlands

Climate change not only alters average weather conditions, but it will also lead to more frequent and more severe natural shocks. Hurricanes are likely to become more intense, as are droughts and floods associated with El Niño events (see Table 6.1, and for more information see Emanuel 2005; IPCC Working Group 1 2001; Trenberth 2005). Intense precipitation events will most likely occur more often, causing increased soil erosion and damage due to flood, landslide, avalanche and mudslides (IPCC Working Group 1 2001). Those who are least able to cope will unfortunately have to bear most of the burden: 85% of people exposed to earthquakes, hurricanes, floods and droughts live in countries with medium or low levels of human development (UNDP 2008), and natural catastrophes between 1985 and 1999 resulted in losses of 13.4% of the gross domestic product in developing countries, compared to only 2.5% in industrialised countries (Munich Re 2006).

Not only are losses relatively greater for developing countries, but these countries are also less able to withstand the initial disaster shock. As a result, a disaster of a similar relative magnitude will cause a much bigger shock to the economy in a poorer country (Noy 2009). The short-term constraints for recovery can subsequently cause poverty traps and result in a reduction of long-term macroeconomic growth rates (Hallegatte and Dumas 2009; Hallegatte 2007).

As developing countries have a limited institutional capacity, private risk management and coping strategies are extremely important. Not surprisingly, poor people are often less able to cope with shocks than their richer neighbours and therefore more vulnerable during the response and recovery phase (Masozera *et al.* 2007). In a situation with high risk and subsistence constraints, poorer agents may therefore invest more heavily in low-risk, low-return assets than wealthier households do (Zimmerman and Carter 2003). Similarly, a natural hazard that destroys productive assets could induce people to replace their relatively remunerative livelihoods with more defensive strategies. This will permanently lower their welfare, as the associated low returns make asset recovery extremely different.

While natural hazards thus present a serious threat to rural livelihoods in developing countries, only a few economic studies analyse this aspect in depth. The general rural risk literature focuses on income fluctuations—often associated with drought—and therefore ignores asset losses, an important consequence of many natural shocks (e.g. Del Ninno *et al.* 2003; Hoddinott 2006; Ito and Kurosaki 2009; Kazianga and Udry 2006; Kochar 1999; Owens *et al.* 2003; Paxson 1992). Recent exceptions are Baez and Santos (2007), Gitter and Barham (2007) and Pörtner (2008), who assess human capital responses to hurricane shocks, and Carter *et al.* (2007), who study the recovery of physical assets after drought and hurricane shocks. From a different perspective, Dercon *et al.* (2005) assess the consumption effects of various natural and other shocks, some of which may have caused both asset and income losses.

We add to this literature by analysing the impact of Hurricane Mitch on rural livelihoods in Nicaragua. Hurricane Mitch hit the country in the last days of October 1998 and was one of the most violent hydro-meteorological phenomena to have struck Central America last century. Small agricultural producers were the population group most affected: they lost their crops, their lands were left unusable, and rural roads and bridges were destroyed (FAO 2001).

We use the Nicaragua Living Standard Measurement Survey (LSMS), which provides unique panel data of almost 2,000 rural households just before Mitch hit the country in 1998, just after that event in 1999, and in 2001 and 2005. Moreover, we executed our own survey and a risk experiment in the Pacific province of Chinandega in 2007. The structure of the paper is as follows. Section 6.1 presents the background to the research: the country of Nicaragua and Hurricane Mitch. Section 6.2 summarises the main findings related to vulnerability and adaptation: the short-term consequences of Mitch for household consumption and how these differ between households depending on wealth; the potential adaptation of livelihood strategies; and the possible changes in risk attitudes. Section 6.3 draws policy lessons and links these to (mainly) Dutch and European policies for development and emergency aid. Finally, Section 6.4 presents recommendations for further research.

6.1 Hurricane Mitch in Nicaragua

Nicaragua is the second poorest country in Latin America and a clear example of a developing country facing a diversity of climatic hazards: between 1980 and 2000, it suffered from seven hurricanes, five floods and three major droughts. During the first half of the 1990s, Nicaragua rapidly became a market economy after years of political conflict and civil war. Poverty rates decreased only slightly from 50% to 48% between 1993 and 1998, which was admittedly a bad agricultural year characterised by a drought associated by the El Niño Southern Oscillation (World Bank 2001a). The 1998 poverty estimates were based on data collected from 15 April to 31 August. In November, Hurricane Mitch devastated the country's infrastructure and destroyed a large part of agricultural production.

Although Hurricane Mitch never entered Nicaragua, it passed by close enough for the associated low-pressure front to cause excessive rainfall in the country's western, central and northern regions. This rainfall brought about floods, strong currents and landslides, and therefore caused extensive damage to agricultural production, the environment and infrastructure. About 2,000 people died in the mudflows after the collapse of the crater of the Casita volcano (National Climatic Data Centre 2004), and about the same amount succumbed to the hurricane in other parts of the country (FAO 2001). About 18% of the population was made homeless (ECLAC 1999), and many were affected by disease outbreaks resulting from poor housing, sanitary and health conditions in the aftermath of the catastrophe (FAO 1999). The estimated total costs associated with the hurricane amount to US$988 million—or 45% of the national GDP—of which US$562 million represents direct storm damage (ECLAC 1999).

Most damage to productive sectors occurred in agriculture (ECLAC 1999). The bulk of first-season crops had been harvested when Mitch hit the country, and yields were above average (FAO 2001). Yet the second cereal and pulse crops were still in the field, and about one-third of expected output from these crops was lost. Mitch also destroyed fruit and vegetable crops and important export crops, such as coffee, banana and sugar cane (FAO 2001). Export crop losses totalled 12% of the expected production of US$37 million (ECLAC 1999). In total, the estimated value of crop losses amounted to almost US$90 million (ECLAC 1999). Added to these income losses, farmers suffered an estimated total of US$62 million of asset losses (ECLAC 1999). Approximately 10,000 ha of cultivable land that were flooded, washed away or silted, and large areas of coffee and banana plantations and fruit orchards were lost. The livestock sector also suffered substantial damage, mainly through death of cattle and loss of installations.

Surprisingly, Hurricane Mitch did not affect poverty rates in the short term, with estimates of 48% of the total population estimated to be poor in 1999, just like a year before. Post-Hurricane Mitch investments brought the average poverty rate down to 46% in 2001 (World Bank Central America Department 2003), but this number was about the same four years later, despite moderate economic growth and low inflation (World Bank 2007).

6.2 Vulnerability and adaptation

This section summarises the main findings of our research on Hurricane Mitch and adaptive livelihoods in Nicaragua. First, we assess differences in vulnerability between households by analysing the short-term consequences of Mitch for the consumption of rural households. Second, we explore whether rural households adapt their livelihood strategy in response to the hurricane. And third, we look at the potential impact of Mitch and other natural shocks on risk attitudes.

6.2.1 Short-term consumption responses

The income of the rural poor fluctuates substantially due to natural, market and institutional risks. They adapt to this situation and generally have smoother consumption than income and smoother income than risk-neutral households would have (Deaton 1992; Morduch 1995; Paxson 1992; Townsend 1994). When markets are perfect, people will smooth consumption using financial and insurance markets, and consumption will be independent of stochastic income fluctuations. Yet financial markets in developing countries are extremely limited, and the poor will have to rely at least partly on their own portfolio of assets for smoothing consumption. They can use liquid assets as buffer stocks to sell when income is low and buy when income is high. (Deaton 1991; Kinsey *et al.* 1998; Rosenzweig and Wolpin 1993). Repeated setbacks may, however, put an end to this strategy (Deaton 1991: Ersado *et al.* 2003). Moreover, if asset sales reduce the productivity of the remaining assets—which will be the case if returns are endogenously increasing—recovery will be difficult or even impossible (Carter and Barrett 2006; Dercon *et al.* 2005; Hoddinott 2006; Jalan and Ravallion 2001; Zimmerman and Carter 2003). Hence, selling assets today may imply permanently lowering future consumption, especially for poorer households. They may therefore choose to cut consumption to smooth assets, instead of the other way around (Dercon *et al.* 2005; Hoddinott 2006; Jalan and Ravallion 2001; Zimmerman and Carter 2003).

We analysed the consumption reactions of rural households in Nicaragua to Hurricane Mitch, using the LSMS from 1998 and 1999. The sample from the 1998 survey covered a representative sample of approximately 2,000 rural households. The household questionnaire included data on characteristics of the household, asset endowments, income and consumption. In May 1999, households located in the area of the country affected by the hurricane were revisited using a similar questionnaire with additional questions about the effects of the hurricane (World Bank 2001b). We use the farm households from the 1998–99 panel for our analysis, as they represent the most affected group and vulnerable to both asset and income shocks.

When asked directly about strategies to cope with El Niño losses, almost one-third of the farm households interviewed mentioned consumption cuts. A second important strategy mentioned by 10% of those interviewed was to work longer hours to increase income. Other strategies, in particular those relying on non-household resources such as financial markets and aid, were mentioned by only a few respondents. While the rare mentioning of financial markets could have been expected because of the low level of market development, the unimportance of emergency aid may come as a surprise considering the large inflow of aid resources. More detailed analysis of the survey data, however, indicates that whereas many households received some kind of gift after Mitch, the value of these gifts was generally low.

More detailed data analysis confirms the importance of consumption cuts in coping with Mitch-related losses. We regressed the change in household consumption between 1998 and 1999 on income and asset shocks, allowing different responses for poor and wealthy households (see Table 6.1, and for more information see Van den Berg and Burger 2008). While shocks did not affect food consumption, total consumption was responsive for both poor and wealthy households, although to a different degree.

As expected, poor households did not perfectly smooth consumption. A negative income shock to the household of C$1 (1 Nicaragua Córdoba Oro or US$0.09) would have resulted in a decrease of C$0.07 in consumption per capita, or C$0.45 in total household consumption. These numbers are similar to the results of Kurosaki (2006) and Paxson (1992) for Pakistan and Thailand, respectively. Wealthy households were not affected by income shocks, confirming our hypothesis that these households would be better able to smooth consumption. Both groups of households lower consumption in response to asset losses, which was expected as these affect their future income-generating capacity.

Given these findings, it may come as a surprise that Mitch did not significantly change Nicaragua's overall poverty profile (World Bank 2001a). The answer lies in the sequence of two adverse natural phenomena: the El Niño Southern Oscillation of 1997, affecting 1998 consumption, and Hurricane Mitch in late 1998, affecting 1999 consumption. Not all households were equally affected by these episodes, and some had at least in the short run incurred more hardship from El Niño than from the infamous Hurricane Mitch. For example, for 48% of the rural farm households in the survey, crop income was higher for the year covering Mitch than for the previous El Niño year. When comparing 1998 and 1999 data, consumption cuts in 1998 could easily be mistaken for consumption growth in 1999. This, however, does not affect the analysis presented above, as our income shock variable measures the income difference between the two years.

6.2.2 Livelihood strategies

While at the aggregate level, the consumption consequences of Mitch did not exceed those of the drought related to El Niño the year before, the associated asset losses may have had persistent effects on the livelihoods of those affected. A livelihood involves the activities that people undertake in pursuit of income, security, well-being and other goals (Adato and Meinzen-Dick 2002; Carney 1999; Chambers and Conway 1991; Ellis 1998). The endowment of capital or assets determines the

Table 6.1 **Effect of asset and income losses on changes in household consumption (N=292)**

	Δ total consumption per capita		
	Mean	**Wealthy**	**Poor**
Change in transitory income (C$)	-195.37	-0.02	0.07**
	(3742.14)	(0.06)	(0.02)
Asset losses			
• Land (mz)	0.23	156.61	-143.24**
	(1.51)	(94.24)	(34.97)
• Livestock (1,000 C$)	1.17	46.00	33.75
	(3.00)	(39.18)	(32.45)
• Draft animals (y = 1)	0.01	-1,540.87*	-3,594.13**
	(0.08)	(752.50)	(619.95)
• Physical agricultural assets (y = 1)	0.12	-1,546.68**	948.61**
	(0.31)	(480.65)	(199.74)
• Non-farm assets (y = 1)	0.03	-273.43	-77.61
	(0.18)	(578.97)	(204.78)
Non-farm enterprise pre-Mitch (y = 1)	0.16	1,468.44*	-145.71
	(0.38)	(591.78)	(152.58)
Constant		-467.53	127.75
		(362.65)	(71.32)
R^2		0.07	0.27

Notes: Grouping results from endogenously switching regression based on asset endowments.
We accounted for arbitrary autocorrelation between clustered households.
Standard deviations in parentheses for means; standard errors in parentheses for regressions.
All values in 1998 prices corrected for regional differences.
** $p < 0.01$; * $p < 0.05$.

Source: Van den Berg and Burger 2008

possibilities people have and thus lies at the basis of the livelihood strategy. In a situation with full information and competitive markets, returns across livelihood strategies are equalised. However, when market constraints exclude particular households from specific activities, this results in differences in returns between activities and consequently in standards of living between households.

Rural markets in Nicaragua are highly imperfect. Only 2% of the lowest wealth quintile received formal loans in 1999, and even for the highest wealth quintile only 14% reported having a formal loan. Moreover, 80% of the poorest quintile and 30% of the wealthiest quintile indicated that they had wanted a loan but lacked resources to back up their demand (Boucher *et al.* 2005). Similarly, land sales markets are thin, and only about 2% of producers included in the 1998 LSMS reported having bought land in any of the five years before the survey (Deininger 2003). These and other market imperfections are likely to exclude the poor from certain profitable activities, and through destruction of productive assets, a hurricane or another natural hazard could induce people with remunerative livelihoods to choose more defensive strategies, which allow them to survive but at a permanently lower welfare level than before (Barrett 2005; Carter *et al.* 2006).

Using the LSMS panel data, we therefore clustered households into different livelihood strategies for the years 1998, 2001 and 2005 (Van den Berg 2009) (see Table 6.2). We quantified livelihood strategies using hierarchical agglomerative clustering and k-means cluster analysis (Hair *et al.* 1992). As defining variables we used the shares of land allocated to annual crops, perennial crops and pastures (ranching); and the shares of labour allocation to farm production, non-farm self-employment, farm wage employment, and non-farm wage employment.

Using stochastic dominance analysis, we classified the strategies based on the associated consumption levels. Annual farming, farm wage employment, and the combination of the two activities resulted in low incomes, whereas non-farm wage employment and ranching were associated with relatively high incomes. Perennial farming, non-farm self-employment, and annual cropping with non-farm employment take an intermediate position. High welfare strategies were associated with high levels of infrastructural, human and, to a smaller extent. natural capital (Van den Berg 2009). A relatively constant distribution of households over the various strategies would suggest that households following low welfare strategies were trapped in poverty.

Yet many households moved actively between strategies of different welfare levels. Even for the most stable livelihood strategies—non-farm wage employment, ranching and perennial farming—between 37% and 64% of households switched to another strategy between survey rounds (Van den Berg 2009). This large-scale mobility is not merely a construct of our clustering procedure: many households actually moved in and out of profitable activities such as non-farm wage employment, and all correlation coefficients between the resource allocation variables used for clustering are less than 0.5 between years.

The question is what role Hurricane Mitch played in the mobility between livelihood strategies. It is possible that the relative stability in the overall distribution of households across livelihood strategies is the result of downward mobility of Mitch-affected households combined with horizontal mobility and upward mobility of other households. Yet whether or not a household lived in a Mitch-affected area did not influence upward or downward mobility between 1998 and 2001 (Table 6.3), nor did it affect the stability of the overall distribution of strategies (Van den Berg 2009). Hence, our results do not provide support for the hypothesis that Hurricane Mitch pushed people into lower-welfare livelihood strategies and thus into chronic poverty.

Table 6.2 **Rural livelihood strategies and poverty**

		High-income strategies		Medium-income strategies			Low-income strategies		
	All	Non-farm wages		Non-annuals farming	Non-farm self-employment	Non-farm with annuals	Farm wages with annuals	Annual cropping	Farm wages
1998 (N = 1,755)									
Share of households	1	0.16		0.07	0.09	0.11	0.10	0.36	0.11
Consumption (C$/cap)	4,033	5,248		4,702	6,148	3,983	3,235	3,324	3,150
	(202)	(370)		(490)	(1,160)	(229)	(274)	(123)	(319)
Poor (%)	68	52		59	48	68	77	77	81
Extremely poor (%)	30	13		23	11	26	39	38	45
			ranching	perennial cropping					
2001 (N = 1,793)									
Share of households	1	0.14	0.04	0.08	0.08	0.12	0.11	0.32	0.10
Consumption (C$/cap)	3,877	5,402	5,149	4,098	4,819	4,159	2,815	3,297	2,795
	(107)	(332)	(394)	(332)	(278)	(181)	(153)	(103)	(178)
Poor (%)	68	46	46	64	49	64	83	78	86
Extremely poor (%)	28	9	14	25	17	19	44	34	46
2005 (N = 3,352)									
Share of households	1	0.14	0.07	0.07	0.09	0.13	0.14	0.27	0.08
Consumption (C$/cap)	4,032	5,386	4,939	5,155	5,592	4,229	2,972	3,172	3,107
	(92)	(253)	(219)	(825)	(297)	(149)	(91)	(81)	(158)
Poor (%)	71	48	54	67	49	62	86	85	82
Extremely poor (%)	31	12	13	23	13	26	46	41	45

Notes: The estimates account for within cluster correlation and sampling weights.

Standard deviations in parentheses.

Source: Van den Berg 2009

Table 6.3 **Upward and downward mobility between livelihood strategies, 1998–2001 (N = 1,157)**

	Ordered logit	Ordered probit
Mitch area (yes = 1)	-0.085	-0.039
	(0.48)	(0.39)
Human capital		
• Household size	-0.001	-0.001
	(0.03)	(0.03)
• Percentage of female adults	-0.174	-0.108
	(0.53)	(0.57)
• Dependent/total members	0.003	0.002
	(0.05)	(0.06)
• Female-headed household (yes = 1)	-0.325	-0.184
	(1.74)	(1.77)
• Age of household head	0.005	0.002
	(0.81)	(0.78)
• Head can read and write (yes = 1)	-0.065	-0.046
	(0.42)	(0.53)
• Head finished high school (yes = 1)	-0.075	-0.040
	(0.21)	(0.21)
Natural capital		
• Land owned (mz)	-0.001	-0.000
	(0.32)	(0.42)
Location capital		
• Home accessed by paved road	0.049	0.012
	(0.22)	(0.11)
• Travel time to nearest health centre	-0.000	-0.000
	(0.31)	(0.250)
• Pacific	-0.100	-0.066
	(0.53)	(0.63)
• Atlantic	0.054	0.041
	(0.27)	(0.34)
Cut 1	-1.758**	-0.986**
	(0.348)	(0.194)
Cut 2	1.051**	0.811**
	(0.341)	(0.195)

Notes: No mobility = 1, upward mobility = 2, downward mobility = 0.
The estimates account for within cluster correlation and sampling weights.
t statistics in parentheses for coefficients, standard deviation for means.
* significant at 5%; ** significant at 1%.

Source: Van den Berg 2009

6.2.3 Risk attitudes

Even if we did not find evidence of an impact of Hurricane Mitch on livelihood strategies, this does not mean that Mitch did not affect economic decisions at all: especially within the self-employment and diversified strategies, many decisions are made that affect the level and stability of income. The more risk-averse people are, the more they will be inclined to forgo income for stability and the less able they will be to recover from possible asset losses. We hypothesise that the occurrence and intensity of shocks affect household risk aversion and test this for Hurricane Mitch and more recent natural shocks in Nicaragua.

Economists mostly see this risk attitude as a relatively stable, innate characteristic, although possibly correlated with gender, wealth and age. There is, however, evidence that risk attitudes are state-contingent. In games-playing, past experience in previous games in the same experiment affects preferences: people who had consistently won had a greater tendency to choose more risk-seeking alternatives (Binswanger 1980; Yesuf and Bluffstone 2009). What is more, risk aversion has been shown to increase when there is greater concern about personal finances or a larger perceived general vulnerability (Anderson and Mellor 2009; Mosley and Verschoor 2005). This is in line with psychological research, which also shows that worry leads to more risk-averse choices (Lerner and Keltner 2001; Raghunathan and Pham 1999). As psychologists have, moreover, shown that experience of hazard makes people more worried and fearful (Cutchin *et al.* 2008; Weinstein 1989), natural hazards could be expected to increase risk aversion. If so, they may affect livelihoods not only directly by lowering income and destroying assets, but also indirectly through increasing the degree of risk aversion, thereby lowering the speed of recovery or even inhibiting recovery altogether.

As the LSMS did not collect data on risk attitudes, we arranged our own fieldwork in the Pacific department of Chinandega, which was hit hard by Hurricane Mitch. In early 2007, we interviewed 222 randomly selected farm households in different communities on general household characteristics and their exposure to Mitch and other shocks. Of these households, 131 lived in communities that were heavily affected by Mitch, 51 lived in communities that were only mildly affected by Mitch, and 40 households were relocated to a resettlement area. Besides the interviews, we randomly selected almost half of the people interviewed in each of the different zones to take part in a small risk experiment with real payoffs. These people were asked to imagine themselves having to choose between Project A, with profits determined by a 50–50 chance of success or failure, and Project B, with profits fixed and secure. The two possible payoffs of Project A were linked to the toss of a coin. The highest payoff of Project A was set at C$80(US$5.33), approximately double the daily agricultural wage rate, and the lowest payoff at C$32(US$2.13). The fixed return for Project B started at the low payoff for Project A and increased in 13 steps to the high payoff. At the start of the session, subjects were informed that one of the games would be randomly selected to determine their final payoff. The payoff at which they switch from the gamble to the fixed return option reveals the value of the respondent's certainty equivalent to the gamble and can easily be converted to the coefficient of absolute risk aversion ρ (Van den Berg *et al.* 2009).

As expected, the subjects generally exhibited risk-averse behaviour in the experiments. The choices of 76% of participants revealed risk aversion, and the average value of ρ was 0.33, which—given that mean income from the experiment equals US$3.73—is equivalent to a coefficient of relative risk aversion of 1.2 and is thus in line with earlier findings (Van den Berg *et al.* 2009).

When regressing the coefficient of absolute risk aversion from the experiments on indicators for the impact of Mitch and other natural hazards, and a set of control variables including personal characteristics and household wealth, we find strong support for the hypothesis that experiencing natural shocks makes people more risk-averse (Table 6.4). Risk aversion was substantially higher for farmers who experienced more damage. This suggests that disasters not only change the asset base of the affected population, but also the nature of their preferences and the weighing of alternative survival strategies. Put differently, risk management and coping strategies and policies that *ex ante* seemed optimal may not necessarily be so after a major disaster has taken place.

6.3 Policy relevance

The considerable impact of natural hazards on development has increasingly been recognised. Our results provide a warning that the resulting risk management and humanitarian aid efforts should not just be directed to incidental shocks with high media coverage but also to less spectacular, often more frequently occurring, shocks: for many rural households in Nicaragua the short-term consumption effects of the 'routine' 1997 El Niño drought were worse than those of the spectacular 1998 Hurricane Mitch (Van den Berg and Burger 2008).

International donors actively integrate natural disasters into their policies. In August 2009, the Inter-American Development Bank approved a project to support the preparation of a Program of Natural Disasters Prevention for Nicaragua, which will focus on the effects of climate change.[1] In addition, the European Union, which accounts for more than 50% of total aid to Nicaragua, recognises the importance of climate change and natural hazards. The prime objective of European development policy according to the European Consensus on Development (EC 2006) is 'the eradication of poverty in the context of sustainable development, including pursuit of the Millennium Development Goals (MDGs), together with the promotion of democracy, good governance and respect for human rights'. Rural areas have a higher proportion of poor people; these people have little political power and therefore require special attention. The European Consensus recognises that rural

1 www.iadb.org/projects/project.cfm?id=NI-T1103&lang=en, accessed 13 March 2010.

Table 6.4 **Determinants of risk aversion (ρ) in Nicaragua (N = 85)**

	Losses and shocks	Losses, shocks and controls	
	Coefficient	Coefficient	% change
Mitch-related losses			
• Home severely damaged/lost (yes = 1)	0.134*	0.166**	33
	(0.075)	(0.074)	
• Share of land lost (mz/mz)	0.090	0.114*	5
	(0.068)	(0.065)	
• Share of animals lost ($/$)	0.173*	0.228**	18
	(0.082)	(0.086)	
• Equipment or farm buildings lost (yes = 1)	-0.058	-0.076	
	(0.081)	(0.063)	
• Crops lost (yes = 1)	0.526**	0.464*	93
	(0.238)	(0.241)	
Occurrence of post-Mitch shocks (yes = 1)			
• Volcanic gases/ashes	0.039	0.057	
	(0.065)	(0.066)	
• Flood	-0.064	-0.113	
	(0.182)	(0.165)	
• Excessive rains (without flood)	0.052	0.074	
	(0.068)	(0.083)	
• Landslide	0.033	0.112	
	(0.163)	(0.106)	
• Drought	0.301***	0.307**	61
	(0.102)	(0.112)	
Heavily affected or resettlement area[a]	-0.198**	-0.184*	-37
	(0.084)	(0.088)	
Constant	-0.402	-1.156***	
	(0.278)	(0.344)	
Controls for personal characteristics and wealth	No	Yes	
Adjusted R-squared	0.36	0.44	

Notes: Standard errors in parentheses.

% changes are computed at median ρ for increasing the independent variable by one sd for continuous variables and shares or a unitary change for dummies.

a Wald tests do not reject equality of coefficients for the two location types.

*** p <0.01; ** p <0.05; * p <0.1.

This table differs slightly from the results presented in Van den Berg *et al.* 2009: we converted all monetary values to US$, merged the two location dummies, and added a regression without control variables. The results have changed only marginally.

poverty is a multi-dimensional problem that includes inequity in access to production factors, low health and education standards, degradation of natural resources and vulnerability to natural disasters. Rural development strategies should tackle all these issues, and risk management has to be mainstreamed. Central to the latter effort is the European Commission humanitarian aid department (ECHO)'s Disaster Preparedness programme (DIPECHO),[2] which targets vulnerable people living in the main disaster-prone regions of the world including Central America. DIPECHO tries to reduce risks by ensuring prior preparedness for the most vulnerable populations by strengthening local physical and human resources.

Our research provides a potentially important lesson for the DIPECHO programme and related organisations and initiatives. We find that experiencing natural shocks makes people more risk-averse. Put differently, natural disasters change the preferences of the affected population and the weighing of alternative survival strategies. As a consequence, risk strategies that *ex ante* seemed optimal may no longer be so after a major disaster has taken place. External support may stimulate people to take additional risk management measures, which *ex post* they would have opted for independently.

While the Netherlands, which alone is approximately the ninth largest donor to Nicaragua,[3] has made the impact of climate change a general policy focus (Ministry of Foreign Affairs 2007), Dutch development cooperation in Nicaragua does not explicitly tackle climate change and natural-hazard related issues. This is surprising, given that Nicaragua is an extremely disaster-prone country. Humanitarian aid has been provided after specific disasters, but this is a separate policy tool independent of development cooperation.[4] Our research suggests that this separation of development policy and humanitarian aid is artificial and could be counterproductive: disasters are not just short-term events; they also affect long-term growth and development. Development aid can decrease the vulnerability of the population for natural, and other, hazards and thus decrease the consequences of shocks and the human suffering in the wake of disasters. Similarly, humanitarian aid may increase the speed of recovery and limit the long-term growth and poverty consequences of natural hazards. This intertwining should be better recognised and exploited while not fully integrating the two types of policy, which could have the result that development funds are siphoned in case of large disasters.

Yet our research also provides support for some of the focal areas of Dutch development cooperation in Nicaragua, namely health, education and the development of small and micro-enterprises and commodity chains. Our research provides interesting insights into possible pathways out of poverty and policies that could support these pathways (Van den Berg 2010). Agricultural wage workers, farmers producing annual crops and those combining these two activities, are equally poor, and the

2 ec.europa.eu/echo/aid/dipecho_en.htm, accessed 13 March 2010.
3 www.minbuza.nl.
4 cccd.minbuza.nl/en/The_Ministry/Organisational_Structure/Policy_theme_departments, accessed 21 July 2010.

transition from agricultural worker to producer is therefore insufficient to escape poverty. Finding a full-time job in the non-farm sector seems a successful strategy to escape poverty. Alternatively, farming households could become engaged in the non-farm sector alongside annual cropping and possibly drop farming altogether later on. Yet these strategies seem open mainly to educated people in locations with a thriving non-agricultural sector. For others, adopting perennial crops or purchasing livestock could increase income. No formal education is required for these activities, although literacy and education do seem to facilitate them. These results provide strong support for the EU–Dutch policy focus on education, not only as a goal in itself but also as an instrument to decrease poverty. The results similarly support the health focus, as illness or death of an educated family member, resulting in the loss of a non-farm job or business, is the most obvious cause of downward mobility to a livelihood strategy with low remuneration. Finally, although in the recent past moderate economic growth has not led to a reduction in poverty, our results suggest that stimulation of the rural non-farm economy could facilitate growth out of poverty and thus support a somewhat targeted policy focus on economic development and private enterprises.

6.4 Further research

Our research provides a number of interesting findings, but much more work needs to be done to understand the complex relationship between natural hazards and poverty. Below, we give two examples of important gaps related to the *ex ante* effects of hurricane risk and the consequences of natural hazards for asset recovery and investment.

First, it is important to realise that we considered only the *ex post* consequences of a natural disaster. Knowing that they live in an area with high disaster risk, people may refrain from investing in more profitable activities even without a disaster actually occurring. Artur and Hilhorst (Chapter 7, this volume), for example, found that cyclical flooding in Mozambique has discouraged investment in productive assets that could be lost during flooding. Our research indicates that this is because of differences not only in risk perception but also in risk aversion. On the other hand, Pörtner (2008) suggests that higher hurricane risk means higher education in Guatemala, possibly because educated people are better able to deal with adverse situations or because human capital is less prone to destruction than physical capital. According to our results, this additional investment in education could have strong positive effects on welfare. It would therefore be worthwhile to analyse the *ex ante* effects of hurricane risk on welfare in more detail to balance possible positive and negative effects.

Second, more research is needed into the consequences of natural disasters for asset recovery and investment. As well as income, Mitch caused households to lose productive assets, and even though this apparently did not force them into livelihood strategies with lower remuneration (Van den Berg 2009), these losses affected and may still affect income and consumption (Van den Berg and Burger 2008). In a related study, Carter *et al.* (2007) find that in Honduras the medium-term effects of Hurricane Mitch differ according to initial household wealth. Their estimates suggest that households that begin beneath (or fall below) a low-asset threshold gravitate to a low-level equilibrium, much lower than that of their better-off neighbours. However, these estimates are based on a large number of strict assumptions, and we find no confirmation of their application for Nicaragua. But our research suggests that Mitch and other natural shocks could hinder recovery by making people more risk-averse (Van den Berg *et al.* 2008). However, whether the higher risk aversion that we measured in gambles in our experiments is associated with more risk-averse choices in real life remains a hypothesis to be tested. Hence, the overall effects of Mitch and other disasters for asset recovery warrant further scrutiny.

References

Adato, M., and R. Meinzen-Dick (2002) *Assessing the Impact of Agricultural Research on Poverty Using the Sustainable Livelihoods Framework* (Washington DC: International Food Policy Research Institute).

Anderson, L., and J. Mellor (2009) 'Are Risk Preferences Stable? Comparing an Experimental Measure with a Validated Survey-Based Measure', *Journal of Risk and Uncertainty* 39.2: 137-60.

Baez, J.E., and I.V. Santos (2007) *Children's Vulnerability to Weather Shocks: A Natural Disaster as a Natural Experiment* (New York: Social Science Research Network).

Barrett, C.B. (2005) 'Rural Poverty Dynamics: Development Policy Implications', *Agricultural Economics* 32.s1: 45-60.

Binswanger, H.P. (1980) 'Attitudes toward Risk: Experimental Measurement in Rural India', *American Journal of Agricultural Economics* 62.3: 395-407.

Boucher, S.R., B.L. Barham and M.R. Carter (2005) 'The Impact of "Market-Friendly" Reforms on Credit and Land Markets in Honduras and Nicaragua', *World Development* 33.1: 107-28.

Carney, D., M. Drinkwater, T. Rusinow, K. Neefjes, S. Wanmali, N. Singh (1999) *Livelihood Approaches Compared: A Brief Comparison of the Livelihoods Approaches of the UK Department for International Development (DFID), CARE, Oxfam and the UNDP* (London: DFID).

Carter, M.R., and C.B. Barrett (2006) 'The Economics of Poverty Traps and Persistent Poverty: An Asset-Based Approach', *Journal of Development Studies* 42.2: 178-99.

——, P.D. Little, T. Mogues and W. Negatu (2006) *Shocks, Sensitivity and Resilience: Tracking the Economic Impacts of Environmental Disaster on Assets in Ethiopia and Honduras* (Washington DC: International Food Policy Research Institute).

——, P.D. Little, T. Mogues and W. Negatu (2007) 'Poverty Traps and Natural Disasters in Ethiopia and Honduras', *World Development* 35.5: 835-56.

Chambers, R., and G. Conway (1991) *Sustainable Rural Livelihoods: Practical Concepts for the 21st Century* (Brighton, UK: Institute of Development Studies).

Cutchin, M.P., K.R. Martin, S.V. Owen and J.S. Goodwin (2008) 'Concern About Petrochemical Health Risk before and after a Refinery Explosion', *Risk Analysis* 28.3: 589-601.

Deaton, A. (1991) 'Saving and Liquidity Constraints', *Econometrica* 59.5: 1,221-48.

—— (1992) *Understanding Consumption* (Oxford, UK: Oxford University Press).

Deininger, K. (2003) 'Land Markets in Developing and Transition Economies: Impact of Liberalization and Implications for Future Reform', *American Journal of Agricultural Economics* 85.5: 1,217-22.

Del Ninno, C., P.A. Dorosh and L.C. Smith (2003) 'Public Policy, Markets and Household Coping Strategies in Bangladesh: Avoiding a Food Security Crisis Following the 1998 Floods', *World Development* 31.7: 1,221-38.

Dercon, S., J. Hoddinott and T. Woldehanna (2005) 'Shocks and Consumption in 15 Ethiopian Villages, 1999–2004', *Journal of African Economies* 14.4: 559-85.

EC (European Communities) (2006) 'The European Consensus on Development' (2006/C 46/01; Luxembourg: European Communities).

ECLAC (Economic Commission for Latin America and the Caribbean) (1999) 'Nicaragua: Assessment of the Damage Caused by Hurricane Mitch, 1998. Implications for Social Development and for the Environment' (Mexico City: United Nations ECLAC).

Ellis, F. (1998) 'Household Strategies and Rural Livelihood Diversification', *Journal of Development Studies* 35.1: 1-38.

Emanuel, K. (2005) 'Increasing Destructiveness of Tropical Cyclones over the Past 30 Years', *Nature* 436: 686-88.

Ersado, L., H. Alderman and J. Alwang (2003) 'Changes in Consumption and Saving Behavior before and after Economic Shocks: Evidence from Zimbabwe', *Economic Development and Cultural Change* 52.1: 187-215.

FAO (Food and Agriculture Organization) (1999) 'FAO/WFP Crop and Food Supply Assessment Mission to Nicaragua'. Special Report; www.fao.org/docrep/004/x1101e/x1101e00.htm, accessed 21 July 2010.

—— (2001) 'Analysis of the Medium-Term Effects of Hurricane Mitch on Food Security in Central America' (Rome: FAO).

Gitter, S.R., and B.L. Barham (2007) 'Credit, Natural Disasters, Coffee, and Educational Attainment in Rural Honduras', *World Development* 35.3: 498-511.

Hair, J.F., R.E. Anderson, R.L. Tatham and W.C. Black (1992) *Multivariate Data Analysis with Readings* (New York: Macmillan).

Hallegatte, S., and P. Dumas (2009) 'Can Natural Disasters Have Positive Consequences? Investigating the Role of Embodied Technical Change', *Ecological Economics* 68.3: 777-86.

——, J.-C. Hourcade and P. Dumas (2007) 'Why Economic Dynamics Matter in Assessing Climate Change Damages: Illustration on Extreme Events', *Ecological Economics* 62.2: 330-40.

Hoddinott, J. (2006) 'Shocks and Their Consequences across and within Households in Rural Zimbabwe', *Journal of Development Studies* 42.2: 301-21.

IPCC (Intergovernmental Panel on Climate Change) Working Group 1 (2001) *Climate Change 2001: The Scientific Basis* (Cambridge: Cambridge University Press).

Ito, T., and T. Kurosaki (2009) 'Weather Risk, Wages in Kind, and the Off-farm Labor Supply of Agricultural Households in a Developing Country', *American Journal of Agricultural Economics* 91.3: 697-710.

Jalan, J., and M. Ravallion (2001) 'Behavioral Responses to Risk in Rural China', *Journal of Development Economics* 66.1: 23-49.

Kazianga, H., and C. Udry (2006) 'Consumption Smoothing? Livestock, Insurance and Drought in Rural Burkina Faso', *Journal of Development Economics* 79.2: 413-46.

Kinsey, B., K. Burger and J.W. Gunning (1998) 'Coping with Drought in Zimbabwe: Survey Evidence on Responses of Rural Households to Risk', *World Development* 26.1: 89-110.

Kochar, A. (1999) 'Smoothing Consumption by Smoothing Income: Hours-of-Work Responses to Idiosyncratic Agricultural Shocks in Rural India', *Review of Economics and Statistics* 81.1: 50-61.

Kurosaki, T. (2006) 'Consumption Vulnerability to Risk in Rural Pakistan', *Journal of Development Studies* 42.1: 70-89.

Lerner, J.S., and D. Keltner (2001) 'Fear, Anger, and Risk', *Journal of Personality and Social Psychology* 81.1: 146-59.

Masozera, M., M. Bailey and C. Kerchner (2007) 'Distribution of Impacts of Natural Disasters across Income Groups: A Case Study of New Orleans', *Ecological Economics* 63.2-3: 299-306.

Ministry of Foreign Affairs (2007) 'Our Common Concern. Investing in Development in a Changing World' (The Hague: Ministry of Foreign Affairs).

—— (1995) 'Income Smoothing and Consumption Smoothing', *Journal of Economic Perspectives* 9.3: 103-14.

Mosley, P., and A. Verschoor (2005) 'Risk Attitudes and the "Vicious Circle of Poverty" ', *European Journal of Development Research* 17.1: 55-88.

Munich Re (2006) *Environmental Report 2005—Perspective: Today's Ideas for Tomorrow's World* (Munich, Germany: Münchener Rückversicherungs-Gesellschaft).

National Climatic Data Centre (2004) 'Mitch: The Deadliest Atlantic Hurricane since 1780', National Climatic Data Centre; www.ncdc.noaa.gov/oa/reports/mitch/mitch.html, accessed 13 March 2010.

Noy, I. (2009) 'The Macroeconomic Consequences of Disasters', *Journal of Development Economics* 88.2: 221-31.

Owens, T., J. Hoddinott and B. Kinsey (2003) 'Ex-Ante Actions and Ex-Post Public Responses to Drought Shocks: Evidence and Simulations from Zimbabwe', *World Development* 31.7: 1,239-55.

Paxson, C.H. (1992) 'Using Weather Variability to Estimate the Response of Savings to Transitory Income in Thailand', *American Economic Review* 82.1: 15-33.

Pörtner, C.C. (2008) *Gone with the Wind? Hurricane Risk, Fertility and Education* (Seattle, WA: Department of Economics, University of Washington).

Raghunathan, R., and M.T. Pham (1999) 'All Negative Moods are not Equal: Motivational Influences of Anxiety and Sadness on Decision Making', *Organizational Behavior and Human Decision Processes* 79.1: 56-77.

Rosenzweig, M.R., and K.I. Wolpin (1993) 'Credit Market Constraints, Consumption Smoothing, and the Accumulation of Durable Production Assets in Low-Income Countries: Investments in Bullocks in India', *Journal of Political Economy* 101.2: 223-44.

Townsend, R.M. (1994) 'Risk and Insurance in Village India', *Econometrica* 62.3: 539-91.

Trenberth, K. (2005) 'Climate: Uncertainty in Hurricanes and Global Warming', *Science* 308.5729: 1,753-54.

UNDP (United Nations Development Programme) (2008) *Reducing Disaster Risk: A Challenge for Development* (New York: UNDP, Bureau for Crisis Prevention and Recovery).

Van den Berg, M. (2009) 'Household Income Strategies and Natural Disasters: Dynamic Livelihoods in Rural Nicaragua', *Ecological Economics* 69: 592-602.

—— and K. Burger (2008) 'Household Consumption and Natural Disasters: The Case of Hurricane Mitch in Nicaragua', paper presented at the *12th EAAE Congress 'People, Food and Environments: Global Trends and European Strategies'*, Ghent, Belgium, 26–29 August 2008.

——, R. Fort and K. Burger (2008) *Natural Hazards and Risk Aversion: Experimental Evidence from Latin America* (Wageningen, Netherlands: Wageningen University).

——, R. Fort and K. Burger (2009) 'Natural Hazards and Risk Aversion: Experimental Evidence from Latin America', paper presented at the *27th Conference of the International Association of Agricultural Economists (IAAE)*, Beijing, 16–22 August 2009.

Weinstein, N.D. (1989) 'Effects of Personal-Experience on Self-protective Behavior', *Psychological Bulletin* 105.1: 31-50.

World Bank (2001a) 'Nicaragua Poverty Assessment: Challenges and Opportunities for Poverty Reduction' (Washington DC: Poverty Reduction and Economics Management Sector Unit, Latin America and the Caribbean Region, World Bank).

—— (2001b) 'Nicaragua Living Standards Measurement Study Survey: Post-Mitch Survey 1999. Supplemental Information Document' (Washington DC: Poverty and Human Resources Development Research Group, World Bank).

—— (2007) 'Nicaragua Country Brief'; go.worldbank.org/AYRB9G1UR0, accessed 13 March 2010.

World Bank Central America Department (2003) 'Nicaragua Poverty Assessment: Raising Welfare and Reducing Vulnerability' (Washington, DC: Central America Department, Latin America and the Caribbean Region, World Bank).

Yesuf, M., and R.A. Bluffstone (2009) 'Poverty, Risk Aversion, and Path Dependence in Low-Income Countries: Experimental Evidence from Ethiopia', *American Journal of Agricultural Economics* 91.4: 1,022-37.

Zimmerman, F.J., and M.R. Carter (2003) 'Asset Smoothing, Consumption Smoothing and the Reproduction of Inequality under Risk and Subsistence Constraints', *Journal of Development Economics* 71.2: 233-60.

7

Climate change adaptation in Mozambique

Luís Artur and Dorothea Hilhorst
Wageningen Disaster Studies

The global policy and scientific climate change communities are increasingly united in bringing the most urgent messages to the world. If the planet is to survive, drastic measures are needed to turn the tide of climate change (NEF and BCAS 2002; Simms *et al.* 2004). Despite the urgency of the matter, it is no easy task to align international governments and forge necessary measures on a global level. This chapter focuses on addressing the question of what happens to the message of urgency as it gets further removed from international meeting grounds.

We will use the case of Mozambique to argue that the message of climate change becomes part of everyday reality, where it derives local meaning, becomes subject to institutional dynamics and politicking and leads to outcomes that may have little resemblance to internationally agreed measures. The chapter concludes with the assertion that climate change adaptation, in order to be successful, must be made consonant with historically grown social and institutional processes and grounded in the adaptive capacities of local people.

This chapter looks specifically at issues of adaptation. There is an emerging consensus that climate change has progressed to such an extent that, along with mitigation, adaptation is necessary. Mitigation deals with the causes of climate change and mainly targets the production and consumption models of the developed countries. Adaptation deals with the consequences of climate change. As it is clear that climate change hits developing countries hardest, adaptation measures have been directed mainly at poor countries. Adaptation refers to complex processes that are more difficult to capture in 'plans for action' than mitigation. Mitigation depends on national governments and international negotiations, whereas adaptation is primarily a matter of individual households and local managers of natural resources in the context of local and regional economies and societies (Tol 2005: 573). Adaptation is largely spontaneous, and the direction in which systems adapt may not lead to more sustainable development that is coherent with mitigation objectives. Even though governments may attempt to improve the capacity to adapt, the mere existence of capacity is not in itself a guarantee that it will be used according to this plan (Burton *et al.* 2002: 150). Furthermore, the capacity to adapt varies considerably between regions, countries and socioeconomic groups and will vary over time (IPCC 2001: 879). Finally, decisions on adaptation to climate change are not isolated from other decisions; they occur in the context of socioeconomic and demographic changes as well as transformations in global governance, social conventions and the globalising flows of capital and labour. This makes it difficult to detach adaptations induced by climate change from actions triggered by other events (Adger *et al.* 2005: 78).

This chapter discusses the responses to climate change by authorities and people in Mozambique. It shows how climate change adaptation becomes subject to politics and social negotiation in the reality of implementation while drawing out policy implications from these findings. It is based on ongoing PhD research by Luís Artur in the context of the Vulnerability, Adaptation and Mitigation programme of the Netherlands Organisation for Scientific Research (NWO), including 18 months of fieldwork in Mozambique (from January 2007 to July 2008). It is informed by preceding research by Dorothea Hilhorst and Suzette Vonhof in 2004 into community-based climate change programmes of the Mozambican Red Cross. After introducing our conceptual approach to climate change adaptation and natural disasters, we will review issues of disasters and climate change in Mozambique. We will then summarise our findings on people's responses and draw conclusions.

7.1 Disasters and climate change

Why focus on natural disasters in a study on climate change adaptation? It has become accepted knowledge that climate change is a factor in the increase of weather extremes. The World Meteorological Organisation reported that 2005 broke numerous weather records all over the world, from drought in Brazil to cold spells in Pakistan and hurricanes in the Atlantic. The economic cost of disasters in 2005 was US$159 billion (IFRC 2005). For property insurers it was the costliest year ever.[1] The year 2005 also marked the worldwide attention for the increase in disaster risks and the potentials of risk reduction. In January, governments and international institutions came together in Kobe, Japan, at the World Conference on Disaster Reduc-

1 Press release, Swiss Re, 24 February 2005.

tion. The Plan of Action that was agreed on, the Hyogo Framework for Disaster Risk Reduction, underlines the importance of addressing underlying risks for disasters. Disasters kill people and livestock, ruin livelihoods, tear communities apart, destroy development efforts and have long-lasting effects on the environment. The Hyogo Framework reflects the growing awareness among civil society organisations, disaster scientists and a growing number of governments such as the United Kingdom that the impact of disaster risks, if well understood, can be reduced (DFID 2006). Disasters are directly undermining the ambitions of the Millennium Development Goals. Disaster risk reduction is important not just for saving lives, but also for protecting infrastructure, natural resources and development investments. This is not a singular effort. It touches on development planning, humanitarian aid, poverty alleviation, adaptation to climate change, sustainable development and the realisation of the Millennium Development Goals.

The number of floods, droughts, earthquakes, hurricanes and other natural hazards has increased over the past decades. During the 1970s, 911 disasters were reported, while in the period 1995–2005 more than 3,000 were reported (IFRC 2005; Guha-Sapir *et al.* 2003). Statistics on the number of people affected by disasters are difficult to interpret owing to a lack of international agreement on the definition of affected. The World Disaster Report nonetheless estimates that numbers of affected people have increased since the 1970s from 55 million to 250 million yearly. Statistics on the socioeconomic costs of disaster indicate an inflation-corrected increase in average yearly costs from US$13 billion in the 1970s to US$73 billion in the last ten years. The economic costs are mainly calculated for industrialised countries. Poor people affected by a disaster have few properties of value, which disappear from view in the statistics. When disaster costs are expressed in terms of percentage of GDP, the 'top ten' lists of affected countries is more or less turned upside down. Despite the increase in frequency and magnitude of disasters, the number of people killed has been reduced from a total of almost one million people in the 1970s to 515,000 people in the 1990s. This can be attributed to improved disaster preparedness, including a more effective delivery of emergency aid.

Disasters do not hit indiscriminately (see also Van den Berg and Burger, Chapter 6 in this book). Vulnerability is a crucial concept in understanding the impact of disaster. People are not equally exposed to natural hazards, because of social, economic and political factors (Wisner *et al.* 2003). Social class, gender, ethnicity, age group, income, health status and citizenship all influence people's vulnerability. The poor, the elderly, women, children and the handicapped are particularly vulnerable to disasters. Vulnerability to disaster increases particularly in less developed countries because of population growth and processes of marginalisation. On average, 13 times more people die per reported disaster in countries of low human development than in countries of high human development (IFRC 2004: 164). Poverty and population growth force people to move to steep slopes with the risk of landslides or to areas prone to floods or earthquakes. Disasters are exacerbated by the HIV/AIDS pandemic: the southern African famine was caused as much by the erosion of peoples' resilience due to AIDS as by the drought itself (De Waal 2002).

Moreover, disasters hit poor people disproportionately; the poor have fewer means to recover from disasters. For them a disaster is often a push back into poverty, which makes them more vulnerable to the next disaster.

7.2 Climate change and adaptation

Discussions on climate change adaptation have started to gain momentum since the Rio conference in 1992 and especially the Kyoto protocol in 1997. Although it is a relatively young concern, we can already distinguish two generations in the conception of climate change adaptation (Burton *et al.* 2002). Early definitions were focused purely on climate change. The IPCC (2001: 881) followed at around the turn of the century the definition of adaptation by Smit *et al.* (1999: 200) as 'the process of adjustment in ecological-social-economic systems in response to actual or expected climatic stimuli, their effects or impacts'. Currently, the IPCC (Adger *et al.* 2007: 720) uses a reformulated definition, which views adaptation as the process of adjustment or change that reduces vulnerability or enhances resilience in response to observed or expected changes in climate and associated extreme weather events.

The first definition fits into an approach that models climate change scenarios, develops impact models and derives from this data information on which regions, countries (and people) are vulnerable to the modelled impacts. This approach looks at vulnerability as an outcome of climate change and leads to responses that single out direct effects of climate change, for instance in water management, agriculture or disaster risk reduction. The second approach starts by looking at vulnerability as a present inability to cope with external pressures or changes, such as the ones posed by climate change. Rather than focusing on climate change scenarios, the second approach focuses on a wider range of contemporary social and economic processes and practices that bring about vulnerability and dampen adaptive capacity. This approach starts from existing policies and regulations to propose policies and measures to improve current adaptation measures (Burton *et al.* 2002: 157). This chapter follows the second approach. It looks comprehensively at the causes of vulnerability and takes a broad perspective of the policy fields that are relevant to climate change adaptation.

7.3 Mozambique: an overview of disasters and climate change

Mozambique is one of the poorest countries in the world, ranking 172 out of 182 on the Human Development Index (UNDP 2009: 145). Natural hazards such as droughts, floods, cyclones and related disasters have all been part of Mozambican history and can be said to have had an impact on shaping the country's poverty and vulnerability situation. Mozambique ranks third in global weather-related damage after Bangladesh and Ethiopia (Buys *et al.* 2007). In 2000, southern Mozambique was hit by a historic flooding that affected 4.5 million people and claimed about 700 lives. In 2001 another flood hit central Mozambique and affected another 500,000 people. From 2007 to 2009, flooding and cyclones have recurrently hit central and northern Mozambique. Some regions in the south are cyclically affected by drought. In a country where most people rely on agriculture, changes in rain patterns may easily turn hazards such as floods and droughts into disasters.

A report by the National Institute for Disaster Management (INGC) (2009) shows the effects of climate change in Mozambique. Over the last 50 years, the average temperature has increased by 1.6°C, the rainfall season has started later and there has been an increase in the length of dry spells. Data from EM-DAT, which is at the basis of most publications on disasters in Mozambique (Christie and Hanlon 2001: 13-14; GoM 1988: 27; INGC *et al.* 2003: 10-12; Negrão 2001: 3-4; World Bank 2005: 29) show that natural hazards have increased in frequency and intensity over the past decade.

Since 1970, Mozambique has been hit by 77 disasters of which 41 (53%) occurred just in the decade 2000–2009 (see Fig. 7.1). There is an increase of nearly 50% in the number of people affected by natural hazards when compared with the previous decade. In 2005, an estimated 94% of the population was affected by natural hazards (Mafambissa 2007: 5). The number of people killed has more than tripled compared to the previous decade. The increase may partly reflect an improved capacity for data collection, and the substantive increase cannot be attributed to climate change alone. Growing population, environmental degradation and limited alternative sources of livelihood tend to increase people's vulnerability to disasters. There is mounting evidence that poverty-reduction programmes have failed to reach the poorest segments of society (Hanlon 2007; James *et al.* 2005; UNICEF 2007).

Droughts have historically been the major hazard affecting nearly 23 million since 1970. Nonetheless, flooding has become the major natural hazard since 1990. This trend may suggest a greater improvement in drought and cyclone management than for floods. The cyclone early warning system has improved owing to a US$4.7 million donation from the US Agency for International Development (USAID). Among other things, irrigation schemes have been expanded for drought management and water reservoirs have been established in a number of drought-prone areas (Marques *et al.* 2002). NGOs such as the Mozambican Red Cross, World Vision and ActionAid have also introduced new technology in drought-prone areas. In comparison, flood management has been more complex, for reasons that will be elaborated below.

Figure 7.1 **Trends of major natural hazards in Mozambique**

Source: Artur (forthcoming) based on EM-DAT

Figure 7.2 **Number of affected people by type of hazard**

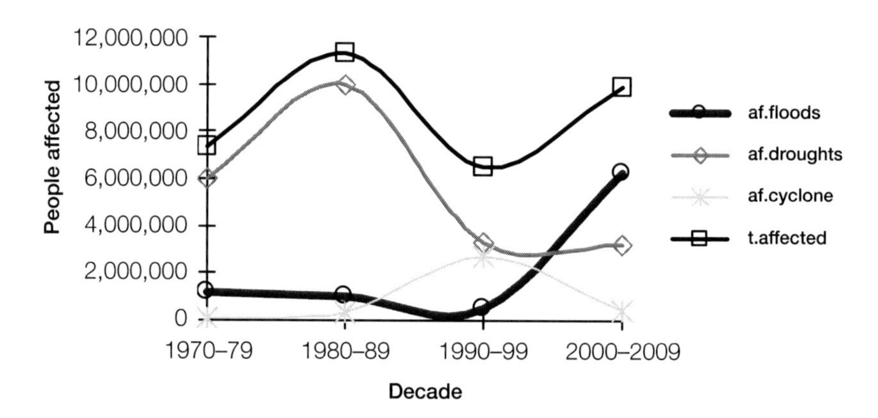

Source: Artur (forthcoming) based on EM-DAT

In response to extreme weather events in Mozambique, a national master plan on disaster management was approved in 2006. The INGC has shifted its policy frame from disaster response to disaster preparedness. The Ministry of Environment has produced the National Action Plan for Adaptation to Climate Change and the Mozambican Meteorological Institute has established an early-warning system to warn vulnerable, isolated people against erratic and extreme weather events. All these measures and further interventions on disaster risk reduction rely on external donor funding and the participation of NGOs and actors at a local level.

7.4 Government response to climate change: between economic growth and environmental concerns

Since the Rio summit in 1992, Mozambique has gradually developed a legal framework to cope with climate change. Under the coordination of the Ministry for Coordination of Environmental Action (MICOA), different working groups and regulatory tools have been established. An inter-institutional working group carried out the first national inventory on greenhouse gases and vulnerability assessment in 1998. This group also led the implementation of the Clean Development Mechanism (CDM) and the National Action Plan for Adaptation (NAPA). Ever since, the country has produced different legal instruments related to climate and sustainable development. These include: (i) Regulation for Environmental Quality Standards and Emissions, (ii) Regulation for Environmental Inspection, (iii) Environmental Strategy for Sustainable Development, (iv) Environmental Law, (v) Environmental Impact Assessment Regulation, (vi) Land Regulation, (vii) Land Legislation, (viii) Water Resources Legislation, (ix) Energy Policy, (x) National Program for Environmental Management, (xi) Regulation for Mine Activities, (xii) Territorial Planning Law.

Despite these legal tools, the overall achievement of environmental sustainability (Millennium Development Goal 7) is unlikely to be achieved by 2015 (GoM and UN 2005). It turns out that, in practice, the government is specifically focusing on Millennium Development Goal 1 (halve the proportion of people living in extreme poverty), which it hopes to achieve in 2010. It appears that economic growth competes with environmental concerns.

A major economic policy has been to encourage private foreign investments through different incentives such as limited taxation. Although some of these incentives seem climate-friendly, such as a 45,000 ha licence for Jatropha production for biofuels and investments in hydropower dams, others will certainly increase greenhouse emissions in the years to come. These include the granting of 300,000 ha for coal exploration to the Australian mining company Riversdale. Similarly, the Brazilian company Vale do Rio Doce will extract about 2.4 billion tons of coal over a renewable period of 25 years. Forests, in the meantime, are being 'depleted' by the exportation of wood to China; what some ironically refer to as the 'Chinese takeaway' (Mackenzie 2006).

Given the extreme poverty and lack of employment and basic infrastructures, it is understandable that the government is preoccupied with development concerns rather than investments in climate change whose returns may only be visible in the long run. The effect is, however, that climate change adaptation becomes a competing claim in the allocation of budgets and investment decisions by the government. The optimistic suggestion that climate change adaptation may lead to economic growth and that economic growth can provide the resources for adaptation may be true, but it cannot be denied that decision-making today often entails negotiation between the two interests.

7.5 Actors and stakes in climate change in Mozambique: power struggles and strategies

In recent years, internationally funded programmes for climate change adaptation have been increasing. Major programmes in Mozambique in 2009 were:

- Joint Programme on Environmental Mainstreaming and Adaptation to Climate Change in Mozambique funded by the Millennium Development Goals Achievement Fund (MDG-F) Spain and the United Nations Development Programme (UNDP)
- Mozambique Poverty and Environment Initiative funded by the government of Ireland
- Joint Programme on Strengthening Disaster Risk Reduction and Emergency Preparedness funded by UNDP
- Coping with Drought and Adaptation to Climate Change funded by the Global Environment Fund (GEF)
- A World Bank study that is expected to result in a fund for climate change adaptation

This has led to competition among government bodies to handle these large programmes. As mentioned earlier, the Ministry for Coordination of Environmental Action (MICOA) was initially the coordinating body for climate change adaptation. Development agencies have questioned MICOA's authority by claiming that climate change is more a development than an environment problem. In 2005, the government of Mozambique created a Ministry for Planning and Development (MPD). In order to mainstream climate change in the national development process, the MPD has started to claim the coordination of climate change interventions. This has created conflicts between MICOA and MPD over leadership, which became apparent when both ministries wanted to handle a World Bank project for integrating small-scale farmers into the market, with a large component for climate change adaptation. Another player is the Ministry for Science and Technology (MCT), which has the mandate to handle interventions regarding science and technology. MCT created a working group on climate change in 2007.

The INGC has also increased its influence on climate change adaptation programmes. It has broadened its mandate by incorporating prevention, vulnerability reduction and reconstruction and development of drought-prone areas. Owing to its strong reputation, nearly all international actors have a preference for working with the INGC. The INGC is handling climate-change-related projects with UNDP, the UK Department for International Development (DFID), Gesellschaft für Technische Zusammenarbeit (GTZ), the World Bank, the Food and Agriculture Organisation (FAO) and the Nordic countries among others.[2] The INGC has also produced

2 Interview with Dr Bonifácio Antonio, Coordinator Office Manager INGC, 28 May 2009.

the first national study on climate change. At the launch of the study in Maputo in May 2009, which Luís Artur attended, the coordination of climate change became a contentious issue, and many guests from competing institutions perceived the initiative as an attempt by the INGC to claim the leadership.

Municipalities are also emerging as actors in the climate change arena. When UN-Habitat launched a project aiming to develop mitigation and adaptation mechanisms in two cities in Mozambique, representatives of the municipalities were keen to defend their lawful duty to develop any intervention in their areas of jurisdiction. Development agencies such as the Mozambican Red Cross, Care International, GTZ and World Vision are also unilaterally or in partnerships implementing projects related to climate change . And lastly, the private sector is stepping into the climate change arena. The Matola Gas Company, Cimentos de Moçambique and the electricity company Electricidade de Moçambique have all asked certificates to access funds under the Clean Development Mechanism.[3]

The emerging characteristic of this unfolding arena is a lack of coherent strategy and leadership for dealing with climate change. This leads to a dispersion of projects, efforts and funds. It is unclear what the outcomes of the dispersed initiatives will be for adapting to climate change at the local level. Actors use the room for manoeuvre created by the lack of coordination to develop and implement climate change interventions according to their own understanding, needs and rules.

7.6 People's responses to climate change

The Zambezi Delta is part of a larger catchment area of the Zambezi River in Africa. The Zambezi is the fourth largest flood plain river in Africa, the largest watercourse in the sub-Saharan region and the largest system flowing into the Indian Ocean (Davies *et al.* 2000: 1; Shela 2000: 65). In Mozambique, the Zambezi Delta covers areas of Mutarara, Mopeia, Caia, Chinde, Marromeu, Inhassunge and Quelimane districts. Based on these districts and using data from the 2007 census (INE 2007), the delta area shelters about 965,859 people, mainly from the Sena ethnic group, who build their livelihoods essentially from agricultural production, fishing and 'petty' trading. These activities are indeed highly influenced by flooding regimes that have been part of the Zambezi Delta history. Records collected by Chidiamassamba and Liesegang (1997) date flooding as far back as 1586 and different authors have recorded more than 21 great floods in the 19th century alone, whose dimension and impacts have remained in people's minds (Taveira 1943; Monteiro 1955; Chidiamassamba and Liesegang 1997; Beilfuss 2005).

The Zambezi Delta has been inhabited for years by people who, despite seasonal flooding, took advantage of the Zambezi River and its ecosystem to build liveli-

3 Interview with Felício Fernando from MICOA, 27 May 2009.

hoods. Changes produced over the past 30 years due to an amalgam of factors such as dams, changes in political ideologies, droughts, civil war, displacement, resettlement and population growth tended altogether to push people to live on the flood plains and to rely heavily on natural resources. In doing so, people became physically exposed to flooding and adaptation to flooding became a crucial element in forging livelihoods. Flooding continues to be the major natural hazard on the delta and over the past ten years the frequency and magnitude of flooding has increased steadily. There were three heavy floods (above 10,000 m^3/s) recorded over the period 2000–2009 with none in the previous decade and only one of that magnitude in the decade 1980–1989. People have developed different strategies over time to adapt to flood risk. This was studied in Cocorico 'community' along the flood plains of the Mopeia district. Luís Artur lived in Cocorico for about six months and for a further year travelled back and forth to Cocorico.

In 2007, Cocorico had about 300 households and nearly 1,500 inhabitants.[4] Due to its location at the junction of the Zambezi River and its tributary Cuacua, Cocorico is the most flood-prone area of the entire Mopeia district and has a long history of flooding. By Western standards Cocorico is a poor community. Most of the households have a house and a granary. There is only one road linking Mopeia main village to Cocorico but it is hardly drivable by car. Livelihoods in Cocorico are forged mainly around agriculture and fishing with a clear division of labour compelling women to agriculture and domestic spheres while men are involved in fishing and trade. Polygamy and multiple households are common. In most cases the wives live distant from each other, and men have a tendency to establish one household on the flood plains where they do their fishing, and one in the higher areas that remains flood-free. All these arrangement are part of historically evolved patterns of household adaptation to flooding.

People in Cocorico have developed adaptation practices to flooding that range from the construction of their houses, to their investment strategies, to all kinds of social arrangement that can be understood as protecting people against floods. Houses in Cocorico are often placed on higher grounds and have thatched roofs. They are made of grass and wood, which is not necessarily a matter of poverty and lack of financial resources, although this may be a relevant factor. In group discussions and individual interviews people suggested that it would be 'irrational' to build conventional houses when there will be a regular flooding that would wash away the investment. They also suggested that grass and wood houses resist flooding events better. Some people learned to build brick houses during a period of forced migration to Malawi and attempted to build some in Cocorico. They made bricks, built the houses and used clay to cement the houses and join the bricks. Their houses were the first to disappear during the 2001 flood.

Savings, investments and insurance are not clearly distinguishable in Cocorico. Canoe and fishing nets are, for instance, investments but also represent savings in a context where saving in currency, jewellery or animals is restrained by flooding

4 Interview with chief of Cocorico, 2 July 2007.

regimes. They can also be used as collateral or insurance for credit, marriage and other circumstances. A survey comparing the flood plain of Cocorico with the upper Mopeia village revealed that households on the flood plains tended to have higher investments in canoes, fishing nets, radios and small-scale poultry (Table 7.1). This shows the relevance of practices that combine everyday livelihoods needs of local people with forms of insurance, saving and preparedness measures for flooding events. Canoes have multiple purposes, including fishing, trading and means of transportation during extreme floods. Another clear example is the investment in radios, which is also a priority investment in the flood plains. The survey found more households owning radios (67%) in lower Cocorico than in the upper Mopeia village (57%). Investment in radio has multiple rewards. As environmental cues for flood forecast become less reliable owing to environmental changes, people tend to rely on radio for information about flooding. In Cocorico, 62% of the households were informed about the 2007 flooding through radio. Listening to the radio is therefore a preparedness measure for floods. But the radio has more everyday functions: to catch up with emerging or profitable markets for fish or agricultural products, and as a socialising tool in a society where festivities and drinking are customary practice. On the other hand, cyclical flooding has tended to discourage investments in productive and non-productive items that would be lost during flooding. There is a tendency to avoid accumulating big animals such as goats and cattle or large furniture such as mattresses, beds and tables. Similarly, items for domestic use such as pots, dishes, cups and spoons are possessed mainly according to their utility. In other words, people's assets are not merely an expression of lack of capital or accumulation, but have a partial meaning as a form of adaptation to recurrent flooding.

People also adapt to flooding through social arrangements that develop over time. They make individual households dependent on each other through marriage, the sharing of gains and responsibilities and, within the household, by the interdependency of females and males, elderly and younger people. Through group interdependency individual households can expect help from others during times of crisis such as extreme flooding. This system was historically guarded by traditional authority, which in the contemporary context is being challenged and re-shaped by new developments. It is also reinforced by among other things the festivities and drinking that are important aspects of social life in the flood plains.

Adaptation to flooding incorporates the physical, natural, political, sociocultural, human, symbolic and economic assets, and a successful adaptation emerges from a delicate balance of these different aspects. However, such a balance is currently not possible owing to rapid changes and different actors' interests along the Zambezi Delta. Increasingly the local physical environment is in degradation and the social fabric is less supportive; as a consequence neither the environment nor the social fabric are able to provide enough protection against disaster-related hazards such as flooding.

It is important to realise that the threats to the lifestyle that people have developed in relation to flooding are manifold. They are environmental, economical, political,

Table 7.1 **Investment priorities on the Cocorico flood plain and Mopeia village**

Group statistics

	Study area	N	Mean	Std. deviation
Radios	Cocorico	83	.8313	.74603
	Mopeia village	114	.6579	.63565
Canoes**	Cocorico	83	.6386	.83488
	Mopeia village	114	.1316	.41033
Plates**	Cocorico	49	5.6327	1.39484
	Mopeia village	105	8.9238	5.40956
Cups**	Cocorico	49	4.2449	1.25051
	Mopeia village	104	5.6923	3.64725
Bikes	Cocorico	82	.6585	.61302
	Mopeia village	110	.6091	.76740
Fishing nets*	Cocorico	82	.7439	.95337
	Mopeia village	109	.3486	.59901
Spoons**	Cocorico	50	3.5800	1.53981
	Mopeia village	71	5.7887	3.35481
Chickens/ducks	Cocorico	81	5.2099	6.61951
	Mopeia village	113	3.6283	4.91389
Goats	Cocorico	81	.4815	1.81046
	Mopeia village	113	.6637	1.94390
Pigs	Cocorico	81	.6296	1.61589
	Mopeia village	113	.5575	1.44505
Cows	Cocorico	81	.0000	.00000[a]
	Mopeia village	113	.0000	.00000[a]
Wives*	Cocorico	83	2.5000	1.06904
	Mopeia village	114	2.0667	.25820
Income per month	Cocorico	80	1.2200	.50669
	Mopeia village	110	1.3000	.56747

a. t cannot be computed because the standard deviations of both groups are 0.

** statistical significant differences at 1%; * statistical significant differences at 5%.

Source: Luís Artur, based on fieldwork data

social and cultural. The physical environment of the delta has been under continuous change. Damming upstream, reduced precipitation, population increases and related needs for crop-cultivation plots and firewood as a source of energy, and timber logging is driving deforestation and accelerating the erosion process. An example of a more sociopolitical element is the erosion of 'traditional' authorities. Traditional authorities tended throughout the history of the delta to administer the group interdependency system that allowed some disaster risk reduction. Ironically, development can also be a factor in undermining resilience to disaster. The manifold development initiatives in the area tend to put local people in contact with external actors so they become aware of new opportunities and lifestyles. As a result, they start to challenge their own perception of social ordering. In group discussions, people suggested that community solidarity is decreasing nowadays as people tend to be more individualistic and concerned with their own affairs. Increasingly people tend to ask for payments for what in the past used to be a social activity. The different factors all contribute to explaining the trend that the number of people affected and killed by floods has increased over the last ten years.

The fieldwork in Cocorico makes it clear that addressing adaptation to climate-related hazards is much more complex than measures directly geared to the environmental reduction of disaster risk. It requires looking at the historical and contemporary processes beyond climate and disaster to analyse how these affect vulnerability to climate-related hazards such as flooding. It is important to study these processes by combining quantitative and qualitative methods. Adaptation is often incorporated in everyday livelihood strategies. Many of the adaptation practices are tacit, and people may be scarcely aware of their adaptive properties.

7.7 Conclusion

We can draw five major conclusions from the analysis presented here:

1. The emergency discourse prevailing around climate change does not resonate with institutional actors in the climate change adaptation arena in Mozambique. While actors experience negative impacts of climate change and take the adaptation agenda seriously to some extent, in everyday practice it gets incorporated into ongoing social processes of ordering and differentiation

2. The stakes of climate change are high in two ways. Climate change is having visible effects on Mozambique and results in increased vulnerability to natural disasters. The stakes are also high in the sense that international attention for climate change has opened up an arena where contestation over ideas and resources on climate change adaptations takes place at all different levels

3. Technocratic responses to the challenge of climate change adaptation are being politicised in practice, as actors endeavour to appropriate them according to their own interpretations and interests
4. For local people, climate change is a reality, but they do not perceive it as a separate factor. It becomes one of the elements of change that they have to deal with
5. Many of the local adaptations to climate change consist of tacit livelihood and sociocultural arrangements that people do not consciously connect to adaptation, yet have a certain functionality in the adaptation process

For both authorities and people, this means that climate change and climate change adaptation become incorporated in the continuity of everyday practice. These conclusions have different policy and research implications.

First, it is important to invest in empirical research facilities, both quantitative and qualitative, in the processes of climate change adaptation. This kind of research is important to provide reality checks on adaptation policies. This is also important because research into the social realities of climate change adaptation can reveal emergent properties that have positive or negative effects for adaptation. These can be used as feedback into policy cycles on adaptation.

Second, the conclusions call for a policy approach to climate change adaptation that is process-oriented instead of technocratic and top-down. Actors negotiate adaptation in their everyday practice. The more explicit these negotiations are, the better they can become subject to policy interventions and agreed change toward adaptations. This means investing in multi-stakeholder governance arrangements around climate change adaptation.

Finally, it is important to invest in monitoring and adjusting capacities for climate change adaptation interventions. In view of the analysis presented here, the distinction between planned and spontaneous adaptation practically dissolves, as actors appropriate planned interventions into their 'spontaneous' strategies. Monitoring for unintended outcomes of programmes, both positive and negative, should enable adjustment of policy to changing contexts and outcomes.

References

Adger, W., S. Agrawala, M. Mirza, C. Conde, K. O'Brien, J. Pulhin, R. Pulwarty, B. Smit and K. Takahashi (2007) 'Assessment of Adaptation Practices, Options, Constrains and Capacity', in M. Parry, O. Canziani, J. Palutikof, P. van de Linden and C. Hanson (eds.), *Climate Change 2007: Impacts, Adaptation and Vulnerability: Contribution of the Working Group II to the Fourth Assessment Report of the Intergovernmental Panel on Climate Change* (Cambridge UK: Cambridge University Press): 717-43.

——, N. Arnell and E. Tompkins (2005) 'Successful Adaptation to Climate Change across Scales', *Global Environmental Change* 15: 77-86.

Artur, L. (forthcoming) 'Continuities in Crisis: Everyday Practices of Disaster Response and Climate Change Adaptation in Mozambique' (PhD thesis, Wageningen University).

Beilfuss, R. (2005) 'Understanding Extreme Floods in the Lower Zambezi River', paper presented at the *Seminar on the Water Management on the Delta of Zambezi*, Maputo, 5–6 September 2005.

Burton, I., S. Huq, B. Lim, O. Pilifosova and E. Schipper (2002) 'From Impact Assessment to Adaptation Priorities: The Shaping of Adaptation Policy', *Climate Policy* 2: 145-59.

Buys, P., U. Deichmann, C. Meisner, T. That and D. Wheeler (2007) *Country Stakes in Climate Change Negotiations: Two Dimensions of Vulnerability* (Washington, DC: World Bank).

Chidiamassamba, C., and G. Liesegang (1997) 'Dados históricos sobre ocorrência e tipos de cheias no vale do Zambeze', paper presented at the *Workshop on Sustainable Use of Cahora Bassa Dam and Zambezi Valley*, Songo, 29 September – 2 October 1997.

Christie, F., and J. Hanlon (2001) *Mozambique and the Great Flood of 2000* (London: Long House Publications).

Davies, B., R. Beilfuss and M. Thoms (2000) 'Cahora Bassa Retrospective, 1974–1997: Effects of Flow Regulation on the Lower Zambezi River', *Verhandlung Internationalen Verein Limnologie* 27: 1-9.

De Waal, A. (2002) 'New Variant Famine: How Aids has Changed the Hunger Equation'; allafrica.com/stories/200211200471.html, accessed 15 March 2010.

DFID (UK Department for International Development) (2006) 'Reducing the Risk of Disasters: Helping to Achieve Sustainable Poverty Reduction in a Vulnerable World: A DFID Policy Paper', 30 March 2006; www.dfid.gov.uk/news/files/disaster-risk-reduction-launch.asp, accessed 13 March 2010.

GoM (Government of Mozambique) (1988) 'Rising to the Challenge. Dealing with Emergency in Mozambique: An Inside View' (Maputo, Mozambique: GoM).

—— and UN (2005) 'Report on the Millennium Development Goals' (Maputo, Mozambique: GOM).

Guha-Sapir, D., D. Hargitt and P. Hoyos (2003) *Thirty Years of Natural Disasters 1974–2003: The Numbers* (Louvain, Belgium: Centre for Research on the Epidemiology of Disasters [CRED]/Université Catholique de Louvain).

Hanlon, J. (2007) 'Is Poverty Decreasing in Mozambique?', Paper no. 14 presented at the *Instituto de Estudos Sociais e Económicos (IESE) Conference*, Maputo, 19 September 2007.

IFRC (International Federation of Red Cross and Red Crescent Societies) (2004) *World Disasters Report* (Geneva: IFRC).

—— (2005) *World Disasters Report* (Geneva: IFRC).

INE (Instituto Nacional de Estatísticas) (2007) 'III Censo Geral da População e Habitação. Resultados preliminares' (Maputo, Mozambique: INE; www.ine.gov.mz, accessed 13 March 2010).

INGC (National Institute for Disaster Management) (2009) *Synthesis Report. INGC Climate Change Report: Study on the Impact of Climate Change on Disaster Risk in Mozambique Synthesis* (ed. B. van Logchem and R. Brito; Maputo, Mozambique: INGC).

——, UEM (University Eduardo Mondlane) and FEWS NET (Famine Early Warning Systems Network) (2003) *Atlas for Disaster Preparedness and Response in the Limpopo Basin* (Maputo, Mozambique: FEWS NET).

IPCC (Intergovernmental Panel on Climate Change) (2001) *Climate Change 2001: Impacts, Adaptations and Vulnerability. Contribution of Working Group II to the Third Assessment Report of the Intergovernmental Panel on Climate Change)* (Cambridge, UK: Cambridge University Press).

James, R., C. Arndt and K. Simler (2005) 'Has Economic Growth in Mozambique Been Pro-Poor?' (FCND Discussion Paper 202; Washington, DC: International Food Policy Research Institute).

Mackenzie, C. (2006) 'Forestry Governance in Zambézia, Mozambique: Chinese Takeaway!' (Report prepared for Forum of NGOs in Zambézia; Quelimane, Mozambique: FONGZA).

Mafambissa, F. (2007) 'Efeitos dos desastres naturais na produção agrícola de culturas alimentares e na segurança alimentar', Paper no. 21 presented at the *Instituto de Estudos Sociais e Económicos (IESE) Conference*, Maputo, 19 September 2007.

Marques, M., M. Vilanculos, J. Mafalacusser and A. Ussivane (2002) 'Levantamento dos regadios existentes no país' (Maputo, Mozambique: Ministério da Agricultura e Desenvolvimento Rural).

Monteiro, G. (1955) *SSE: Sessenta anos de açúcar na Zambézia 1893–1953* (Lisbon).

NEF (New Economics Foundation) and BCAS (Bangladesh Center for Advanced Studies) (2002) *The End of Development? Global Warming, Disasters and the Great Reverse of Human Progress* (London: New Economics Foundation).

Negrão, J. (2001) 'O impacto socioeconómico das cheias: Oração de Sapiência por ocasião da abertura do ano lectivo 2001–2002', Maputo, Mozambique; www.4shared.com/file/35806396/7e92d7a5/O_IMPACTO_SCIO.html, accessed 22 March 2010.

Shela, O. (2000) 'Management of Shared River Basins: The Case of the Zambezi River', *Water Policy* 2: 65-81.

Simms, A., J. Magrath and H. Reid (2004) *Up in Smoke? Threats from and Responses to the Impacts of Global Warming on Human Development* (Working Group on Climate Change and Development; London: new economics foundation/International Institute for Environment and Development).

Smit, B., I. Burton, R. Klein and R. Street (1999) 'The Science of Adaptation: A Framework for Assessment', *Mitigation and Adaptation Strategies for Global Change* 4: 199-213.

Taveira, J. (1943) *Açucar de Moçambique: A SSE, Limitada* (Lourenço Marques, Mozambique: Imprensa Nacional de Moçambique).

Tol, R. (2005) 'Adaptation and Mitigation: Trade-off in Substance and Methods', *Environmental Science and Policy* 8: 572-78.

UNDP (United Nations Development Programme) (2009) *Human Development Report 2009: Overcoming Barriers: Human Mobility and Development* (New York: UNDP).

UNICEF (2007) *Childhood Poverty in Mozambique and Budgetary Allocations* (New York: UNICEF).

Wisner, B., P. Blaikie, T. Cannon and I. Davis (2003) *At Risk: Natural Hazards, People's Vulnerabilities and Disaster* (London: Routledge).

World Bank (2005) 'Learning Lessons from Disaster Recovery: The Case of Mozambique' (Working Paper 12; Washington, DC: World Bank).

8

Adaptation to climate change induced flooding in Dutch municipalities

Maya M. van den Berg, William M. Lafferty and Frans J.H.M. Coenen
Twente Centre for Studies in Technology and Sustainable Development (CSTM), The Netherlands

8.1 Introduction

Based on the most recent data, the Intergovernmental Panel on Climate Change maintains that climate change is an inevitable development. Current observations prove that many natural systems around the globe are already being affected by climate change (Schneider *et al.* 2007). Even if we were able to curb our emissions today, the climate will change owing to the accumulation of greenhouse gases (GHG) emitted in the past. Thus, even from a sceptical point of view, it is now widely recognised that we must begin to initiate adaptation measures now.

The issue of climate change adaptation has thus far been viewed as primarily a 'top-down' initiative, with national governments taking a lead on the issue. In its 2009 climate adaptation white paper, however, the European Commission stresses the crucial role of the local level, where most of the detailed knowledge on local characteristics is available and where civil awareness can be most effectively raised (CEC 2009). The Dutch government has also published its own National Adaptation Strategy (NAS) (VROM 2007a). Known generally as the ARK Programme, the strategy also refers to the importance of local-level government.

In the Netherlands, climate change impacts are primarily a result of excess water entering the Dutch delta.[1] In addition to adjusting for increased precipitation, the NAS focuses on sea-level rise and higher river discharges. Both the national government and the 2008 Delta Commission focus on the most vulnerable areas of flooding,[2] while the impacts from climate change will be felt locally throughout the country. In its report on climate change, the Netherlands Scientific Council for Government Policy deals with these local impacts, such as a changing water system and shifting agricultural activities (Ministerie van Algemene Zaken 2006). The Council states that, given the scale of local impacts, municipalities will have to develop solutions to deal with climate change.

The local level is thus crucial, with numerous decisions affecting vulnerability decided at the local level. Whereas a country as a whole can often be considered resilient, local areas can nonetheless be vulnerable due to their contextual economic structure, geographic situation and infrastructure (Næss *et al.* 2005). As a preliminary step in determining local-level vulnerability, several assessment models were reviewed in an earlier phase of the project discussed in this chapter (Clausen 2007). It was also established in this early phase that local adaptation to climate change was attracting greater research attention (Adger and Vincent 2005; Adger *et al.* 2005; Wall and Marzall 2006; Wilson 2006). There were, however, very few studies that addressed the role of local-level government in adaptation initiatives in a multi-level governance structure; particularly as to how local institutional capacity affects the level of preparedness.

This chapter thus applies a local perspective to adaptation strategies by investigating the effect of institutional capacity on local initiatives within a multi-government context. The goal is an improved knowledge base for both scientists and politicians involved in the development of local adaptation policy.

First, we briefly discuss the terminology related to climate change preparedness, and present our research questions. We then provide an updated overview of Dutch local-level preparedness for climate change impacts (Section 8.2), before proceed-

1 The leading Dutch climate change scenarios—considered as the national standard for adaptation questions—are formulated by the Royal Netherlands Meteorological Institute (KNMI). Apart from variations in expectations on temperature rise (+1°C or +2°C) and air currents (changed or unchanged), the scenarios show similar characteristics of climate change in the Netherlands. According to the KNMI (2006) we can expect the following primary climate change effects: sea-level rise, increased average temperatures, more summer droughts, a greater number of heat waves, increased winter precipitation and greater overall levels of precipitation. (Secondary effects result from these, e.g. higher river discharges due to temperature rise.)

2 This climate change induced flooding risk is problematic since the traditional dykes and water embankments are not designed to withstand sea-level rise or structurally higher water levels. The 2008 Delta Commission—advising the national government on becoming a 'climate-proof' country—recommended to enforce all dykes tenfold (Deltacommissie 2008). Today, the inland river protection system is already being adapted to the needs of the latest climate change scenarios.

ing to a documentation of patterns of local preparedness (Section 8.3). Finally, we draw preliminary conclusions on possible future roles for local governments in a changing climate context (Section 8.4).

8.1.1 Climate change adaptation terminology

The study of preparedness for climate change impacts operates with several key concepts, such as vulnerability and adaptation. In addition the idea of 'risk' is also important, an idea that combines both the magnitude of impact and the probability of its occurrence. As the idea captures uncertainty in the underlying process of climate change exposure, it is essential for decision-makers (Schneider *et al.* 2007). In the Netherlands, many types of risk have been visualised through regional 'risk maps'.[3] Apart from the exceptional risk of flooding, however, other climate-induced risks are generally not treated.

Vulnerability can be seen as the degree to which a system is susceptible and unable to cope with climate change impacts (Klein *et al.* 1998). The Netherlands is considered to be one of the most vulnerable areas in Europe since the majority of the Dutch live below sea level and 70% of Dutch GNP is actually earned below sea level (Kolen *et al.* 2009).

Stemming from evolutionary biology, the term adaptation is relatively new to the climate change field. In the Netherlands, traditional adaptation is reactive, as it has generally involved dyke reinforcement in the aftermath of flooding (e.g. the Delta coastal defence system after the 1953 North Sea Flood). At present the actual planning and implementation of effective adaptation is quite modest in the Netherlands due to limited technical, financial and institutional capacity. Planned adaptation is thus either very limited or very costly (Schneider *et al.* 2007). As a consequence, adaptation is rarely undertaken in response to climate change effects alone. Some success in practical implementation has been seen when measures to address climate change risk are incorporated into existing decision structures. Smit and Wandel (2006) call this the 'mainstreaming' of adaptation to climate change.

8.1.2 Research design and questions

One of the aims of the project was to investigate the level of local governmental preparedness for climate change impacts. A major aim was to explore the actual role of specific local governments in relation to current and future adaptation to climate change impacts. Research question 1 was thus formulated as follows: what is the current and projected role for local-level government within a multi-level governance model for climate adaptation? Further, with an eye to future adaptation initiatives, the second research question focuses on the capacity at the local level: how far does institutional capacity influence the possibilities and limitations for developing local adaptation strategies, and how can this capacity be expanded?

3 Available at www.risicokaart.nl (in Dutch only).

The research design rests on the clear premise that impacts from natural, weather-related events (principally in the present case, flooding) are not new. What is at issue now is the specific 'increment' of future impacts that are specifically attributable to climate change. We thus focus on relating past experience with severe flooding impacts to future probabilities of climate change related impacts. In our case, these 'severe' floods have been recorded in the International Disaster Database EM-DAT. The approach leads to the conceptual fourfold scheme presented in Table 8.1.

For ease of communication and analysis, the logic of the scheme is synthesised in terms of four 'ideal types'. At one end of the continuum, 'spectators' are seen as units that have no serious history of exposure to flooding events and a relatively low probability for climate change impacts in the future. At the opposite end, 'veterans at the front' are seen as units that have both a history of exposure to flooding events and a relatively high probability of climate change impacts in the future. In between these two types, we have 'veterans in reserve' (with an earlier history of flooding impacts, but low probability of impacts from climate change) and 'recruits at the front' (with no significant history of exposure to flooding events, but high probability of future impacts of climate change).

Table 8.1 **Fourfold table for assessing the level of adaptation action in Dutch local government**

		History of exposure to a severe flooding event	
		No	Yes
Projected risk of negative climate change impact	Low	Spectators	Veterans in reserve
	High	Recruits at the front	Veterans at the front

To increase the general scope of relevance of the analysis, we have further differentiated the four types as to 'rural' and 'urban' characteristics. This generates a grid of eight possible conceptual categories, based on three relevant criteria. The rural–urban dimension was largely operationalised according to size of population. The history of exposure dimension was operationalised in relation to the 1953 North Sea Flood and the 1993/1995 high waters,[4] events that could be assumed as 'settled'

4 In 1953, dykes in the south-western parts proved not to be resistant to the combination of spring tide and a north-westerly storm. Around 1,800 people died, comparable to Katrina's death toll in New Orleans in 2005. In the following decades a heavy set of coastal works was carried out to prevent the threat from the sea once and for all. In 1997, the project was finalised with the deliverance of the Maeslant Storm Surge Barrier in the Nieuwe Waterweg, the main entrance to the Port of Rotterdam. The risk of flooding is

in the institutional memory of the relevant cases. And the projected risk dimension was operationalised in relation to a 'regional risk map' for climate-related flooding (see above and reference at footnote 1). A preliminary scoping of relevant municipalities with respect to these three dimensions was then conducted, resulting in the selection of specific units as indicated in Figure 8.1. These eight units thus provide the eight empirical 'cases' for the analysis.

Figure 8.1 **Map of the Netherlands with case selection indicated**

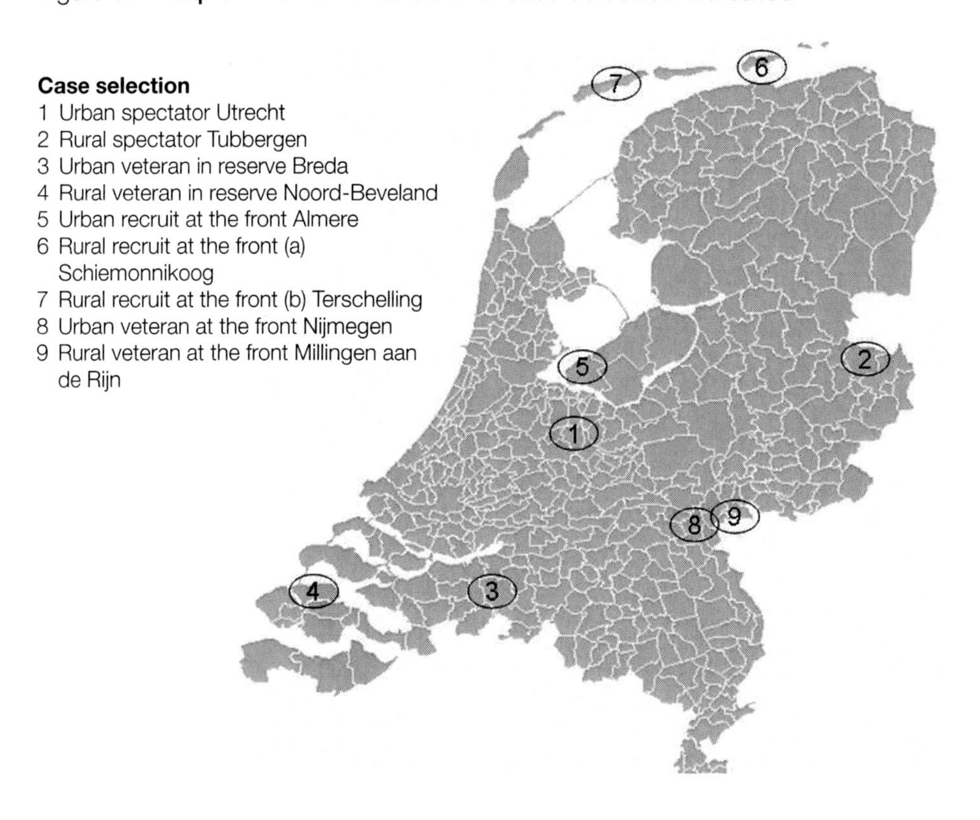

increasing due to sea-level rise.

Four decades later, the high waters of 1993 and 1995 are the most recent climate-related events that were recorded in the EM-DAT. In both cases, dykes proved to be stable in the end. During the 1995 'near-flooding' event, the largest post-war evacuation took place: 250,000 people (and all livestock) were forced to move.

8.2 Local-level climate adaptation in Dutch climate policy

Before discussing local-level climate adaptation in a Dutch context, we briefly deal with the role of local authorities in Dutch climate policy. In the Netherlands, environmental policy has its roots in local-level policy. Until the 1970s, the local level was responsible for environmental policy. At the end of the 1960s, more and more local-level environmental tasks shifted towards the national level owing to the complexity and importance of environmental issues. In the 1980s and '90s the national level tried to improve the local-level environmental duties. Traditionally, the national government considers local environmental policy as a municipal task financed by central funding. A long discussion on who should pay what in local-level environmental policy was only solved by research showing that municipalities had a severe deficit in funding the extension of environmental tasks in the 1970s; from the 1990s onwards local authorities receive earmarked funding to bring their environmental policy to an adequate level (Coenen 2001). Today, local authorities consider climate policy as part of environmental policy.[5]

By signing a climate agreement with the national government, the Dutch municipalities[6] have now joined in the national climate adaptation goals (VROM 2007b). It is here agreed that local authorities will inventory climate adaptation measures fitted to its existing policies. The agreement maintains that, of all governing layers, the local level has the closest connection to civilians and companies and for that reason can set an example. The local authorities also have a specific facilitation and information role *vis-à-vis* the public. The local level is seen as best capable of mobilising people to organise effective actions and acting as a stimulator in placing climate change on the local agenda. In practice, however, our data-gathering showed a more complicated reality. In the following sections we focus on the first research question: what is the current and projected role for local-level government within a multi-level governance model for climate adaptation?

5 VNG, the Association of Netherlands Municipalities, defines 'climate' as one of the components of the environmental policy domain on its website: www.vng.nl (accessed October 21 2009); this also counts for the Ministry of Housing, Spatial Planning and the Environment (Ministerie van Volkshuisvesting, Ruimtelijke Ordening en Milieubeheer—VROM: www.vrom.nl/pagina.html?id=4178 (accessed 24 February 2009).

6 When classifying local government systems, the Netherlands is placed within the 'Napoleonic' system, which is characterised by a relatively high degree of national governmental control (Kok *et al.* 2002). The Dutch 'decentralised unitary state' is based on agreement between all levels of government. Within this co-government system, local authorities have the freedom of initiative as long as they take higher-government legislation into account.

8.2.1 Legal and financial constraints for climate adaptation

Local-level government in the Netherlands is based on the Municipal Law (1851),[7] which prescribes some environmental tasks, such as an annual Environmental Policy Plan, but does not deal with climate-related issues. The Disaster Law (1985)[8] prescribes the municipal tasks involved with disasters and heavy accidents, and it only deals with the possibility of 'regular' extreme events such as flooding or extreme weather events. More latent developments, such as an increasing average temperature, are not covered. Further, the most important law on the environment, the Environmental Conservation Law (1993),[9] does not deal with climatic issues. Besides this legal deficit, the existing legislation is also considered to be insufficient as it proves not to be flexible enough to deal with a changing climate (Verschuuren 2007). Anticipating this, the public campaign Cool Climate urged sympathisers to join in its call for a climate adaptation law.

Several interviewees stressed a lack of tools available to enforce adaptation measurements within their communities. They state that the municipality is unable to implement its adaptation strategies sufficiently, for example, the municipality cannot oblige real estate developers to build climate-proof houses. This also has internal effects: local administrators first focus on the tasks that are obliged by higher-level governments. Local-level voluntary activities such as climate adaptation policy have low priority and funding is minimal.

There were also constraints on funding. Compared to other EU member states, Dutch local government is largely dependent on the national government for its finances. No less than 68% of all income of the lower governments comes from the national government. About half this central funding is earmarked funding; the other half originates from the Municipal Fund, from which budgets are distributed based on criteria such as the number of participants. Another 9% of the lower government's income originates from its own taxes; this is the lowest percentage in the EU, except for Malta, which hardly has a local governmental layer (CBS 2008). The many earmarked funds leave little room for voluntary tasks such as climate adaptation policy. Budgetary constraints thus pose limits on what can be attempted in the way of climate change adaptation.

8.2.2 Institutional involvement in local-level climate adaptation

In spite of the fact that climate change induces many climatic and meteorological effects, Dutch local government commonly 'translates' its impacts into increasing precipitation quantities and an increasing frequency of heatwaves. Therefore, the policy domains involved in climate change adaptation strategies in general are the spatial planning and environment departments. Concerning local water manage-

7 Municipal Law (Gemw) (1851); wetten.overheid.nl, accessed 22 March 2010.

8 Disaster Law (Wrzo) (1985); wetten.overheid.nl, accessed 22 March 2010.

9 Environmental Conservation Law (Wm) (1993); wetten.overheid.nl, accessed 22 March 2010.

ment, municipalities can rely on the regional water-board system, which prevents the Dutch from getting wet feet by maintaining waterways and dykes; it also distributes water equally and is responsible for the purification of waste-water. We will deal with this special governmental level in sub-sections 8.2.3 and 8.2.4.

A broader local-level institutional involvement could also include: disaster management, public health, economic affairs including recreation, economic strategic planning (such as climate change impact assessments on local industries) and institutional adaptation. The case studies showed a remarkable variation in institutional involvement (see Table 8.2).

As well as increasing precipitation quantities and a higher frequency of heatwaves, climate change also affects the domain of public health, which is a municipal responsibility. The municipal healthcare system is organised by the Public Health Services (Gemeentelijke Gezondheidsdienst) (GGD 2009). The GGD obtains its knowledge from the National Institute for Public Health and the Environment (Rijksinstituut voor Volksgezondheid en Milieu) (RIVM 2009),[10] the leading Dutch centre of expertise in the fields of health, nutrition and environmental protection. The RIVM has published several reports on the health effects of climate change on the Netherlands. Owing to its close ties with the RIVM, the GGD appears to be well aware of the effects of climate change on public health, such as an increasing frequency of heatwaves and new diseases entering the country. Yet preventive measurements should be assigned by the municipality itself. In the case studies we found only one case that considered public health to be an issue in its climate adaptation strategy. In addition, a GGD interviewee confirmed that in the region of Twente, no municipality asked the regional GGD for additional measurements.

Climate change also involves an increase of climatic disasters and extreme weather events. Disaster management is an important task for local government: the mayor is responsible for public order and acts as the commanding officer in disaster management. However, the cases showed only a limited institutional involvement from this policy domain. In the case of the 'urban veteran at the front', about 20 civil servants constituted the municipal safety department. Many disaster-management plans are written and revised here. Yet climate change is considered to be 'too big and too slow' to act on. No action is taken since hazardous climate change effects—being extreme weather events and flooding events—are not considered to be 'new' but as already a part of the disaster plans.

In the following section we concentrate on this local-level civil preparedness more thoroughly by describing the current role for local government within a multi-level governance model for climate adaptation. The case studies will be discussed according to the four dimensions distinguished in Table 8.1.

10 www.rivmvoorlichtingscentrum.nl, accessed 22 March 2010.

Table 8.2 **Institutional involvement in local-level climate adaptation**

	Environment (trigger of adaptation file)	**Public space (adjusting public spaces to heat and/ or water storage)**	**Local-level water management (adjustments for water storage and smart discharge)**	**Spatial panning (adjusting new housing development areas to heat and/or water storage)**	**Economic affairs (plans for recreation in the new climate)**	**Public health (involvement of public health issues in adaptation)**	**Internal adaptation measures (awareness rising, cross-department corporation)**	**Disaster management (adjusting planning and training system to new climate conditions)**
Urban spectator	X	X	X	X			X	
Rural spectator			X					
Urban veteran in reserve	X	X	X	X	X	X	X	
Rural veteran in reserve			X					
Urban recruit		X	X	X				
Rural recruit (a)	X		X					
Rural recruit (b)	X		X					
Urban veteran at the front	X	X	X	X			X	
Rural veteran at the front	X		X				X	

8.2.3 The current role of local government in climate adaptation

Spectators

'Spectators' run no increased risk of climate change impacts and have no history of large-scale climate impacts before. This combination of low risk and little or no experience provides a certain baseline picture of local climate adaptation. This baseline offers a starting point to study the concept of climate adaptation and civil preparedness.

In the selected case for 'urban spectators' we found an 'integral approach' to climate change and a proactive way of thinking to adapt to it.[11] The case is in the process of formulating an adaptation policy. Further, the 'urban spectator' is a front-runner in climate change mitigation strategies. On the other hand, the 'rural spectator' case shows a 'minimum-level approach'.[12] The interviewees explain this difference in scope from their size. The 'urban spectator' runs economic and societal risks from climate change (because of its huge economic value and a large population), whereas the 'rural spectator' runs low economic and societal risks. Furthermore, as small municipalities are obliged to fulfil the same tasks as larger municipalities, the larger ones have more capacity to focus on voluntary tasks, such as (currently) climate adaptation strategies.

This general conclusion does not, however, explain the huge differences between the two cases and can perhaps be better understood in context. In its present climate action programme, the 'urban spectator's' municipal board[13] stresses the presence of many research institutes working with climate change. This 'favourable' circumstance is surprising, as we did not anticipate effects from such a 'coincidence' factor. Yet this 'favourable' factor might explain best why the 'urban spectator' is so remarkably active, in addition to its relatively high economic and societal risks. There is also a striking difference in political orientation: the rural case has a right-wing municipal board, whereas the municipal board of the 'urban spectator' (until recently) consisted of left-wing parties with a Green alderman responsible for environmental affairs.[14]

11 The concept of an 'integral approach' towards climate change adaptation is chosen to describe a broad, multi-disciplinary view on climate change preparedness; in Table 8.2, the local-level approach is visualised in the concerned policy domains.

12 A 'minimum-level approach'—as opposed to the 'integral approach'—stands for a narrow view on climate change preparedness. In practice, the 'minimum-level approach' means a nationally stimulated adjusting of the local sewage system. Table 8.2 shows all cases being involved in the 'minimum-level approach'.

13 In the Netherlands, the Municipal Board is the management of the municipal organisation on a daily basis. The Board consists of a centrally appointed mayor and, depending on the number of municipal inhabitants, between two and eight locally elected aldermen.

14 The Dutch political spectrum varies from left-wing to right-wing parties. In general, left-wing parties are progressive and more 'green minded' (environmentally minded) than right-wing, conservative parties. One of the Dutch left-wing parties is GroenLinks (in English: GreenLeft), the Dutch Green political party. Aldermen from this party are called Green aldermen throughout the chapter.

Our 'rural spectator' proved to be preparing for climate change impacts on a minimum level. This was to be expected, as the triggers of experience and increased risk are absent and its rural character implies low capacity for voluntary tasks. More striking was the 'urban spectator', which appeared to be very active on climate adaptation. As indicated, we feel that this can be explained by an interactive favourable context of climate research activity and a Green alderman.

Veterans in reserve

'Veterans in reserve' are cases that have faced weather-related events in the past, but are now believed to be freed from increased risk, including from climate change impacts. The criteria appear to be contradictory, but we did find some relevant cases that fitted into the 'veteran in reserve' category. A sense of urgency was low for both cases, and it was stressed that time is needed to develop an integral organisational awareness. Political support for the 'urban veteran' is ensured by a Green alderman responsible for the environment. For the 'rural veteran' case, with a board to the political right, support was much lower. All in all, the 'urban veteran' emerged as a mitigation front-runner.

The integral approach of the 'urban veteran in reserve' has similarities to the 'urban spectator', whereas the narrower approach of the 'rural veteran in reserve' is roughly comparable to the 'rural spectator'. Similar to the 'urban spectator', the 'urban veteran in reserve' has a Green alderman responsible for environmental policy and a 'favourable' local context in the form of a regional urban network willing to cooperate in progressive climate policy, along with a stimulating provincial actor stressing the need for adaptation. Whereas the 'urban veteran in reserve' stresses the opportunities that climate adaptation brings, the 'rural veteran in reserve' has a more limited view on adaptation. The trigger of experience made no apparent difference for both 'veterans in reserve'. Only in the rural case did we confirm a striking risk perception by an interviewee, not because of memories of the impacts from the 1953 North Sea floods (see footnote 4), but because the national government is considering adjusting the coastal defence works surrounding the island in favour of the densely populated parts to the north. Furthermore, disaster management is described as inadequate because national funding for this is limited because of the thinly populated area.

Climate-related experiences did not seem to have made much of an impact on the local preparedness level of the 'veteran in reserve' cases. We found the 'urban veteran in reserve' to be very active on climate adaptation, but this appears to stem from a stimulating urban network including an active province, also here combined with an enthusiastic Green alderman responsible for the issue. Similarities to the 'urban spectator' are striking. The activities of the 'rural veteran in reserve' are much more limited. Apart from adapting the sewage system to increased precipitation, and a minor adaptation in the municipal mitigation activities, the degree of civil preparedness is at a minimal level.

Recruits at the front

'Recruits at the front' are cases without any experience of climate-related events in the past, but who now face a greater than average risk of climate change impacts in the future. This category thus offers us insights into the impact of projected risk with no major experience of extreme weather impacts. We can thus elaborate on the previous case categories dealing with the impacts of the combinations of 'no risk and no experience' and 'no risk and experience'.

Despite the risk profile of the 'recruit' cases, the urban case interviewees deny running increased risks, mainly because of the national government's promise not to increase the water level of the neighbouring Marker Lake. In a similar manner, the rural case interviewees stressed the national governmental role to protect the population by dykes and dunes. The cases run a much higher risk of flooding compared to municipalities on higher grounds in the eastern or southern parts, but from a local perspective this risk is not perceived as increasing because of climate change.

The 'urban recruit' is involved in a large-scale spatial development project involving the construction of 60,000 new houses. These dwellings must be constructed in less-favourable areas, such as on polders lying below sea level, or on Marker Lake. Heavier precipitation and drought are among the scenarios being considered in the development phase. Besides this project, no adaptation initiatives have been undertaken apart from minimum-level sewage system adjustments. The 'rural recruits' also do not appear very concerned with climate change adaptation. Climate change mitigation is one of the motivations for the joint mitigation project (aiming at 'carbon neutrality' in 2020), but adaptation has no separate role.

The 'recruits' appear to have developed their own perspectives on climate change impacts. While one is mostly interested in climate change effects on nature, the other focuses primarily on climate change effects and tourism.[15] There is, however, no sense of urgency and no new climate-related policies are being pursued. This narrow approach can partly be explained by a lower overall capacity, but a more crucial factor would appear to be the general self-interpretation of the municipality's duties (the 'perceived role'). Both 'rural recruits' emerged as very strict in defining their responsibilities in relation to the responsibilities of other levels of government. Here we see a striking similarity with the 'urban veteran in reserve', which also considers safety and protection from flood as a national governmental task.

Summing up on this dimension, the perception of climate-related risks does not seem to have a significant impact on the cases monitored. In both the urban and rural cases the increased risk for flooding proved to be of less impact at the local level, since the cases feel they already are protected by initiatives from the national government. The rural cases are involved in mitigation and thus interested in cli-

15 ECLAT-COAST, another project within this VAM research programme, primarily focuses on tourism. The project investigates the influence of climate and weather on tourist demand in Europe's coastal zones.

mate change, but adaptation strategies appear to be a 'non-issue'. The islanders apparently do not fear the consequences of either heavier precipitation or sea-level rise. The urban case considers climate change (that is, increased precipitation and heatwaves) in its enlargement plans, but there has always existed a certain level of risk perception because of its location at 5 metres below sea level.

Veterans at the front

As for the 'veterans at the front', both the rural and urban cases experienced climate-related events and face an increased risk of more to come. This maximum effect along both dimensions clearly triggers a maximum of local preparedness. Interestingly, we have seen earlier that the effects of a single trigger experience or increased risk do not emerge as significant in explaining adaptation activities. At this stage, the two front-runners—the 'urban spectator' and the 'urban veteran in reserve'—appear to be influenced by more general internal and external factors, such as a dynamic Green alderman and a motivating network of climate change research and interest.

The 'urban veteran at the front' is very proactive on a broad number of adaptation measurements involving water, public green areas and urban planning. In addition to activity on climate change mitigation, political support is also institutionalised through a Green alderman responsible for environmental affairs. The 'rural veteran at the front' on the other hand is a small-scale front-runner in waste management differentiation. Both 'veterans' are member of the regional environmental and waste network that primarily focuses on mitigation. Because of its large impact, the network is dominated by the 'urban veteran at the front', which actively pursues an integral approach to climate adaptation. The experiences of 1993 and 1995 (see footnote 4) can also explain this pioneering role in addition to the presence of a Green alderman and being a mitigation front-runner. Fresh memories of flood experience have proved to provide a perfect 'window of opportunity' for a progressive water plan written by an enthusiastic Green alderman.

The 'rural veteran' shows clear memories of the 1995 evacuation, but nobody appears to be frightened by the expectations of more water to come. The national government is 'trusted' to provide protection, and is thus considered to be the problem-owner. This also demonstrates a perceived role of local-level dependence on the national government to solve increased risks induced by climate change. We see here a clear similarity to the 'rural veterans in reserve' and the 'rural recruits'. The 'rural veteran at the front' does not take any adaptation measurements apart from the usual sewage system adjustments. An interviewee stressed his concerns for a broader implementation of adaptation and an adjusted safety policy, but both are seen as lacking political support. For the 'rural veteran at the front', therefore, a history of impacts does not appear to lead to markedly increased preparedness.

We have found a very progressive 'urban veteran at the front' and a rather passive 'rural veteran at the front'. The 'urban veteran at the front' is very progressive in its civil preparedness because of political support and a context (the river) that triggers action. This corresponds to a pattern we have already observed in the other categories. Next to the 'urban spectator' and the 'urban veteran in reserve' we can add an 'urban veteran at the front'. The most important triggers in all cases seem to be active support by political leaders and a diversity of local contextual factors that support more active adaptive initiatives (the presence of respected climate research institutions; a well-informed and concerned 'urban' network; and an ever-hazardous river).

8.2.4 Local-level preparedness within a multi-level governance system

The water boards are clearly present to fulfil their duty as regional flooding protectors.[16] Most of the cases are on good 'speaking terms' with the concerned water boards so they can 'fine tune' local strategies, but mostly it is the water board that tells municipalities to meet certain rules. Where primary water dams are involved, the National Executive Authority on Water Management (Rijkswaterstaat) is responsible. Yet, despite its regional presence, the role of Rijkswaterstaat in general is seen as problematic. While local authorities experience cooperation with the water boards as taking place on an equal basis, the Rijkswaterstaat is felt to be rather distant and incommunicative. A clear hierarchal gap is thus present between the national Rijkswaterstaat and the municipalities involved.

When we enquired about a potential future role for the local government in climate adaptation, most changes were expected to occur within their own municipal organisation. This means taking time for local awareness raising or simply 'waiting' for climate change related events to occur. More external demands are not salient at this moment, except for a general wish for more effective coordination at the national level and some demands for funding. The interviews offered a bottom-up perspective on the division of roles in the multi-level governance system (Table 8.3). The roles listed at the bottom of the table are those most often expressed by the local informants.

The national government is considered to be a coordinator and facilitator. Some interviewees indicated a strong need for national recognition for their adaptation efforts ('rural veteran at the front'), while some will only start with their adaptation measurements if national support is forthcoming ('rural veteran in reserve' and 'urban recruit'). Others do not need more national support but expect that the state will upgrade and adapt the dykes ('rural recruits'). One interviewee also suggested a national adaptation 'toolbox' with more specific guidelines ('urban vet-

16 The inhabitants of the Dutch delta have always tried to adjust water patterns by developing new earthworks and other modes of creative water engineering. On a social-administrative level, the water challenge has also led to new forms of social cooperation at an early stage. Already in the middle ages, local communities were involved in local water-management cooperation, and by the 13th century these communities developed into the forerunners of today's water boards.

Table 8.3 **Division of roles in a multi-level government in climate change adaptation from a bottom-up perspective**

Level of government	Desired role from a bottom-up perspective
National level	Facilitating and coordinating role: channelling knowledge and exchange of best practices
Regional level	Pioneering and steering role: crucial in the present context (urban veterans and urban spectator). This level is not necessarily represented by the province. It can also be a water board or another supra-local body such as an urban network
Local level	Executive role: awareness raising, local vulnerability assessments, local mitigation plans. Capturing local diversity to supplement top-down governance

eran in reserve'). Another interviewee stressed the need for national attention to be paid to adaptation since that would motivate local administrators to start working with climate adaptation from the top down ('urban veteran at the front').

The importance of the national government setting an example was frequently mentioned. This could be in an international context, with the Netherlands representing 'best practice' on water protection ('urban spectator'); but it was also viewed as counting within the Netherlands. From another perspective, however, the responsible ministry was viewed as failing in the integration of spatial planning and adaptation, and a wish was expressed that it more actively propagate an interdisciplinary approach ('rural spectator'). Some interviewees experience the national-level government to be 'distant' in both senses of the word ('rural recruits', 'rural veteran in reserve' and 'rural spectator'), while for others a concern over climate change is considered to be part of this distant state ('rural spectator', 'rural recruits').

8.3 Key variables in local climate adaptation

Local-level preparedness for climate change impacts in the Netherlands is influenced by many internal and external factors. As indicated earlier, we have employed the following variables in our case selection: risk (projected risk of negative climate change impacts), experience (history of exposure to a specific type of extreme weather event) and size (urban and rural character). Furthermore, during the data gathering the effect of the local context emerged as an important aspect to consider. We have, therefore, added this factor as our fourth variable. We briefly discuss the four variables below, and then proceed to analyse the second research question about the 'internal' effects of institutional capacity.

8.3.1 The role of local-level government

The application of the dimensions of the fourfold table (Table 8.1) was designed to inductively explore the selected case studies as to the effects of experience, projected risk and size on civil preparedness for climate change adaptation by local-level government.

The effect of risk perception

To determine the effect of risk perception, low- and high-risk cases have been studied. When comparing the four low-risk cases with the five high-risk cases, no clear-cut differences can be noticed between the two groups (Table 8.4).

Table 8.4 **Fourfold table with the results on risk perception and experience**

		History of exposure to a specific type of extreme weather event	
		No	Yes
Projected risk of negative climate change impact	Low	Rural spectator: – – Urban spectator: – +	Rural veteran in reserve: – + Urban veteran in reserve: ++
	High	Rural recruits (a and b) at the front: – + Urban recruit at the front: – +	Rural veteran at the front: – + Urban veteran at the front: ++

Results are reflected from a minimum-level approach (– –) up to an integral-approach (++)

The 'rural spectator' does not feel threatened by climate change impacts (translated into heavier precipitation), while the 'rural veteran in reserve' fears an increased flooding risk but feels protected (and somewhat relieved of responsibility) by higher levels of government. Despite the lack of a clear threat, the low-risk urban cases appeared to be nonetheless very active in the area of climate policy. This can probably be explained by a more favourable political environment due primarily to Green aldermen with designated responsibility for environmental issues. Climate change effects (translated into more precipitation and heatwaves) are considered to have great impact on the city and an institutional sense of urgency contributes to higher levels of awareness, concern and action.

The high-risk cases show different individual characteristics but nonetheless a common institutional awareness of increased risk. Climate change induced risks (interpreted as increased flooding risk) are not perceived as constituting major threats and protection is, once again, expected from the national government. For the high-risk urban cases, both recognise a broad range of climate change effects (heatwaves, increased flooding risks, increased precipitation). The major difference here is the sense of urgency: the 'urban recruit' is quite careless about its risk profile, whereas the 'urban veteran' is much more concerned about its vulnerability. This difference too can be explained from the attitude towards the national government. The 'urban recruit' counts fully on national protection, whereas the 'urban veteran at the front' feels a greater need to act autonomously. This difference can be primarily explained from the experience the 'urban veteran at the front' has.

Risk perception is essential in climate change adaptation. Following Lorenzoni *et al.* (2007) action on adaptation is taken when: (1) risks are known, and (2) resources are available to minimise these risks. However, next to this awareness and the means to take specific measures, a sense of urgency is also crucial. In virtually all cases a certain risk perception is present. All interviewees showed clear awareness of the importance of climate change (generally interpreted as more precipitation, sometimes also increased frequency of heatwaves and increased river discharge), but for only some cases was the lack of adaptive means considered to be the next barrier (the 'urban spectator', the two 'veterans in reserve' and the 'urban veteran at the front').

For this reason, in addition to risk awareness and institutional capacity, we must add a further dimension for action: the 'perceived role' of responsibility and potential for local government units. Even though a certain risk perception is present in all cases, this does not imply that the issue is given high priority on the political agenda. Only when the local authority experiences a high risk and considers climate adaptation to be one of its tasks will actual adaptation initiatives be placed on the political agenda. It is only at that point that a sense of urgency arises because of a lack of resources, which in turn creates an additional barrier to adaptation. Both 'urban veteran' cases express a clear sense of urgency and identify in this light the absence of resources to be the major obstacle for taking action.

The effect of experience

The impact of experience also emerges from a more general comparison. When comparing the five experienced 'veteran' cases with the four non-experienced 'recruit' and 'spectator' cases, the two groups show distinguishing characteristics. Not surprisingly, the 'veteran' cases show a tendency towards greater preparedness than their less-experienced counterparts (Table 8.4).

Experience implies a clear perception of vulnerability in the 'veteran' cases, which clearly can affect views on the perceived role of local-level government in climate change issues. In both 'urban veteran' cases the issue of adaptation has been placed on the political agenda. And in both cases, a Green alderman is responsible for municipal climate programmes that include several adaptation measures. The 'rural veteran' cases, on the other hand, do not show marked political attention. Their risk perception is concentrated on a national government dependency for protection. A lack of resources to minimise risks is considered to be a result of national government rules (for example, poorly developed rescue services for the 'rural veteran in reserve').

The non-experienced 'recruit' and 'spectator' cases show no clear sense of risk, except for the somewhat anomalous 'urban spectator', which is highly involved in preparations for climate change adaptation. The 'urban recruit' shows minimal-level preparations that stem from a low political awareness that can be attributed to a lack of experience. Experience was apparently also an issue for one of the 'rural recruits', but due to the special local character of the event in question (heavy rain showers) no political follow-up ensued. Similar to the high-risk rural cases, both 'rural recruits' and the 'rural spectator' look to the national government to take the appropriate adaptive measurements for protection. In short, they perceive the national government to be the problem-owner. This 'indifferent' attitude can be explained by both a low level of political attention and perceived limited means for climate change initiatives.

Experience also appears more generally in the case comparison to make a difference as it increases the overall effect of risk. If risk is perceived, the factor of experience causes a heightened sense of urgency (for the urban 'veteran in reserve' and both the urban and the rural 'veteran at the front'). Further, if a sense of urgency is present, it appears to lead to a concern over the lack of means. Strikingly, however, both the experienced and non-experienced rural cases do not see a need to take action themselves, but rather trust the national government for protection.

The effect of size

Next to the impacts of risk perception and experience, it appears that an even more determinative factor among the cases is the question of size (rural or urban as measured by population). In our sample, urban communities have invariably large populations and rural communities have, in most cases, very small populations. Relating the size variable to vulnerability, there is obviously a greater risk attached to the more populous units, and, not surprisingly, the case studies show a striking difference between urban and rural cases. The rural cases thus tend to depend more strongly on greater governmental involvement for their security, and, more specifically, to expect more from the national and regional levels of governance to avert impacts from climate change. The larger urban cases, on the other hand, are more generally well informed about the effects that will occur and have greater resources to estimate the impacts. Three of our urban cases thus qualify as 'adaptation front-runners', although they too are definitely not averse to looking to the national government for expert assistance and funding.

The effect of local context

An emergent effect throughout the study has proved to be the effect of specific local contexts. The study reveals that the local situation exerts a significant impact on the climate change problem. It makes clear that all cases have their own particular situation that is determinative for the effects of climate change impacts. For instance, while the 'veterans at the front' cases are used to preparing themselves for climatic events, the 'urban veteran in reserve' appeared to be similarly involved in climate change adaptation without any clear threat. The 'rural recruit' cases appeared to be less active as expected, but one of these cases had a striking awareness of climate change impacts because of a climate-related incident. On the other hand, the 'urban recruit' appeared to be only laterally involved with climate change adaptation, despite being situated in one of the lowest parts of the country.

The factors risk and experience did not prove as decisive as anticipated. After the major impact of size on the local-level preparedness for climate change (as shown in the previous section), the factor of local context proves to be the most decisive. The case narratives reveal that local communities can be clearly more involved in adaptation activities than can be explained by risk and past experience alone. The factor of size proves to have considerable effect, since we have shown that urban municipalities can be much more involved in climate change adaptation then their rural counterparts. We had proposed to explain this from a perspective of general vulnerability. Having said this, it emerges that the most relevant explanations for local-level adaptation initiatives seem to be related to the specific local situation. This is the only explanation for the actual adaptation actions taking place at the local level—or their absence. Of the factors considered, only the local context of the 'urban recruit' can explain why this is the only urban case to be active on a limited level. Having thus shown the relevance of the local context, we can now turn to the more 'internal' factor of institutional capacity.

8.3.2 Institutional capacity and local-level preparedness

In this section, we address the second principal research question: how far does institutional capacity influence the possibilities and limitations for developing local adaptation strategies, and how can this capacity be expanded? 'Institutional capacity' is here considered quite broadly. It refers to governing mechanisms and manpower, but it also includes knowledge. In general, here we have also revealed a major gap between rural and urban communities. Small rural communities obviously have less manpower and more limited means of governance to implement their tasks. It is quite normal, for example, to have only one civil servant with responsibility for climate change adaptation. This places clear constraints on the ability to maintain a relevant network and to improve necessary knowledge skills. This in turn can explain low levels of climate change awareness (or at least concern) and thus a limited sense of urgency to implement adaptation strategies. In terms of both the number of inhabitants affected and the more limited economic resources at stake, the overall vulnerability is perceived as small. Whether or not this is a correct perception is, of course, another matter. Small may be 'beautiful' (Schumacher 1974), but in our study it does not automatically convert to proactive measures for either climate change mitigation or adaptation.

The analysis also demonstrates the importance of institutional capacity in other ways. Increased capacity means more knowledge and awareness of threat, as well as more 'hands' to work on the issue. While the urban cases that were most active on climate adaptation all stressed the limited means to implement measures effectively, the rural cases tended to stress limited resources as a major barrier for working on climate change problems at all.

Knowledge is thus a key element in the development of institutional capacity. This includes a general awareness of climate issues and the growing threat of serious climate-related impacts, as well as more hands-on knowledge of local adaptation challenges and the appropriate solutions for reducing and redressing risks. One can, after all, only expect action if a community both knows what the threats are and has an understanding of how threats can be met—and at what cost to other local priorities. In all of the cases studied, the first level of more general knowledge was more-or-less present. All cases appeared to have their own sources of knowledge of climate change impacts (see Table 8.5). The second level of knowledge of specific challenges was more limited (mainly the 'urban spectator' and 'urban veteran'), and the third level (potential instruments of adaptation) was only present in the 'urban veteran at the front' case—a case that combines both past experience and perceived future risk.

Table 8.5 **Channels of climate change knowledge of local-level civil servants**

	Mitigation front-runner network	Professional networks (Rioned, VNG)	Municipal networks (Wadden, B5, MARN)	EU projects (Building with CaRe, Future Cities)	Higher-level government	Direct scientific contacts
Urban spectator	X				X	
Rural spectator		X			X	
Urban veteran in reserve	X		X		X	
Rural veteran in reserve				X	X	
Urban recruit					X	
Rural recruit (a)			X		X	X
Rural recruit (b)			X		X	
Urban veteran at the front	X		X	X	X	X
Rural veteran at the front			X		X	

The availability of advanced knowledge and preparation is most abundantly present at higher levels of governance, most particularly the active work of water boards to stress the need for more water-storage areas. It has here been pointed out that there is an urgent need for national-level coordination and facilitation. Local communities express a wish to learn from front-runner examples, yet there is hardly any direct connection between the front-runners and others. A major capacity for increasing local adaptation awareness and resources should be available here. More effective means of dissemination and top-down/bottom-up communication are necessary to realise the potential. When institutional capacity (in knowledge, means and expertise) is limited, the municipality must rely on higher levels of governance for backup and protection. Where knowledge and manpower have been made available, the potential for more effective action is clearly present. We have observed in this connection a striking capacity difference between the urban and rural cases. In the smaller communities, one civil servant is usually responsible for environment, spatial planning and housing, while in urban cases entire administrative units can be responsible for this.

There remains, nonetheless, a question as to how decisive capacity can be, since rural communities also tend to have a very different perception of the problems. The role of local administrators and politicians in smaller communities is more limited, with both challenges and solutions usually of a totally different scale (while the 'rural veteran in reserve' plans 60 new houses, the 'urban recruit' builds 60,000!). The less-active role of small communities also exhibits an important social component. Small communities are most often characterised by stronger social solidarity. They tend to trust their community in solving any future problems. This came clearly to the fore in both the 'rural veterans' and 'rural recruits'. These small communities proved to be relatively indifferent to climate change scenarios, and they tended to trust past experience for coping with extreme weather and flooding. Should the problems grow too large and get out of hand, the national government would be there to take charge. Given their greater vulnerability economically, higher population density and (most often in the Netherlands at least) greater experience of natural disasters and impacts, it appears that larger urban communities develop their own particular culture of joint civil protection. The common memory is more oriented towards both adaptive preparation and self-help.

We have differentiated adaptive capacity as to institutional capacity, vulnerability, risk perception and the perceived role. It is the overall combination and interaction of these elements that determines the general quality and effectiveness of local-level adaptive capacity. The degree to which all four elements are present and interactive will strongly influence the degree of local preparedness for climate change impacts. If this preparedness is also combined with the necessary political will to act, then local-level adaptation strategies are generally much more robust (as variously manifest for the 'urban spectator', 'urban veteran in reserve' and 'urban veteran at the front').

8.4 Looking ahead: a future role for local-level climate adaptation

Climate change adaptation is one of the major political challenges we face today. Its context is very complicated, not only because of the many uncertainties associated with the climate change problem itself, but also because we are dealing with a 'premature' policy domain which lacks national focus on implementation, and which is characterised by a lack of uniform understanding and approach. The political–administrative apparatus must learn to deal with these uncertainties. Risk control in society is a basic political responsibility. Identifying threats and adapting to risks has been a driving force of political culture and democratic development for centuries (Van de Donk 2008). Today, however, as first formulated by Ulrich Beck (1992) in his concept of 'the risk society', modern life has become so complex that politicians and governments must increasingly rely on science and research for effective decision-making. There are, moreover, few other areas where this dependency is both more critical and more controversial than in the area of climate change. A recent report for the Netherlands Organisation for Scientific Research (NWO) points out that: 'a much wider set of structural changes will need to be prepared if the dramatic potential of future climate development is to be anticipated' (NWO 2008: 5). Our analysis clearly shows, however, that in the area of adaptation the current channels of knowledge and coordination across different levels of government are fragmented and diffuse. We conclude the chapter, therefore, by broadening our perspective on a future role for local-level climate adaptation in terms of the barriers to more effective initiatives revealed by the project (Table 8.6).

Table 8.6 **Barriers to and motivations for local-level climate change adaptation**

	Internal conditions	External conditions
Barriers	• No internal sense of urgency (rural and urban veterans in reserve, urban spectator, rural veteran at the front) • No means (urban veteran at the front) • Knowledge shortage (urban veteran at the front, urban spectator, rural recruit)	• No obligatory character (urban veteran in reserve, rural veteran at the front) • No public sense of urgency (urban veteran in reserve, rural veteran at the front, rural veteran in reserve, urban spectator) • Scope of the scenarios/lack of visualising the climate change problem (urban spectator, urban veteran at the front)
Motivations	• Mitigation experience (urban veteran in reserve, urban spectator) • Green-minded municipal board (urban spectator, urban veteran at the front, urban veteran in reserve)	• Favourable conditions (the 1993/1995 floods, demands of new residential districts) • Existence of innovative networks (MARN, B5, EU projects, the Wadden Isles cooperation)

8.4.1 Conquering the barriers

Throughout the case studies, most of the barriers mentioned by informants are internal: sceptical colleagues, lack of political support, lack of interest in climate change and difficulties involved in cooperation between different departments and domains. If a certain sense of urgency has already developed at the local level, most interviewees foresee a lack of capacity to implement adaptation actions. This barrier can also account for the fact that several local administrators, with scarce time to devote to their many areas of responsibility, simply do not have the capacity to go deeper into the climate change challenge. Shortages of staff in general are mentioned as a relatively common barrier in the small municipalities, where working on non-legal tasks such as climate adaptation cannot be prioritised.

Where sufficient adaptive capacity and political support for climate-change adaptation is available, several more practical barriers emerge. First, due to their urban situation, cities face major challenges in visualising and risk-scoping the highly complex and interdependent environment constituted by the urban infrastructure. A further complicating factor is uncertainty as to the functionality of the most effective adaptation solutions. And, even if a 'best-possible solution' does emerge, it must then be decided who is going to make the investment. These implementation barriers can only be overcome by more concentrated coordination efforts within and across national, regional and local governmental domains. Our study also shows, however, that best practice at the local level can also be a major knowledge input into these multi-level processes.

Besides these institutional barriers, knowledge gaps are also a major hindrance. This involves climate change knowledge in general and its application at the local level in particular. In general, uncertainties about the 'starting point' of climate change impacts were mentioned by several informants. Many of these viewed a lack of effective instruments as a major barrier in climate adaptation, but very few were able to suggest specific means for addressing the problem. Others simply looked for national funding to support adaptation measures, as this would spur local administrators to greater activity on climate adaptation.

When applying climate change knowledge at the local level, major difficulties also arise when trying to downscale and visualising climate change in combination with the extensive scope of climate change scenarios. Several of the case studies viewed the national scope of the prevailing KNMI as too broad to apply to their adaptation strategies. They feel a need to know the impacts in a manner as detailed as possible in order to prepare as efficiently as possible. Informants also mentioned that predictions about climate change must be reliable and consensual if they are to convince local administrators to act on climate change. Provincial climate scenarios do not fulfil the need for downscaled knowledge as they are based on the national scenarios. It is unlikely, however, that this deficiency can be solved in the near future, since the introductory scoping of downscaled models in the project revealed that the degree of resolution for model predictions will not be specific enough for the Netherlands to accommodate individual variance in local community conditions. This is a technical–methodological issue that clearly requires further multidisciplinary research (Jacques 2006). While most provinces in the Netherlands are already covered by detailed model predictions, the challenge in the future will be to adapt regional predictions to local community conditions. Effective adaptation can only take place through the interaction of 'top-down' (climate-related) and 'bottom-up' (socioeconomic) modelling (Aall and Norland 2005; Clausen 2007).

Limited attention to climate change adaptation is also mentioned as a general problem. More attention is desired both within the organisation as well as throughout society. Some interviewees foresee changes in the course of time due to an increasing chance of events and due to greater awareness in society. Several also feel that the specific issue of climate adaptation could benefit from more focused legislation. Supported by national law, municipalities could then act more forcefully to implement resource-demanding adaptation measures, particularly, for example, in new residential areas.

8.4.2 Drivers for local-level climate adaptation

In addition to the major perceived barriers, we also gained insight into several more positive factors for promoting local climate change adaptation. In a manner similar to earlier studies of Local Agenda 21 in Europe (Lafferty and Coenen 2001), the project indicates that local 'firebrands' are of significant importance in a positive direction. The presence of a Green alderman in one case (and, more generally, specifically 'environmentally oriented' administrators), was strongly reported as crucial to the promotion of climate-related initiatives. The factor usually reflected and enhanced positive institutional support for addressing the climate change issue.

Similarly, we also found that the more 'willing' and positively disposed cases were also active in all sorts of network. This varied from EU projects to urban networks and inter-municipal cooperation. Interviewees actively confirmed that these networks played a key role, as they enabled the local actors to exchange knowledge and best practices, and to share the costs of research and trial projects. Within such stimulating networks, local actors are more motivated to explore climate adaptation efforts that would otherwise be too ambitious (resource-demanding) for a single municipality.

The 'drivers' revealed show similarities to key implementation factors identified by Bulkeley and Betsill (2003) in their analyses of local climate mitigation efforts: (1) a committed individual in a local-level government that (2) manifests a solid climate-protection policy (preventing GHG emissions), (3) has funding available, (4) has power over mitigation-related domains, and (5) perhaps most crucially, has the political will to act. By adjusting factors (2) and (4) from mitigation to adaptation, this list provides a solid baseline for future adaptation initiatives. On the basis of our study, however, we would also add such local contextual factors as: (6) an awareness of the specifics of local climate change impacts (see Table 8.5), (7) hands-on experience with emission-preventing policies, (8) attempts to factor risk assessment into long-term policies, (9) experience with previous extreme-weather events, and (10) the size of the municipality population.

Despite the complex nature of the problem, interviewees in the mitigation front-runner cases express their belief that the problem of climate change adaptation will gradually 'settle' into a more commonly accepted issue. They frequently pointed out that mitigation was also a tough issue, but that it had now been commonly accepted. There is no longer (at least not in our case studies) a perceived need to convince and motivate colleagues within the local administration. Most citizens are seen as being aware of energy conservation, the separation of waste and a need for emissions reductions. While mitigation has become an urgent issue, the challenge of specific adaptation initiatives is new and vague. In nearly every case, there was little sense of urgency in relation to either vulnerability or preparedness. Those informants who are motivated to act on adaptation issues face difficulties convincing their colleagues and administrators on the basis of broad and uncertain scenarios. They see a clear need for translating the threat of impacts into local consequences and for heightening the need for risk assessment. They also feel, however (somewhat naively?), that there is sufficient time to achieve the necessary preparation, and that a sense of urgency will be triggered by either actual experience (!) or increased knowledge.

References

Adger, W.N., and K. Vincent (2005) 'Uncertainty in Adaptive Capacity', *Comptes Rendus Geosciences* 337.4: 399-410.

——, N.W. Arnell and E.L. Tompkins (2005) 'Successful Adaptation to Climate Change across Scales', *Global Environmental Change Part A* 15.2: 77-86.

Aall, C., and I.T. Norland (2005) *Indicators for Local-Scale Climate Vulnerability* (Prosus Report No. 6; Oslo: Programme for Research and Documentation for a Sustainable Society/ University of Oslo).

Beck, U. (1992) *Risk Society: Towards a New Modernity* (London: Sage).

Bulkeley, H., and M.M. Betsill (2003) *Cities and Climate Change: Urban Sustainability and Global Environmental Governance* (Abingdon, UK/New York: Routledge).

CBS (Centraal Bureau voor de Statistiek) (2008) 'Lokale overheid financieel grotendeels afhankelijk van Den Haag', *Webmagazine*, 3 December 2008; www.cbs.nl/nl-nl/menu/themas/overheid-politiek/publicaties/artikelen/archief/2008/2008-2624-wm.htm, accessed 15 March 2010.

CEC (Commission of the European Communities) (2009) *White Paper—Adapting to Climate Change: Towards a European Framework for Action* (SEC(2009) 386, SEC(2009) 387, SEC(2009) 388, COM/2009/0147; Brussels: Commission of the European Communities; ec.europa.eu/environment/climat/adaptation/index_en.htm).

Clausen, G. (2007) 'Assessing Local Vulnerability to Climate Change, Institutional Dimensions and Cross-national Comparisons', paper presented at the *13th Annual International Sustainable Development Research Conference*, Västerås, Sweden, 10–12 June 2007.

Coenen, F.H.J.M. (2001) 'The Netherlands: Probing the Essence of LA21 as an Added-Value Approach to Sustainable Development and Local Democracy', in W.M. Lafferty (ed.), *Sustainable Communities in Europe* (London/Sterling, VA: Earthscan): 153-80.

Deltacommissie (2008) 'Het adviesrapport: Samenvatting en de aanbevelingen' [Advisory Report: Summary and Recommendations]; www.deltacommissie.com.

GGD (Gemeentelijke Gezondheidsdienst) (2009) 'Over GGD' ('On GCD'), retrieved from www.ggd.nl [in Dutch].

Jacques, P. (2006) 'Downscaling Climate Models and Environmental Policy: From Global to Regional Politics', *Journal of Environmental Planning and Management* 49 (March 2006): 301-307.

Klein, R.J., M.J. Smit, H. Goosen and C.H. Hulsbergen (1998) 'Resilience and Vulnerability: Coastal Dynamics or Dutch Dikes?', *Geographical Journal* 164: 259-68.

KNMI (Koninklijk Nederlands Meteorologisch Instituut) (2006) *Klimaat in de 21e eeuw: Vier scenario's voor Nederland* [*Climate in the 21st Century*] (De Bilt, Netherlands: Koninklijk Nederlands Meteorologisch Instituut; www.knmi.nl).

Kok, M.T.J., W. Vermeulen, A. Faaij and D.D. Jager (2002) *Global Warming and Social Innovation: The Challenge of a Climate Neutral Society* (London/Sterling, VA: Earthscan).

Kolen, B., K. Engel, H. van der Most and K. van Ruiten (2009) 'Leven in de Nederlandse Delta: Wachten tot een megacrisis?' ['Living in the Dutch Delta: Waiting for a Mega Crisis?'] *Magazine Nationale Veiligheid en Crisisbeheersing* 7 (August/September 2009): 42-44.

Lafferty, W.M., and F.H.J.M. Coenen (2001) 'Conclusions and Perspectives', in W.M. Lafferty (ed.), *Sustainable Communities in Europe* (London/Sterling VA: Earthscan): 266-304.

Lorenzoni, I., L. Næss, M. Hulme, J. Wolf, D.R. Nelson, A. Wreford, W.N. Adger, S. Dessai and M. Goulden (2007) 'Limits and Barriers to Adaptation, Four Propositions' (Tyndall Centre Briefing Note 20; Norwich, UK: Tyndall Centre for Climate Change Research).

Ministerie van Algemene Zaken (2006). 'Klimaatstrategie: Tussen ambitie en realisme' ['Climate Strategy: Between Ambition and Realism'] (Amsterdam: Amsterdam University Press; www.wrr.nl).

Næss, L.O., G. Bang, S. Eriksen and J. Vevatne (2005) 'Institutional Adaptation to Climate Change: Flood Responses at the Municipal Level in Norway', *Global Environmental Change Part A* 15.2: 125-38.

NWO (Netherlands Organisation for Scientific Research) (2008) 'What If . . . Abrupt and Extreme Climate Change?' (The Hague: NWO).

Schneider, S.H., S. Semenov, A. Patwardhan, I. Burton, C.H.D. Magadza, M. Oppenheimer, A.B. Pittock, A. Rahman, J.B. Smith, A. Suarez and F. Yamin (2007) 'Assessing Key Vulnerabilities and the Risk from Climate Change', in M.L. Parry, O.F. Canziani, J.P. Palutikof, P.J. van der Linden and C.E. Hanson (eds.), *Climate Change 2007: Impacts, Adaptation and Vulnerability Contribution of Working Group II to the Fourth Assessment Report of the Intergovernmental Panel on Climate Change* (Cambridge, UK: Cambridge University Press; www.ipcc.ch/publications_and_data/publications_ipcc_fourth_assessment_report_wg2_report_impacts_adaptation_and_vulnerability.htm, accessed 22 March 2010): 779-810.

Schumacher, E.F. (1974) *Small Is Beautiful: Economics as if People Mattered* (London: Blond & Briggs).

Smit, B., and J. Wandel (2006) 'Adaptation, Adaptive Capacity and Vulnerability', *Global Environmental Change* 16: 282-92.

Van de Donk, W. (2008) 'Politiek als kunst van het onzekere: Omgaan met onzekerheid als politieke kwaliteit' ['Politics as the Art of Uncertainty'], in A.M. de Gier and H.B. Opschoor (eds.), *Onzekerheden en Klimaatverandering* (Amsterdam: Koninklijke Nederlandse Akademie van Wetenschappen, KNAW): 43-55.

Verschuuren, J. (2007) 'Adaptatie aan klimaatverandering vraagt om adaptatie van de wet' ['Adaptation to Climate Change Demands Adaptation of the Law'], *Nederlandsch Juristenblad* 45.

VROM (Ministerie van Volkshuisvesting, Ruimtelijke Ordening en Milieubeheer) (2007a) 'Maak ruimte voor klimaat: Nationale adaptatiestrategie. De interbestuurlijke notitie' ['National Programme on Climate Adaptation and Spatial Planning'] (The Hague: Ministerie van VROM).

—— (2007b) 'Klimaatakkoord gemeenten en Rijk 2007–2011: Samen werken aan een klimaatbestendig en duurzaam Nederland' ['Climate Agreement between Municipalities and State 2007–2011']; www.vrom.nl/pagina.html?id=2706&sp=2&dn=w1005, accessed 17 March 2010.

Wall, E., and K. Marzall (2006) 'Adaptive Capacity for Climate Change in Canadian Rural Communities', *Local Environment* 11.4: 373-97.

Wilson, E. (2006) 'Adapting to Climate Change at the Local Level: The Spatial Planning Response', *Local Environment* 11.6: 609-25.

9

Human responses to climate change

Flooding experiences in the Netherlands

Ruud Zaalberg and Cees J.H. Midden
Department of Industrial Engineering and Innovation Sciences,
Eindhoven University of Technology, The Netherlands

9.1 Climate change: observations, projections and consequences

Since the early 1970s scientists have been debating heatedly about the causes and consequences of global climate change. To date, there seems to be a strong consensus among scientists about continuing and remarkable changes in the global climate over the past 150 years. According to the Intergovernmental Panel on Climate Change (IPCC 2007), the global average surface temperature has increased by almost 1°C in the past 100 years. In general, climatologists and other experts agree that this global warming is partially attributable to the combustion of fossil fuels by humans. Other physical evidence pointing in the direction of global climate change is the global average sea level, which rose by approximately 17 cm during the 20th century. Thermal expansion of sea water and loss of land ice are the main causes of this effect. The IPCC (2007) further predicts that global warming will continue owing to ever-increasing human energy consumption and economic growth. Global average surface temperature will increase by another 2° to 4°C, and the sea level is projected to rise by another 20 to 60 cm during the 21st century.

Global climate change will have far-reaching consequences for delta areas throughout the world. For example, in Europe winter precipitation is projected to increase by 4–14% in the next 40–50 years. As a consequence, the volume of water from the main rivers Rhine and Meuse will increase, while at the same time drainage of water will be hampered because of the sea-level rise predicted by the IPCC (Klijn *et al.* 2004). Around 50% of the Netherlands lies below sea level, resulting in higher probabilities for river flooding (Van Dorland and Jansen 2006; Royal Netherlands Meteorological Institute 2006). Low-lying polders along these Dutch rivers are encircled and protected by so-called dyke rings. However, high-water flood defence by means of dykes may sometimes fall short as a consequence of different failing mechanisms. For example, overflowing can occur when dykes are too low or piping takes place in which dykes are sapped (see Ministry of Transport, Public Works and Water Management 2005). In recent Dutch history two major river floods occurred, caused by extreme river discharges. In 1993 material losses were at least €23 million, and in 1995 more than 250,000 people were evacuated owing to the potential breakdown of river dykes. In contrast, during summer periods the chances of water shortage in the Rhine and the Meuse are also predicted to increase, which may lead to serious economic damage, simply because the transport of goods via these rivers will be hampered (see Chapter 2 in this book).

9.1.1 Governmental responses to climate change

The Dutch are well known for their knowledge and expertise in the prevention of flooding by building flood defence systems such as levees or dykes. However, in 1953, the south-west part of the Netherlands was struck by a catastrophic sea flood, in which 150,000 ha were flooded and 1,835 people died, as well as thousands of farm animals (Allewijn 1983). In response to the 1953 and 1990s floods, Dutch national policy now focuses especially on flood protection by strengthening and heightening the dykes, combined with more recent spatial planning programmes such as 'room for the river' to lower maximum water levels (Ministry of Transport, Public Works and Water Management 2006b). Unfortunately, Dutch national policy continues to stimulate economic investments in flood-prone areas. As a result, these 'flood-sensitive' areas are becoming more and more densely populated, leading to expected economic losses up to €290 billion and up to 6,000 human casualties in the event of a major river flood in the western part of the Netherlands (Ministry of Transport, Public Works and Water Management 2005). Becoming increasingly aware of the impact of the consequences of global climate change for the local community, many European governments are launching information campaigns to inform individuals and companies about their personal vulnerability to future flooding. For example, the Environment Agency in the United Kingdom informs the general public and the economic sector about flooding risks via its website (Environment Agency 2009). UK residents and companies can use so-called flood maps to discover whether they live in potentially threatened areas by simply entering their residential name on this website. Furthermore, residents can find out about what they can do to prepare for flood and what to do to stay safe in a flood. In the Netherlands a similar initiative was taken by the Dutch national government in collaboration with local authorities. Residents can zoom in on risk maps depicting all kinds of natural and technological disaster (Ministry of the Interior and Kingdom Relations n.d.). In other European countries local authorities or research institutes inform residents about flooding risks (for Belgium see County Board East Flanders n.d.; for Germany see Innig n.d.). At the organisational level, the Dutch national government is also trying to improve crisis management plans through Taskforce Management Flooding to streamline the evacuation process if floods do occur (Ministry of Transport, Public Works and Water Management n.d.).

9.1.2 Public information campaigns: are they effective?

An important question is whether these attempts to inform the general public about the local consequences of global warming truly influence public opinion about these issues and will eventually stimulate appropriate adaptive or even mitigating actions against the local consequences of global climate change (see Chapter 8 in this book). In a recent study, the Dutch Institute for Public Opinion and Market Research (2007) concluded that 75% of the Dutch population is concerned about global warming. However, it remains unclear whether the information campaigns mentioned above are responsible for these widespread feelings of concern about global climate change. Even more importantly, at least from a psychological perspective, increasing awareness of global climate change among residents is different from motivating these same residents to effectively cope with local floods. In the end, the effectiveness of information campaigns is solely determined by the extent to which human behaviour is influenced. The ideal information campaign should, therefore, be aimed at motivating residents living in flood-prone areas to cope effectively with imminent floods. When successful, such information campaigns may minimise material losses and loss of life and reduce stress and anxiety among threatened residents (Ministry of Transport, Public Works and Water Management 2006a; see also Klijn *et al.* 2004).

9.1.3 General research aims

The general aims of our research are (1) to increase our understanding of climate and flood risk perceptions, (2) to describe the factors that influence these judgments, and (3) to seek interventions that can contribute to a realistic assessment by laypersons of long-term flooding risks caused by global climate change.

Our research started with an extensive survey among residents of threatened and flooded river areas to study the consequences of various types of flooding experience on the motivation of residents to cope with future flooding risks (Zaalberg *et al.* 2009). Having assessed the significance of direct flooding experiences, the next objective was to explore ways to simulate direct flooding experiences as a means to enhance risk awareness and coping potential for residents who lack direct flooding experiences (Zaalberg and Midden 2009). A major part of our research consisted of the development of a multi-sensory interactive 3D simulation to mimic direct flooding experiences, evoking resultant coping responses without the dangers present in real life.

9.2 Learning from real flooding experiences: conducting survey research

The goal of this research was to identify factors capable of motivating individuals to cope effectively with future flooding. Risk research has shown that disaster experience influences the extent to which people become motivated to cope effectively with future risks (e.g. Siegel *et al.* 2003; Siegrist and Gutscher 2008; Weinstein 1989). Interested in the psychological processes behind this experience–coping link, we placed special emphasis on affective appraisals (e.g. risk-as-feelings hypothesis: Loewenstein *et al.* 2001) and cognitive appraisals (protection motivation theory: Rogers and Prentice-Dunn 1997) as intervening processes.

9.2.1 What we already know

An extensive body of literature in the domain of risks has shown that real disaster experience influences the extent to which people become motivated to cope with future risks. Weinstein (1989) showed that the purchase of flood insurance increased with the severity of past flood damage. Sattler *et al.* (2000) showed that psychological distress due to Hurricane Hugo predicted preparation efforts for the impending Hurricane Emily four years later. However, Siegel *et al.* (2003) showed that El Niño preparation activities were not predicted by property damage due to an earlier earthquake. Instead, residents who indicated that they were emotionally injured by the earthquake had done more to prepare for El Niño three years later than those who were not injured.

9.2.2 Broadening our theoretical knowledge on relevant psychological processes

Lindell and Perry (2000) recommend careful conceptualisation and consistent measurement of the construct 'disaster experience'. Zaalberg *et al.* (2009) therefore distinguish between physical exposures to a threat and resultant subjective experiences, aiming for a stronger consensus about this important construct in risk analysis literature. We compared the subjective experiences of actual flood victims with non-victims who had been exposed to flooding only vicariously. Non-victims were exposed to flooding only indirectly because the major 1993 and 1995 river floods in the Netherlands were broadcast live and published as lead stories in newspapers.

The cognitive approach to behavioural change such as protection motivation theory (PMT) predicts that **people will take coping actions** when (a) perceived vulnerability is appraised as high, (b) perceived consequences are appraised as severe, and (c) coping actions are appraised as effective and feasible. Moreover, PMT predicts **coping thoughts** (e.g. threat denial) when (a) perceived vulnerability is appraised as high, (b) perceived consequences are appraised as severe, and (c) coping actions are appraised as ineffective or unfeasible. However, research investigating the influence of these appraisals on coping actions, other than in the domain of health behaviour, is rare. In the present research survey these cognitive appraisals were investigated in the domain of flooding.

Fear-appeal research stresses the role of negative affect as a motivating force persuading people to act, thereby alleviating their negative affect (Eagly and Chaiken 1993). Loewenstein *et al.* (2001), when discussing their risk-as-feelings hypothesis, talked about the importance of assessing anticipatory emotions (i.e. felt visceral reactions to future risks) in the decision-making process of how to cope with future risks (cf. the affect heuristic: Slovic *et al.* 2004, and experiential processing: Marx *et al.*, 2007). Siegrist and Gutscher (2008) acknowledged the central role of negative affect in motivating residents to take precautionary actions.

9.2.3 Hypotheses and research set-up

The following hypotheses are of central importance in the present study. First, flood victims have stronger intentions than non-victims of taking coping actions against future flooding. Second, flood victims' subjective experiences and appraisals (affective and cognitive) are stronger than those of non-victims. Third, we expect the differences in intentions to be mediated or explained by subjective experiences and appraisals.

We visited 1,597 households living near the rivers Rhine and Meuse. The response rate was 32.3%. Victim locations were chosen in flood-prone areas that had been struck by high-impact floods in 1993 and 1995. Non-victim locations were neither flooded nor evacuated in the past, although they were selected in the vicinity of the victim locations.

First, respondents were questioned about their intentions to engage in coping actions to minimise damage. Respondents judged the likelihood that they would carry out nine different coping actions, rated on rating scales ranging from *chance is very small* (0) to *chance is very big* (4). A distinction was made between actions preventing incoming water (e.g. putting sandbags in front of the house) and actions aimed at adapting to incoming water (e.g. tying up/removing curtains). Second, subjective experiences were rated on 14 rating scales ranging from *not at all* (0) to *extremely* (4). Respondents rated the extent to which they had experienced positive emotions (e.g. sociability), negative emotions (e.g. fear) and social support.[1] Finally, affective appraisals (e.g. worries about possessions) and cognitive appraisals (i.e. threat and coping) were rated on rating scales. Scale points for affective

1 Anchor points for social support ranged from *no support at all* (0) to *extremely supported* (4).

appraisals ranged from *no worries at all* (0) to *extremely worried* (4). Threat appraisals assessed expectancies about the perceived probability of nine different flood outcomes when doing nothing (e.g. home walls become damp) and evaluated the extent to which these flood outcomes were negative.[2] Coping appraisals were assessed by asking respondents for their ideas about perceived action-efficacy, and the perceived self-efficacy of nine different coping actions aimed at minimising damage (e.g. moving furniture to upper floors).[3]

9.2.4 Results

With the help of structural equation modelling (SEM) (Jöreskog and Sörbom 1993)[4] it was shown that flood victims had stronger intentions of taking adaptive actions in the future than non-victims. No differences were found on preventive actions (see Fig. 9.1).

Figure 9.1 **Coping intentions as function of flood victims versus non-victims**

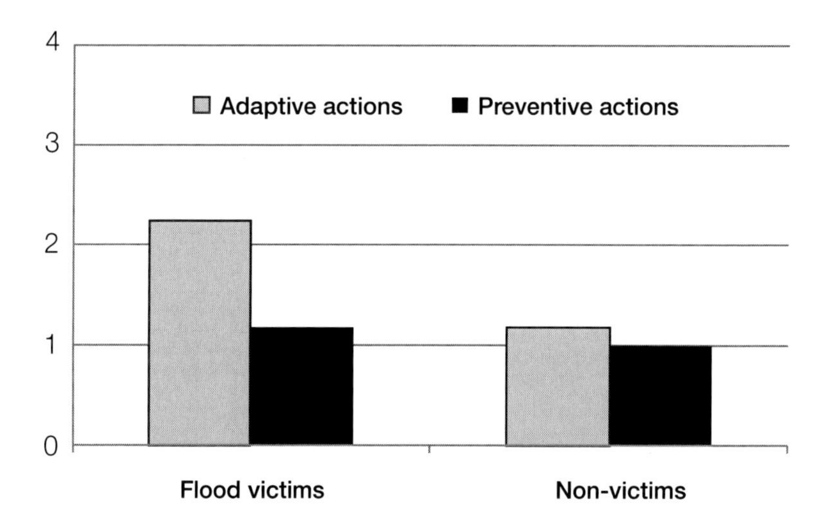

2 Anchor points for perceived probability ranged from *chance is very small* (0) to *chance is very big* (4). Anchor points for perceived flood outcomes ranged from *no damage at all* (0) to *very severe damage* (4).

3 Anchor points for perceived action-efficacy ranged from *not at all useful* (0) to *extremely useful* (4). Anchor points for perceived self-efficacy ranged from *no impediments at all* (0) to *a lot off impediments* (4).

4 Structural equation modelling is a statistical technique for testing and estimating causal relationships using a combination of multivariate regression models and qualitative causal assumptions.

Moreover, flood victims experienced stronger emotions (negative and positive) and received more social support than non-victims (see Fig. 9.2). In addition, flood victims worried more about future flooding, perceived themselves as more vulnerable to future flooding and perceived the outcomes of future flooding as more severe than non-victims. Finally, although not predicted, flood victims judged the effectiveness of adaptive actions as higher than non-victims did (see Fig. 9.3).

Figure 9.2 **Subjective experiences as function of flood victims versus non-victims**

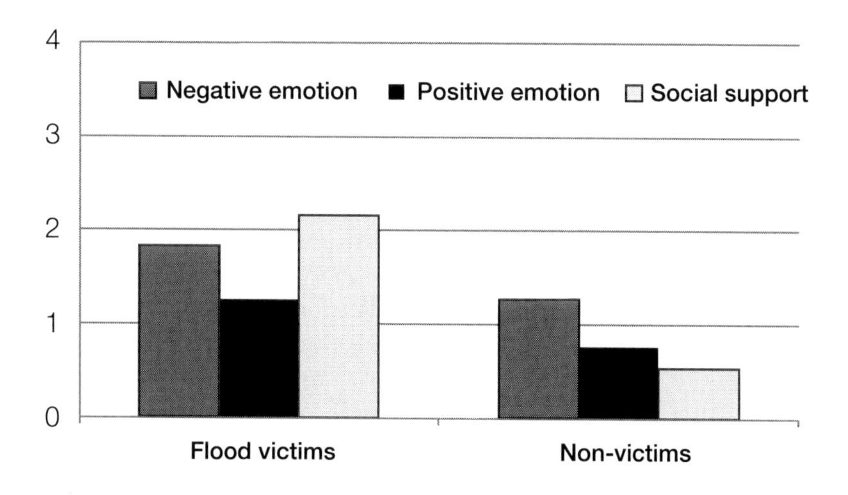

Figure 9.3 **Appraisals as function of flood victims versus non-victims**

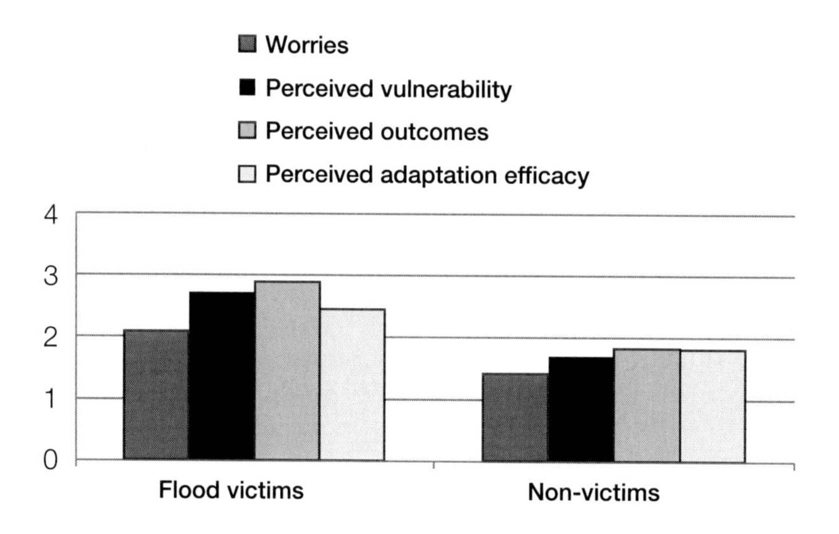

SEM analysis explained the presence of mean differences on adaptive actions in terms of psychological processes. Flood victims (via their higher negative emotions) appraised themselves as more vulnerable to future flooding than non-victims did, leading to stronger intentions to take adaptive actions among flood victims. As a parallel process, flood victims (via their higher negative emotions and greater social support) appraised adaptation as more effective than non-victims did, leading to stronger intentions of taking adaptive actions among flood victims.

9.2.5 Conclusions

Protection motivation theory was partially supported in the domain of flooding (Rogers and Prentice-Dunn 1997). The theoretically predicted causal relationships from threat and coping appraisals to coping actions are empirically supported by our 'manipulation' of some of the key concepts constituting PMT. These are perceived vulnerability and perceived action-efficacy as intervening psychological variables for the experience–coping link.

These findings seem compatible with both affective (cf. experiential processing: Loewenstein *et al.* 2001; affect heuristic: Marx *et al.* 2007; risk-as-feelings: Slovic *et al.* 2004) and cognitive approaches (PMT: Rogers and Prentice-Dunn 1997) to behavioural change. Findings suggest that subjective experiences not only predict adaptive actions via cognitive appraisals, but also form an independent and indispensable part of the mediating process, explaining mean differences in adaptation between flood victims and non-victims. Clearly, both the experiential processing mode based on concrete experiences and the analytical processing mode based on reasoning play a role in decisions on how to deal with future risks.

The results of this survey have practical relevance in developing interventions to inform residents about future flooding risks. Such interventions should preferably be aimed at influencing threat and coping appraisals as important determinants of coping actions. For example, persuading potential flood victims of the urgency and impact of global climate change at the local level, while informing them about efficient and non-efficient coping strategies, should influence coping actions in an appropriate direction.

An alternative and perhaps more powerful approach to achieve behavioural change is the development and use of persuasive technology to simulate real 'disaster experience', thereby going beyond traditional persuasion attempts such as the use of fear-evoking images. A picture may say more than a thousand words, but experiencing a 'virtual' threat may influence the individual more than simply observing images. High-end simulations in a virtual environment (VE) could provide participants with a simulated flood experience at the experiential or sensory level. Ideally, flood simulations should make an attempt not only to simulate physical exposure to real flooding, but preferably also to include the emotional component of real flooding experiences.

9.3 Can real flooding experiences be simulated? Conducting lab research

The next step in our research was to explore the extent to which the immersive quality of different presentation modes influences coping responses to virtual floods. New immersive media technologies such as 3D visualisations (e.g. immersion via media form: Lessiter *et al.* 2001) might support the persuasive quality of future information campaigns communicating real-world flooding risks. Lab research was conducted to investigate the role of multimodal sensory stimulation by means of interactive 3D technology to simulate real flooding experiences, evoking resultant coping responses. We exposed participants to a simulated dyke breach and flooding of their virtual residence positioned in a typical low-lying Dutch polder landscape. The central question was whether multimodal sensory stimulation by means of an interactive 3D simulation facilitated participants' coping responses compared to non-interactive 2D simulations (Zaalberg and Midden 2009). In the next section the theoretical framework will be explained.

9.3.1 Simulating real flooding experiences: what should be evoked?

The following question must be answered when attempting to simulate real flooding experiences realistically: which (psychological) components constitute real disaster experiences? Based on the risk analysis literature four interrelated components can be distinguished. First, Dooley *et al.* (1992) focused on the emotional component of disaster experience (see also Siegel *et al.* 2003; Siegrist and Gutscher 2008). They simply asked their respondents using a yes–no format whether they had ever experienced an earthquake that scared them. Second, Sattler *et al.* (2000) assessed not only the emotional impact (i.e. the amount of distress), but also the amount of property damage caused by Hurricane Hugo (i.e. the material component of disaster experience). Third, Weinstein's (1989) review referred to physical injuries resulting from disasters, that is, the physical component of disaster experience. Additionally, Zaalberg *et al.* (2009) assessed the amount of social support residents had received during a threatening flood: that is, the social component of disaster experience (see also Baan and Klijn 2004). Components constituting real disaster experiences are depicted in Figure 9.4.

As presented above, Zaalberg *et al.* (2009) distinguished between the extent to which residents had been physically exposed to flooding, and investigated its influence on experienced disaster components. Actual flood victims had received more social support and had experienced stronger emotions compared to non-victims who had been exposed to flooding vicariously through different media channels. In other words, a mediated or indirect flooding experience is qualitatively different from an unmediated or direct flooding experience. As stated earlier, Zaalberg *et al.* (2009) confirmed that real disaster experience influenced future coping actions.

Figure 9.4 **Components constituting real disaster experiences**

Emotional component
Material component
Real disaster experiences
Physical component
Social component

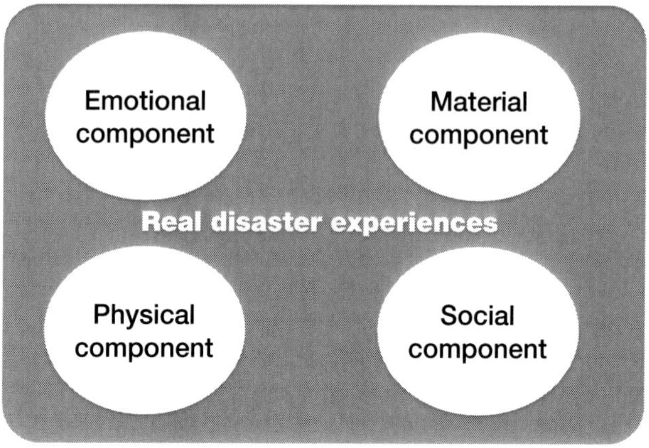

They reported that actual flood victims were more motivated to take adaptive actions in the future compared to equally threatened non-victims. The psychological processes explaining this influence were that victims' intensified social and emotional experiences increased their perceived vulnerability to future flooding and increased their perception of the effectiveness of adaptive actions to deal with these future threats.

In the next section we elaborate on similar and additional psychological processes that could possibly intervene in the relationship between different presentation modes, evoking resultant coping responses to virtual floods. We briefly discuss three theoretical concepts as intervening processes: cognitive appraisals, affective appraisals and presence.

9.3.2 Intervening psychological processes

PMT (Rogers and Prentice-Dunn 1997) is a cognitive theory explaining coping responses under threat. Very briefly, PMT predicts that people will resort to coping responses when threat appraisal is high and coping responses are appraised as effective and feasible.

The risk-as-feelings hypothesis (Loewenstein *et al.* 2001) stresses that anticipatory emotions, that is, felt visceral reactions to future risks, are important in the prediction of coping responses (cf. affect heuristic: Slovic *et al.* 2004; experiential processing: Marx *et al.* 2007). The role of immersive media on affective responses was investigated by Persky and Blascovich (2008). They showed that playing a violent video game leads to more aggressive feelings when played with a head-mounted device compared to desktop playing.

Presence is often defined as the subjective experience or sensation of being located within a mediated environment, despite being present (by definition) in the physical world (for specific definitions see Lessiter *et al.* 2001; Witmer and Singer

1998). Research has revealed the multi-dimensionality of the presence concept. Three sub-dimensions have been reported. First, spatial presence or the sense of being physically present in a VE. Second, realism or the subjective experience of reality in the VE. Third, involvement or attention devoted to the VE (Lessiter *et al.* 2001; Schubert *et al.* 2001). The immersive quality of the presentation mode not only has an influence on presence but also on behaviour. For example, Persky and Blascovich (2008) showed that playing a violent video game leads to more aggressive behaviour when played with a head-mounted device compared to desktop playing, although presence did not mediate this presentation mode effect. However, Nowak *et al.* (2008) presented empirical evidence for the presence concept as a mediator in the relationship between playing a violent video game versus a non-violent control game, and participants' behavioural intentions in response to an aggressive prime. Importantly, presence was manipulated through differences in the (perceived) media content in contrast to the proposed influence on presence via presentation modes in our laboratory research (see also Bouchard *et al.* 2008).

9.3.3 Research set-up and hypotheses

In a laboratory experiment we manipulated the extent to which participants were immersed in a VE by using different presentation modes. One condition consisted of multimodal sensory stimulation by means of an interactive 3D simulation on a 72-inch translucent screen. The remaining conditions consisted of non-interactive 2D simulations. In these conditions participants watched the flood simulation on a 15-inch laptop via film or slides without auditory support. Participants were randomly assigned to one of the three conditions.

In a virtual Dutch polder landscape, participants were exposed to a simulated dyke breach and serious flooding of their virtual residence. In this scenario, a water depth of approximately 300 cm was simulated by the end. Several dependent variables were assessed. At the information processing level, the information search related to preventive evacuation from the threatened area, and to preventive actions to avoid water entering one's residence (e.g. use of sandbags). At the behavioural level, we assessed the evacuation intention, the prevention intention and the intention to buy flood insurance. Cognitive appraisals, affective appraisals and presence were assessed as potential mediators.

The following hypotheses were derived. First, information search and coping intentions are both strengthened in the interactive 3D simulation compared to the non-interactive 2D simulations. Second, presence and affective appraisal are stronger in the interactive 3D simulation compared to the non-interactive 2D simulations. Third, the experimental effects on information search and coping intentions are mediated or explained by presence and affective appraisal. Fourth, the content of the simulation (i.e. the story or media content unfolding in the VE; cf. Bouchard *et al.* 2008; Nowak *et al.* 2008) was kept equal across experimental conditions. Due to the unambiguousness of the simulated risk in all conditions, cognitive appraisals were predicted to be equal across experimental conditions.

9.3.4 Experimental procedure

Fifty-five participants subscribed online and received €10. Participants were seated in front of a 15-inch laptop. A second table with a wireless keyboard and mouse was placed in front of a 72-inch translucent screen, which was audio supported by means of two 150-watt boxes to the left and right of the screen. Participants were seated approximately 250 cm in front of the translucent screen and received a short navigation training on how to walk (with arrow keys) and look around (with mouse) in a virtual 3D maze.

Participants first watched a five-minute introduction film. The purpose of this film was to introduce participants to the causes and consequences of global climate change and to support participants in their understanding of specific simulation events presented later in the simulation. Participants were then asked to portray themselves as residents living in a low-lying polder landscape at the bottom of the dyke and close to a river. Participants were told that they had to go for a walk through the VE once the simulation started. Heavy rainfall was simulated visually and was audio-supported with a real-life recording of rain falling on grass and trees. Participants first climbed the stairs towards the top of the dyke, where they had the opportunity to watch the river rising slowly. A clear contrast was visible between the water rising on the outer side and the low-lying polder on the inner side of the dyke. After approximately 200 m, participants reached a ditch located behind and at the bottom of a dyke. Piping and dyke breach were simulated. The fast-flowing water was audio-supported with a sound recording of a real waterfall. Moreover, the sound was localised in such a way that when participants looked around in the VE, the sound source shifted accordingly from one speaker box to the other. The magnitude of the sound decreased when participants moved away from the sound source, which gave the participants an extra sensory 3D cue in the VE, making it even more realistic and persuasive.

After the dyke breached, participants walked back to watch the consequences for their residence. The maximum water level (approximately 300 cm) had flooded the first floor of their residence. At this point the simulation ended. Once seated behind the laptop, participants were asked to imagine a situation in which they would stay in their virtual residence while the dyke was about to breach. A post-simulation task gave the participants the opportunity to search for information related to different coping responses to protect their health and belongings in the threatening situation of an imminent dyke breach. Participants also answered questions about their coping intentions, cognitive appraisals, affective appraisals and presence.

Participants in the film and slide conditions did not receive the navigation training. Even more importantly, the simulation was shown on the 15-inch laptop in full colour without supportive audio recordings. The film simulation lasted for 288 seconds. The slide show consisted of 26 slides of 11 seconds each.

9.3.5 Assessments

Information search was measured using a so-called information board holding pieces of information that users could choose to read. The information on the board was organised into two rows of seven cells each (using MouselabWEB: see Willemsen and Johnson 2008). One row provided information on preventive evacuation, the other row provided information on preventive actions to avoid water entering the user's residence. Each cell contained a short question related to preventive evacuation or preventive actions, for example, 'How can I leave the threatened area?' Answers to each question were revealed by moving the cursor on top of a specific cell, which then opened, for example, 'Evacuation routes will be signposted.' Total reading times (seconds) were recorded for each cell, and these cell recordings were then summed across each row and used as indicators of the thoroughness with which participants had searched for information.

Coping intentions were assessed with seven items, ranging from *chance is very small* (0) to *chance is very big* (4). Preventive evacuation (e.g. evacuation of family members) and preventive actions (e.g. the use of water pumps) were distinguished. A single item about buying flood insurance to protect the user's belongings in the real world was added to this list.

Cognitive appraisals, such as perceived vulnerability and perceived consequences, were assessed by means of rating scales in relation to three hypothetical flood outcomes. Participants first rated the likelihood that the three flood outcomes would occur in the case of an imaginary dyke breach. Scale anchors ranged from *chance is very small* (0) to *chance is very big* (4). An example item was: 'water damage to your imaginary residence'. Participants were then asked to rate the same three items on perceived consequences in case the water level in the low-lying polder had reached its maximum at the end of the simulation. Scale anchors ranged from *not at all severe* (0) to *very severe* (4). Response-efficacy of the seven coping responses mentioned above was measured using rating scales, ranging from *not at all effective* (0) to *extremely effective* (4).

Affective appraisals were assessed with four rating scales, ranging from *not at all* (0) to *very much* (4). For example, the extent to which participants expected to feel anxious about their imaginary belongings being damaged.

Presence during the simulation was assessed with the Igroup Presence Questionnaire (IPQ) (Schubert *et al.* 2001; and see Carlin *et al.* 1997; Hendrix 1994; Slater and Usoh 1994; Witmer and Singer 1998 for additions to the IPQ). An example item was: 'Somehow I felt that the imaginary world surrounded me'. Scale anchors ranged from *fully disagree* (0) to *fully agree* (4). The IPQ assessed the three presence subdimensions stated earlier (spatial presence, realism and involvement). As a corroborative measure of presence, participants were also asked to rate the extent to which they had lost track of time during the simulation, functioning as a subjective time perception indicator (ITC-Sense of Presence Inventory [ITC-SOPI]: Lessiter *et al.* 2001). High scores on the above-mentioned scale were indicative of a greater sense of presence.

9.3.6 Results

A series of multiple regression analyses were used to test the effects of presentation mode on information search and coping intentions. The indirect, mediated effects via affective appraisals and presence were also tested (Preacher and Hayes 2008).

Controlled for age differences in information processing, reading time related to preventive evacuation was longer for participants in the 3D condition compared to participants' reading time in the film and slides conditions. Mediation was not found. No (mediating) effects were found for reading time related to preventive actions. Reading time for participants in the 3D condition was equal to the reading times for participants in the film and slides conditions (see Fig. 9.5).

Figure 9.5 **Reading time (seconds) of preventive evacuation and preventive actions as function of presentation mode**

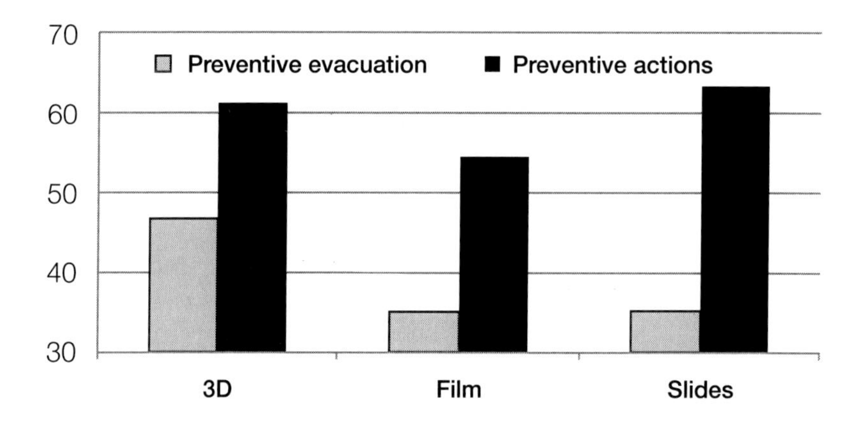

With control for differences in the perceived water level at the end of the simulation, participants in the 3D condition were more willing to evacuate from the virtual polder landscape compared to participants in the slides condition, but not to participants in the film condition (see Fig. 9.6). This significant contrast was reduced when we controlled for subjective time perception. Participants in the 3D condition were more willing to evacuate from the virtual polder landscape than participants in the slides condition, because they experienced a greater sense of presence during the simulation (see Fig. 9.7). Participants' motivation to take preventive actions in the 3D condition was similar to participants' motivation in the film and slides conditions (see Fig. 9.6).

Figure 9.6 **Coping intentions as function of presentation mode**

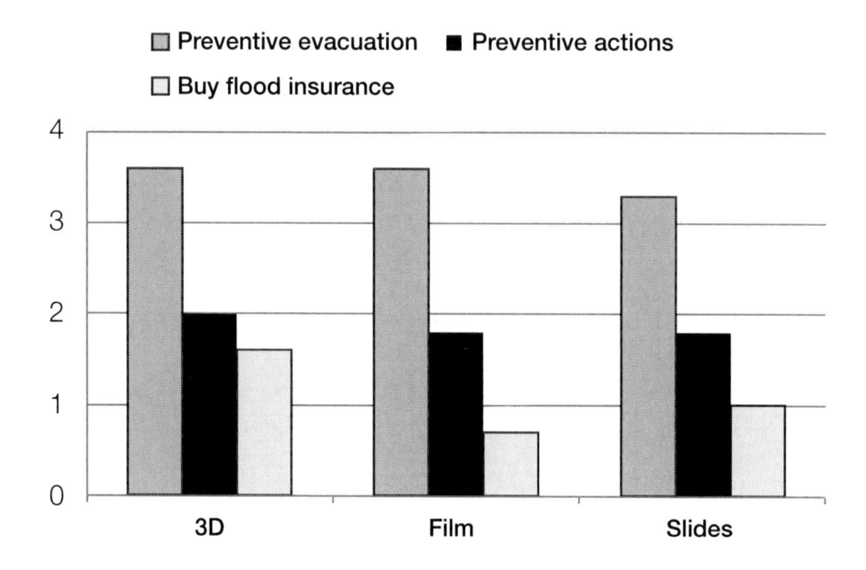

With control for differences in the perceived water level at the end of the simulation, participants in the 3D condition were more willing to buy flood insurance compared to participants in the film condition, but no more willing than participants in the slides condition (see Fig. 9.6). This contrast effect reduced when we controlled for spatial presence. Participants in the 3D condition were more willing to buy flood insurance than participants in the film condition, because they experienced a greater sense of presence during the simulation (see Fig. 9.7).

Figure 9.7 **Presence indicators as function of presentation mode**

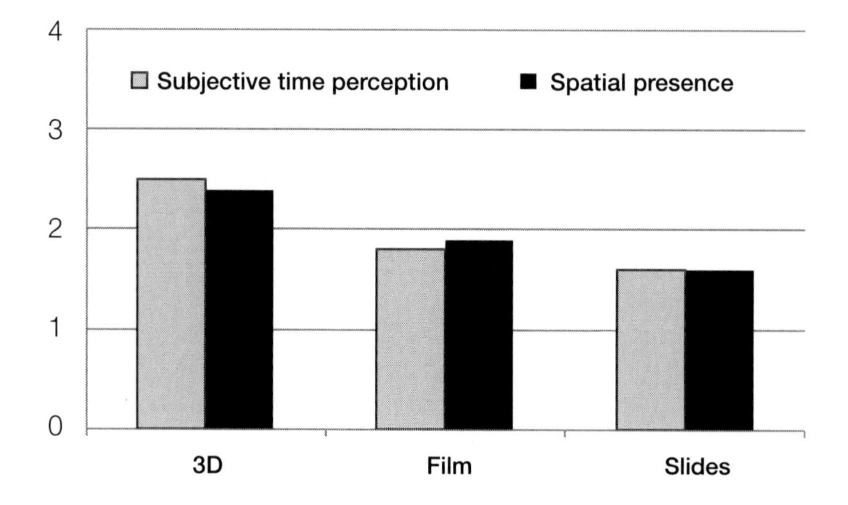

In contrast to our predictions, affective appraisals were not influenced by presentation mode. As predicted, no effects for presentation mode were found for cognitive appraisals. In sum, neither affective nor cognitive appraisals acted as mediating processes in the relationship between presentation mode versus information search and coping intentions.

9.3.7 Conclusions

Multimodal sensory stimulation by means of interactive 3D technology was used to simulate real flooding experiences, evoking resultant coping responses. Participants were exposed to a simulated dyke breach and flooding of their virtual residence positioned in a typical low-lying Dutch polder landscape. The central question was whether multimodal sensory stimulation facilitated participants' coping responses compared to non-interactive 2D simulations. An interactive 3D simulation increased not only the motivation to evacuate from the threatened virtual polder, but also the motivation to buy flood insurance in the real world. Both effects on the behavioural level were mediated by a greater sense of being present in the VE. To put it differently, human intentions change in response to simulated flooding provided that humans feel 'physically' present in the VE. We believe this to be an essential feature of future information campaigns when aiming at changing people's perceptions and coping responses in response to real-world flooding risks. However, the effects of the interactive 3D simulation on the information processing level are probably mediated by processes other than presence. These unknown processes should be investigated in future research.

9.4 Future directions

The 3D simulation enriches the sensory experiences of the participants, but does not increase their emotional experience. Therefore, a mismatch remains between real and simulated flooding experiences. Future research should further optimise the match between direct flooding experiences, which are inherently emotional, and simulated flooding experiences, where emotions have to be brought in explicitly. The following directions may be followed. First, the addition of extra-sensory stimulation such as wind and vibration to the flood simulation will enhance not only the 'sense of being present', but also the affective quality of the flood simulation. We predict that the addition of extra-sensory stimulation will increase the motivation to take appropriate coping actions in the VE. Second, enrichment of the flood simulation with affect-laden content will further stimulate the emotional involvement of the subjects. For example, the psychological relevance of 'flooded property' will increase when subjects first have to invest in their virtual property, which is then flooded. Third, the comparison of samples differing in the extent to which residents have direct flooding experiences may confirm the relevance of triggering the experiential system (i.e. images and associations of past flooding) in the prediction of coping actions in a VE. An additional process that may explain the influence of direct flooding experiences on coping actions is that future flooding risks and coping effectiveness become less ambiguous for residents with direct flooding experiences. The central question then is whether perceived ambiguity regarding the simulated flooding risk and coping effectiveness in the VE influences coping actions.

9.5 Discussion

The negative consequences of global climate change for low-lying countries are realistic and severe, especially when considering the large number of historical records of flooding in the Netherlands and the ever-increasing population density and growth of local economies located near to Dutch dykes (Klijn *et al.* 2004). However, it remains unclear to what extent residents living in flood-prone areas can be motivated to effectively cope with future floods. We present empirical evidence that past flooding experience acts as a motivating force to increase the intention to take adaptive actions against future flooding. Awareness of one's own vulnerability to future flooding due to climate change and insights into the effectiveness of coping actions to deal with these new risks are driven by direct flooding experiences. The use of new immersive 3D technology can be used to provide residents at risk with a simulated flooding experience at the experiential or sensory level, influencing information processing and coping intentions. These go beyond traditional persuasion attempts such as the use of fear-evoking images via film or slides. However, the affective component, inherently present in real flooding experiences and capable of influencing adaptive intentions via cognitive appraisals, was not influenced by presentation mode. Therefore, simulation researchers should feel challenged to transform traditional vicarious experiences via images into a full-blown 3D experience that is perceived as 'real', personally relevant and threatening. Future research should test the effects of technology-induced emotions in a VE on threat and coping appraisals as intervening variables predicting flood protection intentions. Future communication efforts should not only use these new technologies to transfer knowledge on effective coping strategies and flooding risks, but should also be directed especially towards the millions of residents living in flood-prone areas in the Netherlands, and who lack direct flooding experience as their guiding principle. Low resolution and static 2D flood maps, which are already available to the general public via websites of local authorities, may be upgraded into high-resolution 3D pictures of one's local community being flooded, comparable to Google Earth's street view.

The policy relevance of the presented research is high in our view. There are sufficient indications showing that people have growing awareness of climate risks, but that the sense of urgency, as the motivating factor for coping actions, is still rather weak. On top of that, people do not seem to be really aware of the scope of the risks and the nature of the required actions. This lack of concreteness and vividness may also lower the tendency to accept unattractive policy measures, which may be necessary to anticipate enhanced risks of flooding.

The present research delivers practical insights for developing intervention instruments to motivate the large numbers of residents who live in flood-prone areas in the Netherlands to protect themselves effectively in case of an imminent flood. The application of our scientific knowledge on simulated flooding experience in practical intervention strategies is without doubt important, not only from a human point of view in terms of a possible reduction in material and human losses, but also from a policy point of view. In our view the appropriate assessment by laypersons of flooding risks due to climate change will be a backbone for successful climate policy. In that sense our research can be of major significance and the present data has improved our insights. However, we should also acknowledge that the final target of developing effective intervention instruments is highly ambitious and will need more work besides the work already done. As stated earlier, the effectiveness of intervention instruments may be enhanced through the elicitation of emotional responses. We believe it is worth the effort.

References

Allewijn, R. (1983) *Een zee van water: Februarivloed 1953 over de Hoeksche Waard en het Eiland van Dordrecht* [*Rising Tide: The 1953 February Flood in the Netherlands*] (Klaaswaal: Waterschap De Groote Waard).

Baan, P.J.A., and F. Klijn (2004) 'Flood Risk Perception and Implications for Flood Risk Management in the Netherlands', *International Journal of River Basin Management* 2: 113-22.

Bouchard, S., J. St-Jacques, G. Robillard and P. Renaud (2008) 'Anxiety Increases the Feeling of Presence in Virtual Reality', *Presence* 17: 376-91.

Carlin, A.S., H.G. Hoffman and S. Weghorst (1997) 'Virtual Reality and Tactile Augmentation in the Treatment of Spider Phobia: A Case Report', *Behaviour Research and Therapy* 35: 153-58.

County Board East Flanders (n.d.) 'Wat bij een evacuatie?' ['What to do in case of an evacuation?']; www.oost-vlaanderen.be/public/over_provincie/veiligheid/evac/index.cfm, accessed 22 July 2010.

Dooley, D., R. Catalano, S. Mishra and S. Serxner (1992) 'Earthquake Preparedness: Predictors in a Community Survey', *Journal of Applied Social Psychology* 22: 451-70.

Dutch Institute for Public Opinion and Market Research (2007) 'Klimaatverandering? Mensenwerk' ['Climate Change? The Human Factor'] (Amsterdam: Dutch Institute for Public Opinion and Market Research).

Eagly, A.H., and S. Chaiken (1993) *The Psychology of Attitudes* (Fort Worth, TX: Harcourt Brace Jovanovich).

Environment Agency (UK) (2009) 'Creating a Better Place', 2 October 2009; www.environment-agency.gov.uk, accessed 17 March 2010.

Hendrix, C.M. (1994) 'Exploratory Studies on the Sense of Presence in Virtual Environments as a Function of Visual and Auditory Display Parameters' (Master's thesis, University of Washington, Seattle, WA).

Innig (n.d.) 'Willkommen auf der informationsseite "hochwasser" für die bürgerinnen und bürger der stadt Bremen'; www.innig.uni-bremen.de/kurzinfo_plattform.pdf, accessed 23 July 2010.

IPCC (Intergovernmental Panel on Climate Change) (2007) *Climate Change 2007: Synthesis Report* (Fourth Assessment Report; Geneva: IPCC).

Jöreskog, K.G., and D. Sörbom (1993) *LISREL8 User's Reference Guide* (Chicago: Scientific Software International).

Klijn, F., M. van Buuren and S.A.M. van Rooij (2004) 'Flood-risk Management Strategies for an Uncertain Future: Living with Rhine River Floods in the Netherlands?', *Ambio* 33: 141-47.

Lessiter, J., J. Freeman, E. Keogh and J. Davidoff (2001) 'A Cross-media Presence Questionnaire: The ITC-Sense of Presence Inventory', *Presence* 10: 282-97.

Lindell, M.K., and R.W. Perry (2000) 'Household Adjustment to Earthquake Hazard: A Review of Research', *Environment and Behavior* 32: 461-501.

Loewenstein, G.F., E.U. Weber, C.K. Hsee and N. Welch (2001) 'Risk as Feelings', *Psychological Bulletin* 127: 267-86.

Marx, S.M., E.U. Weber, B.S. Orlove, A. Leiserowitz, D.H. Krantz, C. Roncoli and J. Phillips (2007) 'Communication and Mental Processes: Experiential and Analytic Processing of Uncertain Climate Information', *Global Environmental Change* 17: 47-58.

Ministry of the Interior and Kingdom Relations (n.d.) 'Risk map'; www.risicokaart.nl, accessed 17 March 2010.

Ministry of Transport, Public Works and Water Management (2005) 'Veiligheid Nederland in kaart: Hoofdrapport onderzoek overstromingsrisico's' ['Mapping Dutch Safety: Final Research Report on Flooding Risks'] (The Hague: Ministry of Transport, Public Works and Water Management).

——(2006a) 'Questionnaire about Dutch Research Programs and Projects on Flood Risk Management' (Delft, Netherlands: WL/Delft Hydraulics).

—— (2006b) 'Spatial Planning Key Decision "Room for the River"' (The Hague: Ministry of Transport, Public Works and Water Management).

—— (n.d.) 'Water and Safety: Taskforce Management Flooding'; www.verkeerenwaterstaat.nl/onderwerpen/water/water_en_veiligheid/platform_overstromingen, accessed 17 March 2010.

Nowak, K.L., M. Krcmar and K.M. Farrar (2008) 'The Causes and Consequences of Presence: Considering the Influence of Violent Video Games on Presence and Aggression', *Presence* 17: 256-68.

Persky, S., and J. Blascovich (2008) 'Immersive Virtual Video Game Play and Presence: Influences on Aggressive Feelings and Behavior', *Presence* 17: 57-72.

Preacher, K.J., and A.F. Hayes (2008) 'Asymptotic and Resampling Strategies for Assessing and Comparing Indirect Effects in Multiple Mediator Models', *Behavior Research Methods* 40: 879-91.

Rogers, R.W., and S. Prentice-Dunn (1997) 'Protection Motivation Theory', in D.S. Gochman (ed.), *Handbook of Health Behavior Research 1: Personal and Social Determinants* (New York: Plenum Press): 113-32.

Royal Netherlands Meteorological Institute (2006) *Klimaat in de 21e eeuw: Vier scenario's voor Nederland* [*The 21st-Century Climate: Four Scenarios for the Netherlands*] (De Bilt, Netherlands: Royal Netherlands Meteorological Institute).

Sattler, D.N., C.F. Kaiser and J.B. Hittner (2000) 'Disaster Preparedness: Relationships among Prior Experience, Personal Characteristics, and Distress', *Journal of Applied Social Psychology* 30: 1,396-420.

Schubert, T., F. Friedmann and H. Regenbrecht (2001) 'The Experience of Presence: Factor Analytic Insights', *Presence: Teleoperators and Virtual Environments* 10: 266-81.

Siegel, J.M., K.I. Shoaf, A.A. Afifi and L.B. Bourque (2003) 'Surviving Two Disasters: Does Reaction to the First Predict Response to the Second?', *Environment and Behavior* 35: 637-54.

Siegrist, M., and H. Gutscher (2008) 'Natural Hazards and Motivation for Mitigation Behavior: People Cannot Predict the Affect Evoked by a Severe Flood', *Risk Analysis* 28: 771-78.

Slater, M., and M. Usoh (1994) 'Representations Systems, Perceptual Position, and Presence in Immersive Virtual Environments', *Presence* 2: 221-33.

Slovic, P., M.L. Finucane, E. Peters and D.G. MacGregor (2004) 'Risk as Analysis and Risk as Feelings: Some Thoughts about Affect, Reason, Risk, and Rationality', *Risk Analysis* 24: 311-22.

Van Dorland, R., and B. Jansen (2006) *De staat van het klimaat 2006: Actueel onderzoek en beleid nader verklaard* [*Climate State 2006: Further Explanation of Current Research and Policy*] (De Bilt/Wageningen, Netherlands: Platform Communication on Climate Change).

Weinstein, N.D. (1989) 'Effects of Personal Experience on Self-protective Behavior', *Psychological Bulletin* 105: 31-50.

Willemsen, M.C., and E.J. Johnson (2008) 'MouselabWEB: Monitoring Information Acquisition Processes on the Web', 14 August 2008; www.mouselabweb.org, accessed 17 March 2010.

Witmer, B.G., and M.J. Singer (1998) 'Measuring Presence in Virtual Environments: A Presence Questionnaire', *Presence* 7: 225-40.

Zaalberg, R., and C.J.H. Midden (2009) 'Living Behind Dikes: A Simulated Flooding Experience', manuscript submitted for publication, Eindhoven University of Technology.

——, C.J.H. Midden, A.L. Meijnders and L.T. McCalley (2009) 'Prevention, Adaptation, and Threat Denial: Flooding Experiences in the Netherlands', *Risk Analysis* 29: 1,759-78.

10

Interactions between white certificates for energy efficiency and other energy and climate policy instruments

Vlasis Oikonomou
SOM Research Institute, University of Groningen, The Netherlands

Energy is used as a basic factor for producing goods and services in the industrial and commercial sector globally and is also a vital commodity for our daily life, providing heating, cooling, cooking, transportation and the means for other activities. The use of energy is, however, a polluting activity with many externalities that are not taken into account in its final price. In fact, the conversion of primary energy (predominantly fossil fuels) to the various forms of final energy emits carbon dioxide (CO_2), which is one of the greenhouse gases (GHG) responsible for global warming.

The EU and its member states are developing their own policies targeting energy supply, energy demand and environmental goals that are indirectly linked to energy use, and energy efficiency, renewable energy promotion and the reduction of GHG are considered as priorities for many countries. Recent trends in environmental and energy policies tend to support market-oriented schemes, owing to their perceived high efficiency and market acceptance. Within the context of the United Nations Framework on Climate Change Convention (UNFCCC) Kyoto Protocol, several energy and climate policy instruments have evolved including, among many others, the EU Emissions Trading Scheme (EU ETS) (European Union 2004, 2009a, 2009b), Kyoto Protocol Joint Implementation (JI) mechanism and Clean Development Mechanism (CDM), benchmarking, white certificates (WhC) (also known as energy efficiency titles), voluntary agreements (VAs), tradable green certificates (TGCs) for renewable energy (RE) and command-and-control measures. As these policies are designed and implemented in an already policy-crowded environment, interactions between these instruments are taking place. These interactions can take different forms and shapes and in general they can be complementary, competitive or self-exclusive. The complexity of policy interaction, with many policy instruments directly or indirectly interacting with each other under current international and EU climate and energy policies, is presented in Figure 10.1.

Figure 10.1 **Complexity of policy interaction in international and EU policy level: an illustration**

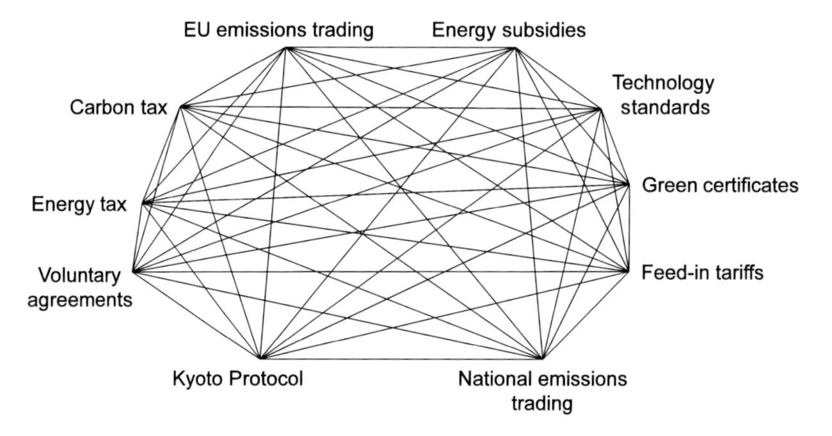

Policy-makers consult studies on policy interaction (PI) that are often based on one single analytical method and do not provide thorough insight covering more details of each policy decision step. Hence there is a substantial information gap between different approaches, which requires input from various sources and can often be time- and money-consuming (for a literature review of PI see Gunningham and Sinclair 1998; Oikonomou and Jepma 2007; Sorrell *et al.* 2003). There are several other cases where conflicts of policies were not taken into account *ex ante* or during their implementation, such as the EU Common Agricultural Policy, where subsidies for agricultural production together with production quotas led to structural and environmental problems and an entire shift of the policy was undertaken. In the energy field specifically, numerous findings in EU countries attest that energy liberalisation objectives in the presence of national financial support schemes have inhibited the main objectives of deregulated markets and led to reconsolidation and oligopolistic tendencies of energy companies. As a result, PI needs special attention when designing and implementing new policies in parallel to the incumbent ones (IPCC 2007).

As a starting point, this chapter demonstrates possible methods of analysing PI while exploring possibilities of synthesising them in a uniform approach. We choose white certificates (WhC), also known as energy efficiency titles, an instrument for energy efficiency improvement in the end-use sectors, and we compare it with other energy and climate policy instruments (for a detailed description of WhC see Bertoldi *et al.* 2005; IEA 2006). WhC is a new and much-debated policy instrument to increase energy efficiency using market-based mechanisms. It belongs to the category of environmental tradable permits and is similar to the JI and emissions trading schemes, with the basic difference that it refers to energy efficiency as the tradable commodity. Its basic principle is that authorities impose energy efficiency obligations on electricity and gas suppliers (i.e. retailers) and distributors, who can then decide whether to implement energy efficiency measures or to purchase WhC, depending on their opportunity costs. WhC have been implemented in Italy, France and Great Britain, while other countries are discussing their implementation (e.g. Poland and the Netherlands). WhC can assist in achieving energy and climate targets, as they are currently pursued in the EU, in the form of 20% energy efficiency improvement and 20% emissions reduction by 2020. WhC is furthermore supported by the EU Directive on energy end-use efficiency and energy services (European Union 2006), which clearly states: 'allowing Member State authorities, inter alia, to tender . . . including systems for White Certificates' and quoting the Green Paper on Energy: 'the Commission considers this to be a possible next step in a few years time and may then come forward with a proposal based on the experiences in some Member States currently developing and implementing such certification schemes.'

The philosophy underlying this system is to combine the guaranteed results of setting obligations (it can also be considered as a smart way of regulation) with the economic efficiency of market-based mechanisms. The effectiveness of the WhC increases if they are bundled with, among others, information campaigns and other means to promote opportunities for energy saving (Farinelli *et al.* 2005) in order to drive consumer demand for energy-efficient measures. In a parallel market of these measures, the introduction of the WhC would reduce the relative price of the energy efficiency measures resulting in an increased demand for these goods. However, the rise in the demand may be limited. The reasons are, first, the rebound effect on consumers' behaviour, which is expected to range from 5% to 50% (for more information see Binswanger 2001; Greening *et al.* 2000; Sorrell 2007). Other studies on the effects of WhC on rebound effects can be found in Oikonomou *et al.* 2007 and Oikonomou *et al.* 2009a. The second reason is that short-term elasticity of energy use, as complemented by durable equipment, is much smaller than long-term elasticity (Velthuijsen and Worrell 1999). As shown in Figure 10.2, the WhC market consists of the following participants: regulatory authority, suppliers and distributors of gas and electricity, energy service companies (ESCOs), households and brokers.

The regulatory authority plays the principal role in distributing the obligations among the participants and issuing the certificates. The participants that can request and trade WhC are:

Figure 10.2 **White certificates energy market**

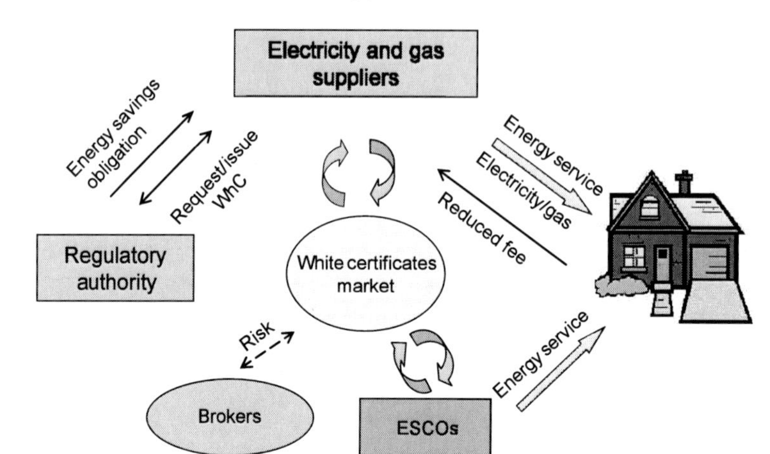

- Suppliers and distributors of gas and electricity, who have an obligation, set by the regulatory authority, to save a certain amount of energy within a specified period. To this end, suppliers have to promote specific energy efficiency projects to end-consumers. Suppliers and distributors receive WhC and can trade them on the market
- Energy service companies (ESCOs), which are companies that offer to reduce a client's energy costs, often by taking a share of such reduced costs as repayment for installing the energy efficiency measure and financing its upgrades. They do not have an obligation, but can participate in the scheme after achieving energy savings and receiving WhC
- 'Other participants', which are entities that do not have an obligation but can purchase and sell WhC, providing thus the necessary liquidity in the market. Examples for such entities are brokers and financing institutions, which facilitate the transactions and reduce the risk of the investments, while speculating on the price of WhC and receiving a commission from the transaction costs. The eligibility and the role of these entities differ among the existing WhC schemes. These entities are included in the UK and French WhC schemes

WhC are implemented in parallel to other policy instruments addressing similar or conflicting targets, which can enhance or inhibit their overall performance. As they are relatively new on the policy agenda, limited or almost no actual *ex post* information is available on interactions of WhC with other policy instruments. In this chapter, our main research areas are: (1) to provide a general explanatory framework for analysing PI in the areas of energy and climate, by employing suitable methods and paving the way for a unified framework, and (2) to draw conclusions for policy-makers when introducing WhC with other policy instruments, stressing the critical aspects and conditions that affect their overall performance. We expect that the methods we employed can provide an analytical framework for PI and generate useful information for policy-makers. The significance of the objectives pursued in this study and the means employed in order to achieve them are driven by the importance of assessment of PI in often-intertwined energy and climate targets. The policy evaluation of WhC with other instruments is also 'crucial in verifying results, withdrawing inefficient policies or providing the corrections necessary to improve the performance of policy instruments in order to resolve the problems and secure policy objectives' (Fischer 1995; Mickwitz 2003; Mundaca 2008).

In the next section we describe four methods that were employed in the study. In Sections 10.2–10.5 we present key results from published articles on policy interactions of WhC with other instruments, based on these four methods. Finally, Section 10.6 gives conclusions from the synthesis of these methods and policy recommendations.

10.1 Research methodology

In order to analyse PI, we refer to policy analysis methods, which are general procedures for producing and transforming policy-relevant information in a wide variety of contexts. When facing multiple policy instruments, various methods can be employed in order to generate meaningful results. Literature provides us with a plethora of methods for assessing single policy problems, which can be also applied to more complex issues of PI (see e.g. Blok 2006; Taylor and Jollands 2007; Vreuls *et al.* 2005). In this chapter we make use of the following four methods:

10.1.1 Institutional policy analysis method

Institutional policy analysis in principle deals with the study of institutional governance and its consequences (Hall 1986; Parsons 2001; Selznick 1949). To study PI, we construct a qualitative method for analysing the impacts of coupling policy instruments, which consists of several steps. The first step defines clearly the type of PI, since different attributes are assigned to different approaches and subsequently policy targets can be specified. In the second step of our method, we break down the design characteristics of instruments and extract conclusions for synergies or conflicts within an integrated scheme. In a PI problem we attempt to identify for each instrument the specific issue it addresses, the implementation network (mix of public and private bodies that implement the policy), the target groups (economic actors that are influenced directly or indirectly by the policy), the outcomes and objectives that reflect the intended or unintended effects of policy, and the context (broader economic, political and cultural context in which the policy operates). Finally, we evaluate the effect that PI has in society and in economic sectors based on six general criteria: effectiveness, efficiency, impacts on energy and market prices, impacts on society, innovation and market competition.

10.1.2 Microeconomic policy analysis method

An analysis of the economic behaviour of an individual company or economic unit can be provided by microeconomic theory. This theory can serve to demonstrate the reactions of market participants, namely consumers and producers, when being challenged with various constraints and policy instruments. The core principle determining the producer's reaction to the market is profit maximisation under production cost constraints. Alternatively, it could be set as cost minimisation target in order to achieve a given profit target. In a policy framework, each instrument acts as an extra cost or cost reduction in the production structure, affecting the value of production (or alternatively taking the form of a technological shift). From a consumer's perspective the main behaviour determinant is the maximisation of the utility through enjoying goods and services (energy services for the given context).

10.1.3 Multi-criteria policy analysis method

An alternative process for analysing PI problems involves the participation of relevant stakeholders that express opinions and own views on the specific problem. Stakeholder participatory methods gain more ground in actual policy-making and are supplementary to other quantitative assessment methods. To this end, when facing options of different combinations of several policy instruments in the energy and climate field, policy-makers often consult views and preferences of directly or indirectly affected market actors in the form of workshops. We develop and apply a decision support tool for analysing energy and climate PI, built on the institutional theory method stated above. Policy-makers can express in order of merit the significance they attribute to the design characteristics of instruments when designing a policy, and select pairs of instruments to consider for implementation. Criteria covering specific objectives for assessing these instruments individually are used and policy-makers can assign weights to them expressing their preferences, in the framework of a multi-criteria analysis (MCA). An overall assessment of combined instruments takes place based on the input from policy-makers (for extensive work on MCA see Belton and Stewart 2002; Gamper and Turcanu 2007; Greening and Bernow 2004; Hajkowicz *et al.* 2000; Konidari and Mavrakis 2007; Leitmann 1976; Rogers and Bruen 1998).

10.1.4 Technoeconomic policy analysis method

A final method we use for analysing PI is quantitative energy models, which assist us in projecting future energy demand and supply, assessing the impacts of different energy systems and finally appraising the energy systems. More specifically, we employ the Energy Technology Systems Analysis Programme (ETSAP) TIMES Integrated Assessment Model (TIAM), a generator of economic equilibrium programming models of energy systems and their time development. The energy system is represented by relations between services and final products on the one hand and environmental and material flows on the other. Each supply/demand technology is characterised by technical economic and environmental parameters and by input commodities or output services. Basic components are processes (conversion technologies from commodities to other commodities), commodities (consisting of energy carriers, energy services, materials, monetary flows and emissions) and commodity flows (as links between processes and commodities). Each flow is attached to a particular process and represents one input and one output of that process (see Loulou and Labriet (2007) for an analytical description of the model).

Each method provides us with various results in terms of blending WhC with various policy instruments in the field of energy or environmental policy. Findings differ as quality of information from each method is substantially different and the scope of each analysis focuses on various actors. For instance, from a top-down policy perspective, implementing WhC with parallel energy tax redemptions might not be optimal from a policy-maker's perspective, because it can induce free-riding behaviour. In contrast, from a company's perspective, as expressed by microeconomic theory, costs of energy efficiency actions can be low and technological innovation can be stimulated, which render the combined scheme desirable.

In the field of energy efficiency, a WhC scheme presents many similarities to other instruments, especially those that include a (tradable or non-tradable) certificate mechanism. When analysing such policy instruments, these similarities can have a positive or negative impact, in terms of lessons learned for the policies' market introduction and implementation. A positive effect is that policy-makers can use experiences from similar instruments and decide which parameters or design characteristics might jeopardise the overall functioning of the scheme; they can therefore adapt these parameters prior to implementing a WhC scheme. A negative side is that policy-makers can often provoke overlaps between instruments, and some instruments that are considered 'flexible' or 'cost-effective' may be redundant when combined with others. In the following sections we demonstrate some examples from interactions of WhC with other policy instruments based on our various methods and finally we conclude with some key findings and policy recommendations.

10.2 Case studies: institutional analysis

Departing from an institutional policy analysis perspective, we employ some examples of combinations of WhC with other instruments, namely TGC, Kyoto Protocol JI mechanism, and VAs for energy efficiency. Initially, a potential implementation of a WhC scheme was tested for the Netherlands, because as a country it has accumulated experience in applying various energy efficiency policies in the market and attempts to play the pioneer in this field (Blok *et al.* 2002; Boonekamp *et al.* 2004; Joosen *et al.* 2004). Taking Dutch households as a case study, we examined the overall effects of a national WhC scheme from 2000 to 2020 (divided into five-year periods), without taking into account policy implementation and other administrative costs (Oikonomou *et al.* 2007). Based on the findings on several energy-efficient technologies we conclude that this scheme can achieve high effectiveness and cost-effectiveness in terms of energy savings. The maximum cost-effective cumulated primary energy savings until 2020 are estimated at 240 petajoules (PJ). The primary energy consumption of Dutch households after the implementation of the WhC can be reduced by 26% (in 2020), compared to a frozen efficiency scenario. These savings stem from the use of new technologies that can be diffused in the market through the application of the scheme. The financial net savings in the last phase of the WhC scheme are three times higher than the initial phase of the scheme. Based on these findings we conclude that WhC is a policy instrument that could be useful for the design of future energy policy in parallel with the existing instruments and policies already applied in the Netherlands.

In a second step in our analysis, by analysing experiences gained from Dutch and Swedish TGC schemes, we extract some policy lessons that can support a successful design of WhC (Oikonomou and Mundaca 2008). Taking into account the lessons learned so far by TGC schemes in the Netherlands and Sweden we identified some crucial policy lessons that policy-makers should take into account when designing similar schemes for energy efficiency. Our analysis indicates key policy suggestions for each specific design characteristic decided by policy-makers in a WhC scheme. To highlight the main points: an energy savings target must be clearly expressed for the long term, a transparent market must be established and standardised procedures for monitoring, and verification of the actual energy savings achieved must be guaranteed.

Furthermore, we identified the feasibility for the Netherlands to allow a WhC scheme where electricity and gas suppliers, under an energy efficiency obligation, can implement energy efficiency projects in buildings in another country under the form of JI (Oikonomou and Van der Gaast 2008). The Netherlands has been selected as a country case study since it is experienced in JI projects and has already implemented many energy efficiency improvement policies (Van der Gaast 2002). In the Netherlands, 33 million tonnes of CO_2-equivalent emissions reduction are to be achieved under JI projects as one of the means to reach the 6% reduction target for the country under the Kyoto Protocol. A WhC scheme for households can be cost-effective and reduce 80 PJ of primary energy in 2012, while cumulative savings up to 2020 can reach a range of 180–240 PJ, which can be translated into 10.3–13.6 Mt CO_2. Therefore, an integrated voluntary JI/WhC scheme, as described in Oikonomou and Van der Gaast 2008, could contribute substantially to achieving national (current and post-2012) targets. A fundamental outcome is that an integrated scheme of WhC with JI for energy efficiency projects in the built environment can be complementary, given that the former remains a national scheme for the Netherlands. Both instruments refer to the same policy context, which facilitates a common design in terms of target setting, and both can be used as WhC trading mechanisms. Furthermore, a voluntary integrated WhC/JI scheme (maintaining the obligatory element of WhC) can be complementary and improve energy efficiency at national and international levels. More specifically, in terms of objectives a hybrid scheme can facilitate electricity and gas suppliers to achieve their obligations through implementing cost-effective energy efficiency projects in another country. Nonetheless, it must be noted that given the present state of international climate negotiations, there is little guidance for investors on whether post-2012 reductions could be accounted, under what conditions and at what prices.

Nevertheless, two aspects must be taken into account that determine the overall performance of the WhC/JI scheme: methods for defining baselines in energy savings and ways of converting different tradable commodities. In order to reduce bottlenecks in monitoring and verification of energy savings of each instrument, harmonised approaches for individual schemes must be used in an integrated WhC/JI policy. Conversion of credits from JI to WhC should be allowed only if the former have been proven to originate from a realised energy efficiency project and under a steady conversion rate.

Based on an *ex ante* assessment, a WhC/JI scheme can be effective, since an obligation to energy suppliers under WhC guarantees the achievement of minimum energy savings targets. Furthermore, it can contribute to the security of energy supply in the JI host countries and provide incentives for long-term spillover effects, provided that free riders and rebound effects are dealt with. In terms of cost-effectiveness, the WhC/JI scheme can achieve both reduction of GHG emissions and energy efficiency improvement at relatively low costs due to flexible options provided to energy suppliers. The range of total costs will be determined by transaction and administrative costs, which can increase if policy-makers opt for a complicated design. Effects on innovation and diffusion of new energy-efficient technologies can be basically positive through suppliers' and ESCOs' efforts to increase market shares under a WhC scheme. In general though, the effects of a hybrid WhC/JI scheme on innovation depend mainly on (local and foreign) market demand, competitiveness between technologies, existing energy saving potential, and transaction costs of the policies. A WhC/JI scheme under our proposed design options can have minimal negative (or almost neutral) impacts on society as a whole while in some cases some positive effects can be present, namely in terms of employment and increase in environmental awareness. Finally, such a scheme can be promising in terms of market effects given the market-oriented character of both policy instruments, because it can be compatible with energy market liberalisation and increase the competitiveness of specific 'cleaner' technologies.

Nonetheless, despite the positive outcome of the feasibility assessment of the hybrid scheme, it is noted that contrary to the even mixing of GHG in the atmosphere, which makes the location of GHG abatement measures irrelevant for their effectiveness, energy efficiency measures have mainly an effect on the site where they are carried out. This implies that energy efficiency measures carried out abroad and translated into WhC for use in the Netherlands replace domestic energy efficiency improvements that would have been carried out with domestic action only. This issue requires political consideration and could lead to a full conversion of JI credits into WhC (as assumed in this chapter) or a limited/discounted conversion, where energy efficiency JI projects are only partially credited as WhC, so that some further energy efficiency investments within the Netherlands would still be needed to comply with energy efficiency commitments.

As a third step, we examined how WhC (as a market mechanism) could eventually function together with VAs for energy efficiency (Bertoldi and Rezessy 2007; Rietbergen *et al.* 2002) in order to stimulate investments beyond the cheapest energy efficiency solutions (Oikonomou *et al.* 2009b). It is worth noting that a basic bottleneck that needs to be resolved in WhC implementation is that suppliers and distributors generally target 'low-hanging fruits' in energy efficiency investments. We made use of a VAs/WhC scheme that addresses energy efficiency improvement in services and residential sectors (indirectly addressed as a recipient of final energy efficiency investments) in the Netherlands. An integrated scheme of WhC and VAs (maintaining the obligatory element of WhC and the voluntary character of participation in VAs) can be complementary and enhance energy efficiency improvement. In this scheme the targets pursued could be double: (1) energy efficiency improvement with installation of energy saving measures in the residential and service sector (hence reduction of energy intensity in these sectors), and (2) diffusion of CFLs (compact fluorescent lamps) and energy-efficient product development for the manufacturing sector. Energy-efficient project packages could address white and brown appliances, the building envelope, small appliances, innovative goods and energy management (for instance, behavioural change). In the aforementioned study we present a list of energy efficiency measures that are eligible for certification in the WhC/VAs scheme, under the WhC obligation, providing also flexibility for all remaining available energy efficiency actions undertaken by participants.

Furthermore, we demonstrated the functioning of this scheme and all possible activities that key players need to undertake as specified by both policy instruments. Two accounting issues that are crucial for the integration of WhC and VAs are recognised: the transaction costs originating from the policy life-cycle, and methods to calculate the baseline beyond which energy savings start counting against targets set by both instruments. Concerning the former, the life-cycle of WhC consists of planning, implementation, measurement and verification, issuance, trading and redemption. Depending on the design of a WhC scheme, each step includes different transaction costs for market players and authorities. In a hybrid scheme most transaction costs are identical to the ones that WhC and VAs as stand-alone instruments entail (market research, projects implementation, reporting, monitoring, issuing and trading WhC). Concerning the second point, participation in VAs requires that companies undertake energy saving investments with a longer payback period, triggering investments in more expensive energy efficiency measures. In the case of WhC, all energy efficiency actions on the part of suppliers are rewarded with certificates. In order to simplify the measuring of savings, some technologies and actions can be *ex ante* pre-approved in a specific list, which determines the amount of savings spread over the technology's lifetime. Market participants under VAs and WhC can select some of these measures based on their financial capacity and on the range of deviation from their energy savings target. In merit order, as economic theory dictates, market players should first exploit all the least expensive measures to fulfil their obligation (and evade a financial penalty set under WhC). As may be expected, the level of the penalty will also determine the selection of investments in technologies.

From an *ex ante* assessment based on several criteria of the proposed WhC/VAs scheme we found that it can enable the realisation of energy security targets. This scheme can also reduce the risk of free-riding behaviour, as all relevant market players have primarily to fulfil their targets and in doing so they can opt in for voluntary actions with a specific reward (in the form of financial assistance, advertisement and others). Compliance costs are expected to decrease in a hybrid scheme, because market players can make use of the cost flexibility option under the agreements and reduce their overall costs in the scheme. Administration costs are usually not substantial under VAs and they are estimated to be lower than under a regulatory policy. The WhC/VAs policy can tackle the information asymmetry on behalf of the authorities so that increased administrative costs can be avoided, but transaction costs can be slightly higher. In the proposed integrated scheme, one could expect that a moderate stimulation of innovation and knowledge exchange would be achieved. In principle, VAs are considered to be quite favourable for entities that undertake obligations for energy efficiency improvements. Their link with a WhC scheme can reduce social opposition to the implementation of the latter, because VAs can serve as a 'cushion' that potentially reduces the costs of abatement and as a means to improve companies' green image in the market.

This comparative analysis of the three examples from an institutional policy analysis perspective mentioned above reveals that in general WhC can be linked to various policy instruments and considerably increase their effectiveness in achieving the desired targets of energy efficiency improvement.

10.3 Case studies: microeconomic theory

By employing microeconomic policy theory, Oikonomou *et al.* (2008) and Giraudet and Quirion (2008) tested interactions of WhC with energy taxation. Because of the multiplicity of energy taxes (Zhang and Baranzini 2004), we confined ourselves to a tax on fossil fuels as an input for electricity production (carbon tax) and a tax on electricity sales (electricity tax). These policy instruments differ in their objectives and final impacts on the price of electricity. The cases examined in Oikonomou *et al.* (2008) consist of electricity producers with and without a carbon tax, electricity suppliers with or without a sales tax, and with WhC obligations. A parallel implementation of WhC for electricity suppliers with a carbon tax on electricity producers and a sales tax with WhC obligations to electricity suppliers is also presented. In all these cases, we demonstrated the differences in optimisation behaviour of producers and suppliers and how the optimal decisions of electricity producers are incorporated in the decisions of electricity suppliers. Given also that WhC refer to the undertaking of energy efficiency actions, we presented possible trade-offs for energy suppliers between reducing electricity supply costs and implementing energy saving projects. According to a general evaluation of a couple of cases of WhC with carbon and electricity taxes, various positive and negative effects of both schemes in target achievement and cost-effectiveness are present, which can lead to an added value of such schemes in the policy mix, although with some uncertainties of outcomes.

According to our evaluation, both combined schemes presented positive and negative effects on the criteria studied. In terms of effectiveness, a WhC/carbon tax scheme is slightly superior to a WhC /electricity tax especially on the grounds of reduction of GHG emissions and technological innovation, while it can be less effective in guaranteeing security of energy supply. Achievement of various targets can nevertheless be ensured by both schemes in comparison to the case where these policy instruments stand alone. As far as cost minimisation is concerned, neither scheme is highly efficient, given the incurred administration and transaction costs required to implement these policies. Innovation towards more energy-efficient technologies can be enhanced through a WhC/carbon tax scheme given the extra incentives for such investments from market actors, while existing technologies can be diffused more easily in both schemes in contrast to stand-alone policies. Both schemes can influence various impacts on society positively or negatively, given that uncertainties are quite high also in this case. We estimate that equity and fairness for directly and indirectly affected market parties remain the same in both combined schemes, and they can both face political opposition due to the mandatory character of taxation and energy efficiency obligations. Furthermore, an increase in employment opportunities is present more in a WhC/electricity tax scheme, while an increase of environmental awareness is stimulated more from a WhC/carbon tax scheme. Finally, both combined schemes generate positive market effects in terms of trading and business opportunities, competition in energy markets and market competitiveness. A WhC/carbon tax can be highly significant in creating further trading possibilities for participants by opening new energy efficiency markets. In contrast, the integrated scheme does not present any added value for the process of market liberalisation and competition, as compared to the case where both instruments stand alone in the market.

A basic finding in policy interaction from WhC and energy taxation is that several parameters can increase final electricity prices: demand for electricity and electricity supply costs on a large scale and followed by the level of sales tax, level of obligation for energy saving, and price of WhC (representing the marginal costs of energy saving projects). Some parameters that influence positively (in the sense of price increases) the price of WhC are the price of electricity, the level of demand and the level of fuel tax. In contrast, parameters that influence negatively the WhC price are the level of obligation, supply cost and the level of electricity sales tax. Finally, energy savings (or quantities of WhC traded) are affected positively by the level of obligation, cost of electricity supply and level of sales target, and negatively by the price of WhC, the price of electricity, the level of electricity demand and the level of carbon tax. The magnitude of impacts of all these parameters depends on the values chosen and on the initial position of suppliers (e.g. if their actual behaviour deviates from full compliance with targets).

Similar findings appear also in the case of an oligopolistic energy market. We argued that in an oligopolistic EU market, as it is currently structured, one leader (e.g. ENEL in Italy and EDF in France) shares the market with a number of firms that have access to the same technology but have to take their decisions under the constraint of the first-mover advantage of the leader (Oikonomou *et al.* 2009b). In general, the leader supplies larger quantities and lower prices in both markets and this leads to higher consumer surplus in terms of electricity consumption and installation of energy saving technologies. Profits for the leader in both energy and energy efficiency markets are higher than those for followers. As a result, at the margin smaller companies can be pushed out of the market. Furthermore, the introduction of electricity and carbon taxes with WhC in such a market generates various results. An electricity tax decreases the quantities of electricity supplied and increases the electricity price to consumers. In combination with WhC, energy saving investments by both large and smaller companies increase, with a potential effect of lowering WhC prices. The level of tax determines the outcome of revenues of taxation in the WhC/electricity tax scheme. Finally, a carbon tax reduces the quantity of electricity and the leader firm is most affected. The price of electricity subsequently increases, which drives the leader firm to invest in energy efficiency, covering the main part of the energy efficiency market. Smaller firms, on the contrary, operating at the fringe market reduce such actions as the carbon tax increases.

10.4 Case studies: multi-criteria analysis

In the third step, we made use of multi-criteria analysis and of the Energy and Climate Policy Interactions (ECPI) decisions support tool that we developed (Oikonomou *et al.* 2008). We presented a couple of illustrative examples of integrating feed-in tariffs for RE with TGC, and feed-in tariffs for RE with carbon taxes, based on stakeholders' judgements and criteria valuation. As demonstrated in the study, integrating a TGC scheme with a quota obligation to producers with feed-in tariffs for energy suppliers might not be the optimal solution compared to where both instruments are implemented separately and their interaction lies only in the final energy price for consumers. This can take place if TGC addresses specific RE technologies, while feed-in tariff finances the most expensive actions, as financing 'low-hanging fruits' in RE can lead to market inefficiencies and lock-out of many technologies. The application has demonstrated that, based on the preferences of policy-makers, integrating a TGC scheme with traditional feed-in tariff schemes might not be optimal. Instead, letting instruments separate in the market is preferred (e.g. by addressing different technologies) in order to eliminate policy overlaps. We conclude that, like the case of carbon taxes with feed-in tariffs for RE, by incorporating policy-makers' weighting preferences on criteria and targets, these two instruments can function more effectively if they are isolated. This outcome does not exclude any interaction between them. On the contrary, it states that due to their interaction in the market level, their integration is not optimal.

10.5 Case studies: technoeconomic analysis

The fourth method we employed was based on technoeconomic analysis, where we made use of the well-established model TIAM, which arises from the energy and environmental system analysis standpoint. We focused on the Western European energy system and all its components and explored the potential impacts that the parallel implementation of several market-based instruments in the fields of energy efficiency and GHG emissions reduction might induce. We tested seven base scenarios reflecting different targets and policy instruments and seven separate scenarios depicting interactions of these targets and instruments (see Table 10.1).

For the energy efficiency policy instruments we further assume that supportive information campaigns can assist in overcoming market barriers to energy efficiency investments, and this is modelled with an apparently higher discount rate than the financial market one. This discount rate attempts to simulate the displacement of the system from the economic optimum from the base case scenario. Indeed, Ruderman *et al.* (1987) point out that an 'implicit real discount rate' used in investment decisions in the residential-household sector goes from 35% to 70%. In our case, we select an apparent discount rate of about 30%, which approaches the bottom level given by Ruderman *et al.* (1987).

Marginal energy saving costs for all scenarios are different based on a variety of financially cost-effective policy options. We can ascertain that of the climate policy instruments tested, Kyoto and EU ETS with energy efficiency subsidies and Kyoto

Table 10.1 **Policy scenarios and targets**

Scenario	Instruments and targets	Sectors
Base	Policy-free environment	All
Base 1	Kyoto commitments and International Emissions Trading Scheme	All
Base 2	EU Emissions Trading Scheme	Energy production and industry
Base 3	20% emissions reduction	All
Base 4	20% energy efficiency improvement and white certificates	All
Base 5	Energy Services Directive (9%) and white certificates	Residential, Commercial, Transport
Base 6	Tax on CO_2 emissions	Energy production and industry
Base 7	Subsidy for energy efficiency	Residential and Commercial
A12	Climate I: Kyoto Commitments and EU Emissions Trading Scheme	
A123	Climate II: Climate I and EU Climate objective	
A1234	Climate I and EU energy efficiency objective with WhC	
A1235	Climate II and EU Energy Savings Directive with WhC	
A1237	Climate II and energy efficiency subsidies	
A1245	Climate I, Energy efficiency objective and EU Energy Savings Directive with WhC	
A126	Climate I and carbon tax	

and EU ETS with carbon taxes are cost-effective, while almost half of the target of Kyoto are covered. Climate policy scenarios alone do not appear to increase substantially the total costs, always in comparison to the business as usual trend, as the targets pursued are relatively moderate (Fig. 10.3).

In order to extract the potentials for financial cost-effectiveness, we employ as a benchmark the average fuel prices in the EU, which are estimated at 0.5 €/KWh. Based on this value, we can ascertain that the EU ETS, Climate II (all climate targets), Climate I with energy efficiency subsidies and Climate I with carbon taxes are met cost-effectively, while almost half of the target of Climate I (Kyoto and EU ETS targets) are covered. The remaining scenarios are above the average fuel price, and hence this signifies that although the potential can be covered, still the costs inhibit implementation of these combined policies.

Another indicator we employ is the absolute energy savings, calculated as the difference of each scenario with the baseline. There, WhC in an enlarged scheme for

Figure 10.3 **Marginal costs of policy scenarios**

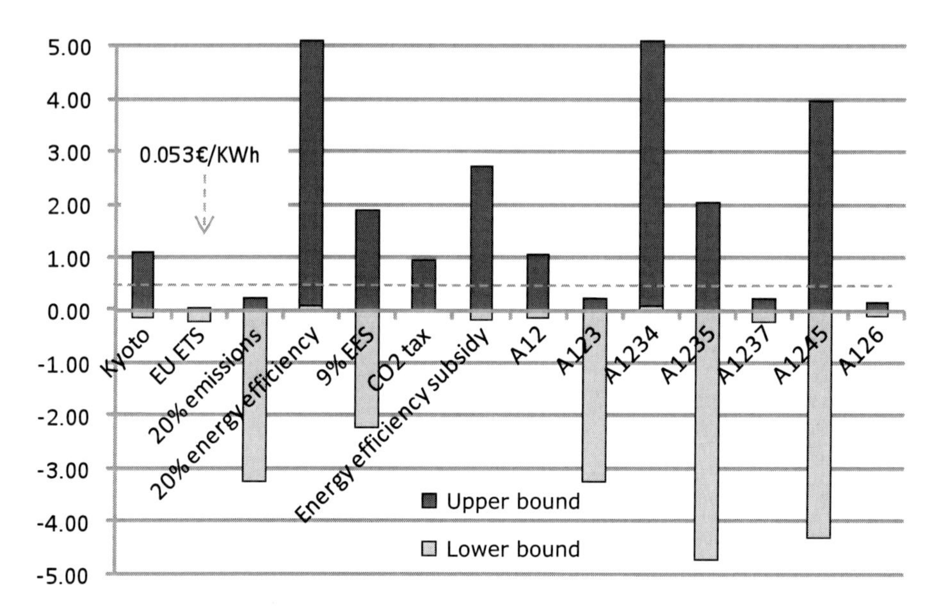

achieving energy efficiency objectives of 20% can generate enormous energy savings, higher than all the remaining policy cases. In contrast, a limited WhC scheme supporting the EES Directive, addressing only end-use sectors, is not sufficient for reducing energy in the entire economy (Fig. 10.4).

Figure 10.4 **Total final energy savings**

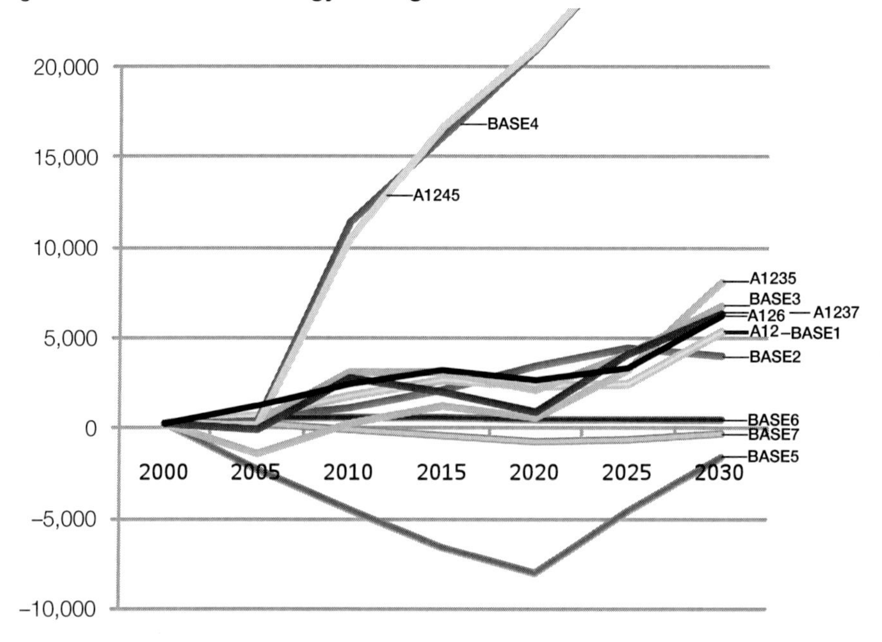

The majority of policies and targets tested in the model successfully reduce CO_2 emissions at different levels. An exception lies with the EES Directive and a limited WhC scheme, and with subsidies for energy efficiency improvement, when left to stand alone in the market, which do not generate substantial GHG reductions. In terms of total CO_2 emissions, the limited WhC scheme to residential and commercial sectors does not substantially reduce emissions, in contrast to all other policies and their combinations. As far as energy effectiveness is concerned, we distinguish between primary energy supply and energy intensity of the entire economy. As a starting point, we identify different patterns in primary energy supply in each policy scenario, which are in line with the emission trends presented. The highest reduction of fossil fuel share is present when climate policies are implemented in parallel with WhC. The highest increase of the shares of renewable sources is due to climate policy primarily and to a much lesser extent to energy efficiency targets. A significant influence in the decrease of energy intensity is attributed to implementation of an enlarged WhC scheme with stringent efficiency targets and when combined with climate targets. The largest improvement in fossil fuel intensity is triggered by the combinations of Kyoto policies and EU ETS with WhC schemes. In general though, all scenarios achieve to a certain extent a reduction of fossil fuel intensity in the economy. An exception to that is the limited WhC scheme, which can be explained by the fact that it covers limited sectors, while fossil fuel intensity refers to the entire economy.

The framework incorporated in each method in this chapter provides various results that could feed in the other methods and generate a meaningful result. As explained above, each method on its own is often not sufficient to demonstrate the added value of introducing a policy instrument next to one already implemented in the market and can hence lead to a limited insight into what is necessary for policy decision and debate. To this end, policy-makers often require more information from different perspectives than each stand-alone method can provide, and this may lead to inconsistent policy decisions. This pitfall can be clearly demonstrated in Table 10.2, where we use an example of the implementation of a WhC on energy suppliers with a carbon tax on energy producers. In our multi-method approach, several independent routes were used to address the analysis of policy interactions of WhC schemes.

Based on the various evaluation outcomes of combinations of WhC with other policy schemes, a basic lesson to be drawn is that a flexible and multi-dimensional evaluation framework for energy efficiency and climate policy needs to be used in similar assessments. Departing from the assessment of the methods of analysing PI, a possible 'synthesis' of these methods can be useful for extracting policy conclusions. In order to account for each method's strengths and weaknesses we propose the efficient use of all mentioned methods in a stepwise approach in order to provide integrated results from different scopes and by using diverse necessary information.

Table 10.2 **Policy interactions information from different methods**

Example: WhC for energy suppliers with a carbon tax for energy producers			
Institutional analysis	**Microeconomic analysis**	**Multi-criteria analysis**	**Technoeconomic analysis**
• Carbon tax and WhC could be necessary in the existing policy framework • Determining level of interactions (national, no fungibility), more complementarities than overlaps • Options of characteristics, targets, energy suppliers and producers, administrative bodies required • Qualitative evaluation (effectiveness increases, cost-effectiveness does not have added value, innovation stimulated)	• Behaviour of supplier with WhC obligation and producer with carbon tax in competitive and oligopolistic markets • Trade-off between paying for carbon or energy efficiency from both producers and suppliers • Electricity cost rises for end-users so more opportunities for energy efficiency • Suppliers face more options, by performing energy efficiency investments, buying certificates, or paying the penalty • Producers can only transfer the increased price to suppliers, either from paying the tax or from performing investments on site	• WhC are complementary to a carbon tax, while targeting CO_2 reduction and energy efficiency improvement • Effectiveness, cost-effectiveness, socioeconomic effects and innovation are determined by decision-makers and weighted • Policy-makers evaluate *ex post* or *ex ante* experiences of carbon tax and WhC • Policy-makers can negotiate on the introduction of both instruments by sharing the same information and overcoming initial biases • Policy-makers can evaluate *ex post* this interaction by determining the effect of both instruments on the previously selected criteria	• Long-term effects of carbon tax and WhC • Static and dynamic effects of combination on electricity, gas, fuels prices for suppliers, end-users and industry • Static and dynamic effects of combination on abatement costs, energy efficiency actions and overall market capacity • Both environmental and economic targets can be achieved simultaneously • Costs for the energy system are increased • Penetration of new low-carbon technologies in the electricity production and energy-efficient ones in end-users

Missing info	Missing info	Missing info	Missing info
• What is the behaviour of producers or suppliers facing carbon taxes and WhC in competitive and oligopolistic markets? • How much energy is saved? At what cost? • What is the effect of both instruments on electricity price? Are these effects static? • Is cost-effectiveness or effectiveness the only criterion to evaluate the added value of the integrated scheme? • Can such a combination be politically accepted? • What are the total energy system costs?	• Is the electricity price effect solely dependent on the behaviour of producer or supplier? • How much energy finally is saved in the economy? At what cost? • Is cost-effectiveness or effectiveness the only criterion to evaluate the added value of the integrated scheme? • Can such a combination be politically accepted? • What are the total energy system costs?	• What is the behaviour of producers or suppliers facing carbon taxes and WhC in competitive and oligopolistic markets? • How much energy is saved? At what cost? • What is the effect of both instruments on electricity price? • Are these effects static? • What are the total energy system costs?	• What is the behaviour of producers or suppliers facing carbon taxes and WhC in competitive and oligopolistic markets? • Is cost-effectiveness or effectiveness the only criterion to evaluate the added value of the integrated scheme? • Can such a combination be politically accepted?
Policy information	**Policy information**	**Policy information**	**Policy information**
• Necessity of instruments (+) • Design modalities (+) • Market performance (–) • Costs (–) • Technological innovation (–) • Market acceptability (–) • Social criteria (+)	• Necessity of instruments (–) • Design modalities (–) • Market performance (+) • Costs (–) • Technological innovation (–) • Market acceptability (–) • Social criteria (+)	• Necessity of instruments (–) • Design modalities (+) • Market performance (–) • Costs (–) • Technological innovation (–) • Market acceptability (+) • Social criteria (+)	• Necessity of instruments (–) • Design modalities (–) • Market performance (+) • Costs (+) • Technological innovation (+) • Market acceptability (–) • Social criteria (–)

Note: (+) = provision of adequate information for proceeding in the decision-making process; (–) = inadequate or no information at all.

10.6 Conclusions

In general, based on the various evaluation outcomes from integrating WhC with other policy schemes, a basic lesson to be drawn is that a consistent evaluation framework for energy efficiency policy needs to be used in similar assessments. Such a framework, although it is complex and time and resource intensive, can enhance the policy learning processes of policy-makers and facilitate policy designing processes in the future.

The key message to policy-makers is that introducing WhC or other policy instruments demands detailed information from various aspects of economy and market entities. The degree of complexity of energy and climate policy problems, with their strongly intertwined objectives cannot be dealt with merely from an economic or technological approach, but rather from a synthesis of methods. To this end, we consider that the necessary steps to be taken, in terms of choice of methods, should be in accordance with a decision tree we propose, which incorporates all four methods stated in this chapter (Fig. 10.5). Each step in this decision-making process depends mainly on the information availability of the PI problem. More specifically, for a given PI examination, we suggest that policy-makers should follow these steps: (1) initiate the policy research from institutional method, (2) microeconomic analysis in order to identify the potential behaviour of market actors in given market conditions, (3) technoeconomic analysis for extracting actual costs of the policies at stake, and (4) multi-criteria analysis for the decision-making process, where the acceptability of PI will be tested with stakeholders.

Figure 10.5 **Decision tree for PI analysis**

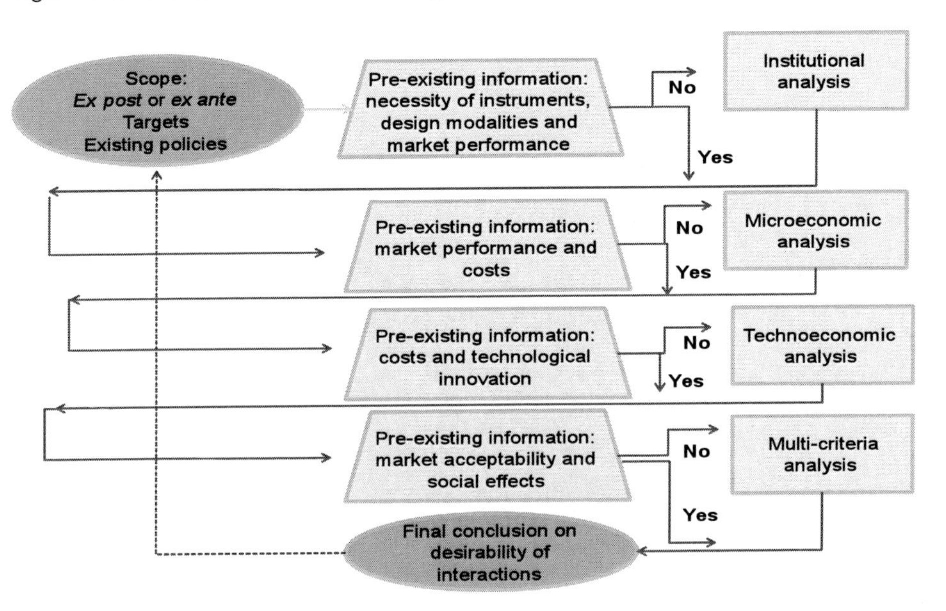

On the whole, this chapter demonstrates that WhC schemes can adapt and function in several market conditions and generate energy savings, but they should not be considered as a panacea, as other complementary instruments are deemed necessary in order to guarantee their effectiveness. All things considered, parameters that affect the performance of WhC schemes when fused with other policy instruments are their design characteristics, their scope, the market conditions in which they operate, and their ability to hold their own in an effective portfolio of policy instruments addressing energy efficiency or similar energy and climate targets. To conclude, we argue that WhC can be a sound instrument from its basic principles, as it makes use of market forces and can assist in overcoming market barriers to energy efficiency. Furthermore, we expect that WhC can be integrated with other policy instruments and achieve cost-effectively multiple environmental objectives. We argue that in order to render these interactions economically viable and effective, some preconditions on the designing phase of WhC schemes must be fulfilled:

- A binding, long-term target must be clearly set within a WhC policy time frame—reducing regulatory uncertainties for market actors
- Standardised common procedures must be employed for energy savings calculations (e.g. baseline setting, 'deadweight' and others). Furthermore, standardised full cost accounting systems, *ex ante* monitoring and verification of energy savings, streamlined procedures and standardised trading contracts must be employed in order to reduce transaction costs
- A proper market must be established, thereby ensuring the participation of numerous actors (e.g. avoiding oligopolistic market conditions, increasing liquidity, etc.)
- In order to keep compliance costs low, market conditions should be such that the tradability of WhC is guaranteed
- A concrete penalty should be set and publicly known before the implementation of a WhC scheme, in order to provide in advance correct market signals and let obliged parties develop their investments plans and further market strategies
- Transparent and fair cost-recovery mechanisms and effective enforcement by the authorities are crucial
- WhC schemes should be as technology neutral as possible, so that they can create competition among different energy efficiency technologies and avoid lock-in/out market situations. The energy efficiency target should not address only the 'low-hanging fruits' that could also be diffused in the market without an extra policy. Innovative technologies can also be stimulated parallel to WhC through additional stimuli from existing instruments; however, clear definition about additionality is required

Despite the tendency of policy-makers in the global energy and environmental community to propose a series of new policy instruments for combating climate change and securing energy supply, several aspects are often overlooked or neglected. The most important of them is the evaluation of potential interactions of these instruments, which can inhibit the achievement of both targets. Similar remarks apply to WhC schemes, which have attracted growing interest from policy-makers, but with little evaluation of their effects when implemented with incumbent instruments, namely financial policies. This research has highlighted some main characteristics of WhC schemes and their interactions with other policy instruments and provided an insight into methods for identifying these interactions.

Departing from the methods we presented for analysing PI as such, a methodological issue arises that demands extra attention: the selection of appropriate evaluation methods. Parallel to these, there can be several methods employed for similar analysis, suitable for providing answers to core questions for policy-makers. As a result, policy-makers must consult findings from various methods, covering several aspects of market actors and policy implementation, in order to promote policy instruments to achieve societal energy and environmental targets. A possible added value in this sense from further research could be the identification of a link between all these methods and a formulation of a tool that can integrate the assumptions and results of each separate method. Such a tool could be suitable to answer questions posed for policy interactions at each step of the policy process.

References

Belton, V., and T.J. Stewart (2002) *Multiple Criteria Decision Analysis* (Dordrecht, Netherlands: Kluwer Academic Publishers).

Bertoldi, P., and S. Rezessy (2007) 'Voluntary Agreements for Energy Efficiency: Review and Results of European Experiences', *Energy & Environment* 18.1: 37-73.

——, S. Rezessy, O. Langniss and M. Voogt (2005) 'White, Green and Brown Certificates: How to Make the Most of Them?' (Paper 7203; Brussels: European Commission Joint Research Centre).

Binswanger, M. (2001) 'Technological Progress and Sustainable Development: What About the Rebound Effect?', *Ecological Economics* 36.1: 119-32.

Blok, K. (2006) *Introduction to Energy Analysis* (Amsterdam: Techne Press).

——, L.F. de Groot, E. Luiten and M. Rietbergen (2002) *The Effectiveness of Policy Instruments for Energy Efficiency Improvement in Firms: The Dutch Experience* (Boston/Dordrecht/London: Kluwer Academic Publishers).

Boonekamp, P.G.M., B.W. Daniels, A.W.N. Dril, P. van Kroon, J.R. Ybema and R.A. van den Wijngaart (2004) 'Sectoral CO_2 Emissions in the Netherlands up to 2010: Update of the Reference Projection for Policy-making on Indicative Targets' (ECN Policy Studies, ECN-C--04-029; Petten, Netherlands: Energieonderzoek Centrum Nederland).

European Union (2004) 'Directive 2004/101/EC of the European Parliament and of the Council amending Directive 2003/87/EC establishing a scheme for greenhouse gas emission allowance trading within the Community, in respect of the Kyoto Protocol's project mechanisms', *Official Journal of the European Union* 13.11.2004.
—— (2006) 'Directive 2006/32/EC of the European Parliament and of the Council on energy end-use efficiency and energy services and repealing Council Directive 93/76/EEC,' *Official Journal of the European Union* 27.4.2006.
—— (2009a) 'Decision No 406/2009/EC of the European Parliament and of the Council on the effort of Member States to reduce their greenhouse gas emissions to meet the Community's greenhouse gas emission reduction commitments up to 2020', *Official Journal of the European Union* 5.6.2009.
—— (2009b) 'Directive of the European Parliament and of the Council of 23 April 2009 amending Directive 2003/87/EC so as to improve and extend the greenhouse gas emission allowance trading scheme of the Community', *Official Journal of the European Union* 5.6.2009.
Farinelli, U., T. Johansson, K. McCormick, L. Mundaca, V. Oikonomou, M. Örtenvik, M. Patel and F. Santi (2005) ' "White and Green": Comparison of Market-Based Instruments to Promote Energy Efficiency', *Journal of Cleaner Production* 13.10–11: 1,015-26.
Fischer, F. (1995) *Evaluating Public Policy* (Belmont, CA: Wadsworth Group).
Gamper, C.D., and C. Turcanu (2007) 'On the Governmental Use of Multi-criteria Analysis', *Ecological Economics* 62.2: 298-307.
Greening, L.A., and S. Bernow (2004) 'Design of Coordinated Energy and Environmental Policies: Use of Multi-criteria Decision-making', *Energy Policy* 32.6: 721-35.
——, D.L. Greene and C. Difiglio (2000) 'Energy Efficiency and Consumption—the Rebound Effect: A Survey', *Energy Policy* 28.6–7: 389-401.
Giraudet, L.G., and P. Quirion (2008) 'Efficiency and Distributional Impacts of Tradable White Certificates Compared to Taxes, Subsidies and Regulations', *Revue d'economie politique* 119.6: 885-914.
Gunningham, N., and D. Sinclair (1998) *Designing Smart Regulation* (Paris: OECD/International Energy Agency).
Hajkowicz, S., M. Young, S. Wheeler, D.H. MacDonald and D. Young (2000) *Supporting Decisions, Understanding Natural Resource Management Assessment Techniques: A Report to the Land and Water Resources Research and Development Corporation* (Clayton, Australia: Commonwealth Scientific and Industrial Research Organisation, Land and Water).
Hall, P.A. (1986) *Governing the Economy: The Politics of State Intervention in Britain and France* (Cambridge, UK: Polity Press).
IEA (International Energy Agency) and A. Capozza (2006) *Market Mechanisms for White Certificates Trading Task XIV Final Report: Implementing Agreement on Demand-Side Management Technologies and Programmes* (Paris: OECD/IEA).
IPCC (Intergovernmental Panel on Climate Change) (2007) *Climate Change 2007. Mitigation of Climate Change: Contribution of Working Group III to the Fourth Assessment Report of the Intergovernmental Panel on Climate Change* (Geneva: IPCC).
Joosen, S., M. Harmelink and K. Blok (2004) *Evaluatie van het klimaatbeleid in de gebouwde omgeving 1995–2002* (EEP03007; Utrecht, Netherlands: Ecofys).
Konidari, P., and D. Mavrakis (2007) 'A Multi-criteria Evaluation Method for Climate Change Mitigation Policy Instruments', *Energy Policy* 35.12: 6,235-57.
Leitmann, G. (1976) *Multicriteria Decision Making and Differential Games* (New York/London: Plenum Press).
Loulou, R., and M. Labriet (2007) 'ETSAP-TIAM: The TIMES Integrated Assessment Model Part I: Model Structure', *Computational Management Science* 5.1–2: 7-40.

Mickwitz, P. (2003) *Environmental Policy Evaluation: Concepts and Practice* (Vaajakoski, Finland: Finnish Society of Sciences and Letters).

Mundaca, L. (2008) 'Markets for Energy Efficiency: Exploring the Implications of an EU-wide Tradable White Certificate Scheme', *Energy Economics* 30.6: 3,016-43.

Oikonomou, V., and C. Jepma (2007) 'A Framework on Interactions of Climate and Energy Policy Instruments', *Journal of Mitigation and Adaptation Strategies for Global Change* 13.2: 131-56.

—— and L. Mundaca (2008) 'Tradable White Certificates: What Can We Learn from Tradable Green Certificates?', *Energy Efficiency* 1.3: 211-32.

—— and W. van der Gaast (2008) 'Integrating Joint Implementation Projects for Energy Efficiency on the Built Environment with White Certificates in the Netherlands', *Journal of Mitigation and Adaptation Strategies for Global Change* 13.1: 61-85.

——, M. Rietbergen and M. Patel (2007) 'An Ex-ante Evaluation of a White Certificates Scheme in the Netherlands: A Case Study for the Household Sector', *Energy Policy* 35.2: 1,147-63.

——, C. Jepma, F. Becchis and D. Russolillo (2008) 'White Certificates for Energy Efficiency Improvement with Energy Taxes: A Theoretical Economic Model', *Energy Economics* 30.6: 3,044-62.

——, F. Becchis, L. Steg and D. Russolillo (2009a) 'Microeconomics of Energy Saving and Energy Efficiency as a Tool to Design Effective Policy Options', *Energy Policy* 37.11: 4787-96.

——, W. van der Gaast, M. Rietbergen and M. Patel (2009b) 'Voluntary Agreements with White Certificates for Energy Efficiency Improvement as a Hybrid Policy Instrument', *Energy Policy* 37: 1,970-82.

——, M. di Giacomo, D. Russolillo and F. Becchis (forthcoming) 'White Certificates in the Italian Energy Oligopoly Market', *Energy Sources Part B: Economics, Planning and Policy.*

Parsons, W. (2001) *Public Policy: An Introduction to the Theory and Practice of Policy Analysis* (Cheltenham, UK: Edward Elgar).

Rietbergen, M.G., J.C.M. Farla and K. Blok (2002) 'Do Agreements Enhance Energy Efficiency Improvement? Analysing the Actual Outcome of Long-Term Agreements on Industrial Energy Efficiency Improvement in The Netherlands', *Journal of Cleaner Production* 10.2: 153-63.

Rogers, M., and M. Bruen (1998) 'A New System for Weighting Environmental Criteria for Use Within ELECTRE III', *European Journal of Operational Research* 107: 552-63.

Ruderman, H., M. Levine and J. McMahon (1987) 'The Behavior of the Market for Energy Efficiency in Residential Appliances Including Heating and Cooling Equipment', *The Energy Journal* 8.1: 101-24.

Selznick, P. (1949) *TVA and the Grass Roots* (Berkeley, CA: University of California Press).

Sorrell, S. (2007) *The Rebound Effect: An Assessment of the Evidence for Economy-wide Energy Saving from Improved Energy Efficiency* (London: UK Energy Research Centre).

——, A. Smith, R. Betz, R. Waltz, C. Boemare, P. Quirion, J. Sijm, D. Mavrakis, P. Konidari, S. Vassos, D. Haralampopoulos and C. Pilinis (2003) *Interaction in EU Climate Policy. Final Report to the European Commission* (Brussels: European Commission).

Taylor, P., and N. Jollands (2007) *Evaluation of Energy Efficiency Policy: Gaps and Conclusions* (Paris: International Energy Agency).

Van der Gaast, W. (2002) 'The Scope for Joint Implementation in the EU Candidate Countries', *International Environmental Agreements: Politics, Law and Economics* 2.3: 275-90.

Velthuijsen, J-W., and E. Worrell (1999) 'The Economics of Energy', in J.C.J.M. van den Bergh (ed.), *Handbook of Environmental and Resource Economics* (Cheltenham, UK: Edward Elgar).

Vreuls, H. (2005) *Evaluating Energy Efficiency Policy Measures and DSM Programmes* (The Hague: International Energy Agency/SenterNovem).

Zhang, Z., and A. Baranzini (2004) 'What Do We Know About Carbon Taxes? An Inquiry into their Impacts on Competitiveness and Distribution of Income', *Energy Policy* 32.4: 507-18.

11

Distributional choices in EU climate change policy seen through the lens of legal principles

Javier de Cendra de Larragán and Marjan Peeters
Law Faculty, Metro Institute, Maastricht University, The Netherlands

In the course of designing and implementing climate change policies, one of the core dilemmas that faces states is how to distribute the costs (and benefits) of those policies among the responsible greenhouse gas (GHG)-emitting sources. The effective implementation of intensive emission reduction targets will require substantial efforts from most sources across all sectors of society. The need to implement such targets has been recognised in the Kyoto Protocol for developed countries and is increasingly seen as a responsibility of developing countries, especially of the most developed ones.

Choices for the distribution of costs and benefits (here referred to as distributional choices) are expressed in laws, regulations and administrative decisions. Law is necessary to impose emission reduction obligations on states, business and citizens. The question therefore arises as to which legal criteria should guide the design and implementation of those distributional choices.

This chapter summarises the main conclusions of a PhD project, financed by the Netherlands Organisation for Scientific Research (NWO) Vulnerability, Adaptation and Mitigation (VAM) project, which has examined distributional choices included in the climate law and policy of the European Union from the specific perspective of legal principles. This very concise summary is of course meant as an invitation to the interested reader to turn to the full PhD study, which will be published in book form in 2010.[1]

The main research question concerned the role of legal principles in making, assessing and testing distributional choices. In order to answer that question, the research developed a framework for assessing distributional choices from the perspective of legal principles, thereby contributing to a further understanding of the legal dimensions of the serious distributional dilemma implied in climate policy-making.

The research focused on EU climate change law and policy. This choice is justified on the basis of two observations: first, the EU has for a long time sought to play a leading role in the development of international climate change policy. During all these years, the EU has developed an elaborate internal climate change legal framework to support its leadership position at the international level. This legal framework lends itself very well to an exploration of the role that legal principles might have had in its development. Second, the EU legal system is crowned by the Treaties, which include a large number of legal principles. Both secondary legislation and administrative decisions need to respect those principles. In addition, the EU legal system is based on a number of general principles of law that have been identified and elaborated on over time by the European Court of Justice (ECJ), and which can be relevant to distributional choices.

The precise questions formulated in this research were:

1. Are distributional choices made in EU law in line with the requirements imposed by the relevant legal principles, and, if so, to what extent?
2. Are there remaining tensions between the distributional choices taken (or proposed to be taken) and legal principles of Community law, and, if so, could those choices be challenged before the European courts?
3. Is it possible to detect in law, policy documents and relevant literature the emergence of new legal principles that could provide additional guidance when making, evaluating and testing distributional choices in climate change law?
4. Can an analysis based on legal principles be used to provide recommendations to policy-makers on the content and mode of adoption of distributional choices? Could the same analysis be of use for courts when reflecting on the legality of distributional choices, and for scholars to structure their thinking on them?

In order to present some of the main findings of the research, the chapter is organised as follows: Section 11.1 identifies several types of distributional choice in the context of climate change policy and presents the legal methodology that has

1 The project has been conducted in the period 2005–2009 by Javier de Cendra de Larragán under the supervision of Michael Faure and Marjan Peeters (Law Faculty, Maastricht University).

been elaborated to study them. Section 11.2 presents a number of (selected) conclusions arising from the analysis of distributional choices from the perspective of legal principles. The conclusions highlighted here are only those that indicate tensions between legal principles and distributional choices. Section 11.3 considers whether new legal principles are emerging that might be relevant in this context. In Section 11.4, a different policy approach based on the emerging per capita concept will be elaborated. Section 11.5 concludes the chapter.

11.1 The possible role of legal principles for climate policies

Distributional choices are seen as those decisions made by the legislator and the executive (and sometimes by courts as a result of their decisions in particular cases brought before them) that aim primarily at shaping the distribution of burdens and benefits, and of environmental damages and benefits, stemming from climate change policies between generations, between countries and within countries. Five types of distributional choice in the context of climate change policy can be identified:

1. Distribution between generations
2. Distribution between countries at international level
3. Impacts of internal climate policies on foreign countries
4. Distribution of the burden between EU Member States
5. Distribution of the burden between sources within the EU

The difference between choices (2) and (3) is as follows: whereas (2) refers to choices regarding rules for burden sharing among states, (3) refers to collateral distributional impacts of internal choices on foreign countries.[2] The difference between choices (4) and (5) is to be found in the subjects of the distribution. Whereas (4) refers to EU Member States, (5) refers to private sources located within those Member States. This distinction is necessary because there is EU law directly regulating burden sharing among private sources located within the EU.

The legal perspective adopted to examine these distributional choices is built on the view that law can be understood as a search for a rational proportion along three dimensions:

2 For instance, principles for burden sharing such as responsibility, capacity and need would fall under choice (2), while policies such as targets on biofuels, which may have negative impacts abroad, would fall under choice (3).

- In the relation between ends and means
- In the distribution of benefits and burdens
- In the participation of interested members of society in the law-making process

If law is understood as a search for a rational proportion along these three dimensions, then those dimensions can be understood to form part of what can be termed the 'meta-principle of proportionality'. The meta-principle of proportionality would include all the principles of practical rationality that form part of a given legal system. Given that these principles are not only norms of practical reasoning (Thomas 2005), but also the fundamental pillars of law and legal systems (Tridimas 2006), they can play a role not only in guiding the activity of courts, but also the activity of law-makers and scholars. The scheme that is presented in Table 11.1 shows an overview of the meta-principle of proportionality and the way in which existing legal principles fit into it.

The research placed a large number of principles of EU law within the meta-principle of proportionality and explored their content and application by courts in detail. Two main conclusions emerged from the analysis:

First, it was found that four legal principles contain a relatively well-defined core: these are the principles of proportionality, equality, legitimate expectations and legal certainty.[3] All the other principles examined are either too vague or lack a clear core except for, perhaps, the precautionary principle (de Sadeleer 2006; Vilaça 2004),[4] and, following a recent strand of case-law from the ECJ and several opinions of Advocate General (AG) Kokott,[5] the polluter pays principle. This ambiguity applies in particular to environmental principles such as integration and rectifica-

3 However, even the structure and mode of application of these principles is not completely settled. European courts do not always apply the principle of proportionality in the same manner; sometimes they distinguish between the three tests of suitability, necessity and proportionality in the narrow sense, while on other occasions they combine the first two tests, and this may have impacts on the outcome of the decision. Likewise, the principle of equality can be interpreted in different ways depending on the type of case, particularly whether it deals with the internal market or with fundamental rights. Other principles that have also been developed with a remarkable degree of sophistication by the European courts are legal certainty and legitimate expectations.

4 Nevertheless, even this principle is subject to strong controversies regarding its meaning and legal consequences in different settings. For an analysis of the role of the precautionary principle in determining liability of private parties for climate change damages, see Chapter 12 in this book by Miriam Haritz.

5 The most important opinion in this regard is the opinion of AG Kokott in Case C-254/08, *Futura Immobiliare srl Hotel Futura, Meeting Hotel, Hotel Blanc, Hotel Clyton, Business srl v. Comune di Casoria*, delivered on 23 April 2009. The ECJ issued its ruling on 16 July 2009. Case C-254/08 *Futura Immobiliare srl Hotel Futura, Meeting Hotel, Hotel Blanc, Hotel Clyton, Business srl v. Comune di Casoria*, 16 July 2009, not yet reported. See also Case C-188/07 *Commune de Mesquer v. Total France SA, Total International Ltd*, 24 June 2008.

Table 11.1 **The meta-principle of proportionality**

Position of legal principles within the meta-principle of proportionality			
First dimension of the meta-principle of proportionality	Suitability	Intensity	Focus on one single area, constant monitoring, adequacy to reality
		Quality	Sustainable development, integration, precaution, prevention, polluter pays principle, rectification at source
		Probability	Legal certainty, legitimate expectations, penalties, effectiveness (in the sense of adequate enforcement)
	Necessity	Principle of attribution	
		Principle of subsidiarity	
		Principles of environmental and cost-effectiveness	
Second dimension of the meta-principle of proportionality	Proportionality between countries	Principles of common but differentiated responsibilities, solidarity, loyal cooperation	
	Proportionality between private parties	Principle of intergenerational equity Principle of proportionality (suitability, necessity, proportionality in the narrow sense Fundamental rights and fundamental freedoms Principle of equality	
Third dimension of the meta-principle of proportionality	Access to information		
	Public participation		
	Access to justice		
	Environmental justice		

tion at source, and suggests the limitations that many legal principles have when used as tools to test distributional choices. Nevertheless, it was also found that most principles have a sufficiently defined core that enables them to usefully guide choices made by policy-makers and to structure academic debate.

Second, because the ECJ affords a very large margin of appreciation to the European Community (EC) policy-maker when making complex policy choices, it is very unlikely that the Court will declare that EC climate change policy is in breach of principles of EC law. Nevertheless, it is more likely that the ECJ may decide that certain decisions taken by the European Commission are in breach of one or more legal principles. In fact, a number of ECJ judgements have already annulled Commission decisions taken in this context for being in breach of legal principles.

11.2 Are distributional choices made in EU climate policy in line with the relevant legal principles?

The research has carried out an extensive assessment of the fit between the principles within the different dimensions of the meta-principle of proportionality and each of the distributional choices identified above. From that assessment, a large number of observations have arisen, which have led to some policy recommendations. However, for reasons of space, this section will present only the most relevant observations before enumerating policy recommendations.

11.2.1 Burden sharing between generations

The research found that the Kyoto target adopted by the EU does not bear any logical relation with the (non-legally binding) long-term target endorsed by the EU. This long-term target seeks to avoid an increase in global average temperature above 2°C in comparison with pre-industrial levels. Nevertheless, the EU has recently adopted a legally binding mid-term target aiming at reducing its 1990 emissions by 20% by 2020, and has offered to reduce them by 30% in comparison with the same base year if other developed countries make comparable commitments and if developing countries—excluding Least Developed Countries (LDCs)—commit to meaningful climate change policies. The European Parliament has sought to adopt a legally binding obligation of reducing emissions annually in accordance with a linear path towards the long-term target, but this has not been accepted by the Council. Instead, a non-legally binding reference has been made to the need to move the EU into a highly energy-efficient and low GHG-emitting economy.[6]

6 See Decision No 406/2009/EC of the European Parliament and of the Council of 23 April 2009 on the effort of Member States to reduce their GHG emissions to meet the EC's GHG emission reduction commitments up to 2020, OJ L 140, 5.6.2009, Recital 4.

From the perspective of the polluter pays principle, one can ask whether the EU contribution towards the long-term target is proportional to its responsibility for creating the problem. The research shows that the EU has justified its contribution on the basis of a number of principles in addition to the polluter pays principle, such as capability to pay, mitigation potential, early action and population trends, which leads to a lower burden for the EU than would arise if the only principle followed was the polluter pays principle (European Commission 2009c: 4).

The EU has sought to justify the adoption of the long-term target on the basis of a cost–benefit analysis concluding that for higher increases in global mean temperature costs will outweigh benefits, both at global level and within the EU. However, the EU has neither made explicit the assumptions made therein nor the rationale, nor discussed alternative approaches that could have led to very different results. In particular, the Commission has not systematically discussed the substantial body of literature on economic modelling of climate change; therefore, the analysis underpinning the adoption of the long-term target appears to be incomplete, if not defective. Hence, the final choice—regardless of its material correctness—does not appear to have been based on a rational balancing of all the available scientific evidence (Tol 2007). Moreover, considerations of an ethical nature have not been discussed in any detail. For instance, while considerations of the rights of future generations are clearly very relevant in justifying an ethical duty to adopt such a target, they have not played any explicit role. The Commission has sought to offer an additional justification based on public support, by noting that the majority of participants involved in some workshops on international climate policy support the target of 2°C. While that is true, those workshops took place a number of years after the initial target was adopted.

Currently, the rights of future generations are not expressly recognised in EU law, although it can be argued that they are implicit in the concept of sustainable development. The Lisbon Treaty includes as one of the tasks of the EU the promotion of solidarity between generations, but there are no specific mechanisms to ensure that the interests of future generations are represented in the law-making process. The Preamble of the Charter of Fundamental Rights says that 'the enjoyment of these rights entails responsibilities and duties with regard to other persons, to the human community and to future generations'. The question was raised of whether establishing legal mechanisms to protect the rights of future generations could add anything of value to the principle of sustainable development. In particular, introducing a provision dealing explicitly with obligations to future generations, establishing organs that are entrusted with producing assessments of the long-term sustainability of certain policies, and granting some people or organisations the right—on an ad hoc or a permanent basis—to challenge regulatory choices when they feel that the interests of future generations in sustainability are being ignored, could add value to the actual content of the principle of sustainable development. Introducing these or similar clauses in constitutions would allow (constitutional) courts to assess whether particular laws sufficiently respect the rights of future generations.

11.2.2 The EU position on burden sharing between countries at international level

The position of the EU on burden sharing at the international level can best be assessed by considering its approach to the principle of common but differentiated responsibilities (CBDR). In fact, the EU understanding of the CBDR principle has over time become more complex, nuanced and constructive. The EU is moving away from a focus on burden sharing to one of effort sharing.[7] The difference is not merely a semantic one. Effort sharing is based on solidarity among countries and generations and on promoting mutual advantage rather than focusing on shared losses.[8] While the EU still advocates the need for developed countries to adopt absolute mitigation commitments, it now places a stronger focus on the need to establish objective criteria to ensure comparability of efforts among developed countries and developing countries, in particular the most advanced ones.[9] Moreover, all developing countries, except for LDCs, need to commit to low-carbon development strategies. These strategies should set out a credible pathway to limit those countries' emissions, and should identify the external support needed to implement actions that are too expensive for the countries themselves. The credibility of those strategies should be determined by means of independent technical analyses. Furthermore, the EU has also proposed, in line with the CBDR principle and with the value of solidarity, that all developed countries should contribute, in accordance with the polluter pays principle, to the funding of adaptation policies in developing countries.

The EU does not accept the view put forward by some developing countries[10] that historical responsibility should play a central role in burden sharing. This view is based on the observation that the process of industrialisation that has taken place in the last centuries in developed countries has been the main contributor to climate change. The EU holds the view that the United Nations Framework Convention on Climate Change (UNFCCC) does not endorse the principle of historical responsibility. Other developing countries consider that the main principle for burden sharing should be equal allocation of emission rights per capita. The per

7 'European Parliament legislative resolution of 17 December 2008 on the proposal for a decision of the European Parliament and of the Council on the effort of Member States to reduce their greenhouse gas emissions to meet the Community's greenhouse gas emission reduction commitments up to 2020' (COM(2008)0017 – C6-0041/2008 – 2008/0014(COD); www.europarl.europa.eu/sides/getDoc.do?pubRef=-//EP//TEXT+TA+P6-TA-2008-0611+0+DOC+XML+V0//EN&language=EN#BKMD-16, accessed 19 March 2010.

8 Nevertheless this chapter will use the term 'burden sharing' instead of 'effort sharing' for convenience, except in the context of burden sharing between EU Member States, where the distinction shall be kept.

9 The Council has considered that developing countries as a group need to achieve a substantial and quantifiable deviation below the currently predicted emissions growth rate. See Conclusions of the Council of the European Union, Brussels 3 March 2009; www.register.consilium.europa.eu. See also European Commission 2009c: 5.

10 Including Brazil, China and Saudi Arabia.

capita approach is based on the idea of sharing fairly the absorptive capacity of the atmosphere. Its application would lead to a huge redistribution of the emission rights between developed and developing countries, and would force the former to purchase from developing countries most of the entitlements they need to maintain current production and consumption patterns. Not surprisingly, the EU has not endorsed the principle of equal allocation of emission rights per capita either. Rather, the EU position resembles to some extent the contraction and convergence (C&C) approach, which requires long-term convergence of per capita emissions, while affording countries with per capita emissions below the global average the right to increase further their emissions before reducing them in line with the required global average. Nevertheless, the EU approach does not completely reflect the spirit of C&C for two reasons. First, the EU rejects the equal per capita dimension on which C&C is founded. While the EU position would lead to some global redistribution of public funds, it would still fall short of the large redistributions that would be required under the equal per capita approach. Second, while equal per capita allocation and C&C would give rise to an unconditioned redistribution of emission rights, the EU position favours a conditional redistribution subject to developing countries adopting monitorable, reportable and verifiable (MRV) low-carbon development strategies. But a more fundamental difference exists between the approach of the EU and of developing countries to burden sharing (Müller 2009). The EU has adopted a burden-sharing paradigm according to which all countries need to engage in mitigation activities, according to their responsibility and capacity. According to this paradigm, it would be possible to exempt some parties from contributing to the policy objectives if they have no or very little capacity to contribute. But what would not seem acceptable within this paradigm is that some countries may profit from doing nothing to mitigate their emissions. Hence, having 'hot air' would not acceptable. Moreover, it would be even less acceptable to benefit from selling 'hot air' in the global carbon market.[11] But, as some have pointed out, the burden-sharing paradigm is not the only possible one and, in fact, many developing countries support a very different paradigm, based on the allocation of (property rights to) a natural resource among countries—the 'right to natural resources' paradigm.

Under the 'right to natural resources' paradigm, concepts such as 'ecological

11 It should nevertheless be noted here that several Member States of the EU have purchased 'hot air' in the international carbon market in order to comply with their mitigation targets under the Kyoto Protocol. For instance, according to the environmental data service ENDS, the Czech Republic has agreed to sell 8.5 million AAUs (Assigned Amount Units) to Spain (5 million) and Austria (3.5 million). Emissions in the Czech Republic are expected to be about 17 per cent below the country's target. The total amount of the Czech Republic's AAUs available for sale is 100 million, including those already sold. See ENDS Europe Daily (2009) and press releases from the Ministry of the Environment of the Czech Republic (2009a, 2009b) In both cases, the AAUs sold are linked to a green investment scheme, a mechanism through which proceeds from the sale of AAUs are invested by Czech Republic in activities that ensure GHG emissions reductions.

space' and equal per capita allocation become meaningful principles to guide the allocation of the absorbing capacity of the atmosphere, because within that paradigm it is possible to defend the position that all human beings have a right to a sufficient share of the absorbing capacity of the atmosphere. Given the current overuse of that absorbing capacity, some citizens, regardless of where they live in the world, should reduce their emissions to a sufficient extent so that others may increase theirs while remaining within 'safe' aggregate levels. (Hayward 2006). The paradigms of burden sharing and the right to natural resources are not only very different, but also very entrenched within the negotiating positions of different parties to the international negotiations.

When considering the findings discussed in the previous paragraph from the perspective of legal principles, and in particular from the perspective of the CBDR principle, the conclusion seems sobering. As international law currently stands, burden sharing at international level remains essentially a political exercise, informed by a number of elements including self-interest and normative views. The CBDR would be at most a conglomerate of an indeterminate number of principles of distributive justice, but neither their meaning nor the precise combination that should be implemented to determine individual contributions enjoys widespread agreement among states. Fundamentally, the position of the EU itself does not seem to be entirely consistent with any particular set of principles: it can best be described as a rather balanced and pragmatic position, and in that sense it may offer some grounds to build a future agreement on the meaning and consequences of the CBDR. This is because the EU position, at its core, reflects an acceptance of the role of many principles, some of which have the effect of increasing the EU burden. Nevertheless, it should be noted that the EU's proposed contribution to burden sharing remains short of what would be required if the polluter pays principle (based on historical responsibility for past emissions), or if the principle of equal per capita allocation were to be used as the main principles for burden sharing at international level.

Several particular elements of EU climate change law and policy that may have negative distributional impacts beyond the EU borders, and particularly among the most vulnerable individuals and communities, have been examined. These are (1) the use of flexible mechanisms and reduced emissions from deforestation and forest degradation plus conservation (REDD-plus), (2) the use of biofuels, and (3) the issue of carbon leakage. The analysis of each of these issues indicated some important tensions.

Regarding the use of credits from flexible mechanisms, it was found that while the EU has sought to reduce its reliance on credits that fail to ensure additionality, environmental integrity and sustainable development, it still relies on them to a significant extent to ensure compliance with its mitigation target. Regarding the use of biofuels, the EU legislator has justified the adoption of a legally binding target of 10% for the use of biofuels in transport, which can come either from the EU or from other countries, on the basis that it is necessary to reduce emissions in the short term. The use of biofuels is contentious, with many saying that it does not contribute to mitigation and has negative impacts on water, biodiversity and food production. The use of imported biofuels in the EU is not subject to quantitative limits, even if it is difficult to imagine how the EU will be able to control the social and environmental impacts of imported biofuels in foreign countries. The EU has required that biofuels, in order to be used for the purpose of complying with the legally binding target, must be produced in a sustainable manner. One relevant question is whether this requirement can be enforced for imported biofuels, particularly considering the difficulties in inspecting production conditions abroad. The EU proposes a dual solution to this problem: first, to accept only biofuels fulfilling sustainability criteria; second, to assess *ex post* potential negative impacts of biofuels. Two criticisms seem in order: first, the EU legislator has paid more attention to the potentially negative environmental impacts than to the social ones; second, the question of whether imported biofuels should be allowed at all and whether there are other alternatives that would achieve a better proportion between means and ends does not seem to have been seriously considered by the EU legislator.

Third, leakage remains one of the most important hurdles for the adoption of stringent climate policy. Whether it is a real issue or not is difficult to prove empirically, but the EU legislator is making a huge effort to ensure that *ex post* empirical analyses do not need to be conducted by substantially reducing the burden upon companies subject to international competition. Moreover, the EU's approach to burden sharing in the international negotiations is inconsistent with its approach during the negotiations of the first EU burden-sharing agreement. A consistent approach would have required considering, also at international level, equal per capita allocation as a guiding principle for the distribution of the efforts within sectors not exposed to international competition.

11.2.3 Burden sharing among EU Member States

The EU has adopted two consecutive burden-sharing agreements in order to share among its Member States the Kyoto target[12] and the unilateral 2020 mitigation commitment.[13] The second burden-sharing agreement was in fact termed 'effort sharing', in an effort to place more emphasis on the opportunities that arise from climate change policy, and less on the costs. The second 'effort- sharing' agreement is a more rational instrument than the first in respect of all dimensions of the meta-principle of proportionality. In particular, it achieves a better proportion between

12 Council Decision 2002/358/EC of 25 April 2002 concerning the approval, on behalf of the European Community, of the Kyoto Protocol to the United Nations Framework Convention on Climate Change and the joint fulfilment of commitments thereunder, OJ L130, 15.5.2002.

13 Decision No 406/2009/EC of the European Parliament and of the Council of 23 April 2009 on the effort of Member States to reduce their greenhouse gas emissions to meet the Community's greenhouse gas emission reduction commitments up to 2020, OJ L140, 5.6.2009.

ends and means, by scoring higher on the three tests of intensity, quality and probability. It is, moreover, well justified under the principle of necessity and provides a central role for the principles of solidarity and common but differentiated responsibilities. Nevertheless, some tensions do remain:

First, the EU targets are not yet in line with the ultimate objective of EU climate change policy, since a direct link between the mid-term and the long-term targets is still missing.

Second, effort sharing within the EU seems to be subject to power relations to the extent that certain countries have clearly benefited from past inaction. That outcome goes against the core of the polluter pays principle, and could weaken the credibility of the EU in the international negotiations, particularly because the EU is pressing 'advanced' developing countries to take on targets as soon as possible. If Member States that miss their targets are not to be punished under EC law, developing countries may get the impression that lack of compliance with their own targets—if they accept them at all—should not carry any negative consequences at international level.

Third, the principle of loyal cooperation could turn out—*a posteriori*—to be very relevant for effort sharing among Member States. In particular, it could mean that Member States that overcomply with their targets are forced to place excess AAUs in the hands of the EU, or otherwise purchase AAUs from other countries to ensure the compliance of the EU with its own target.

Last but not least, it must be noted that neither the burden-sharing agreement nor the effort-sharing agreement mention the size and distribution of adaptation costs within the EU, although it is clear that they will be considerable and very asymmetrically distributed. Adaptation has only recently been tackled within EU climate change policy[14] but the argument can be made that effort sharing should also apply to it, just as the EU has proposed to distribute among developed countries the costs of financing adaptation in developing countries. While the EU proposed to use the polluter pays principle and capacity to pay as the main guiding principles for sharing those costs among developed countries, it has not determined yet which principles will be used to share the financial commitment for developing countries among Member States. It would be logical to think that the same principles proposed for effort sharing in mitigation would be used in this context, in addition to giving attention to the vulnerability of each country to climate change impacts.

11.2.4 Burden sharing within the internal market

The test of quality asks to what extent distributional choices are in line with the integration principle, the polluter pays principle and the principle of rectification at

14 The European Commission elaborated a green concluding chapter on adaptation to climate change in 2007 (European Commission 2007). This led to the publication of a white paper (European Commission 2009a).

source. EU climate change policy is progressing towards a fuller application of these three principles. Integration can be seen particularly in the fact that the EU institutions have declared that, from 2013 onwards, all sectors and gases will be covered in one way or another by EU climate change policy. The polluter pays principle has been advanced particularly with the move from grandfathering to auctioning as the core allocation method in the EU Emissions Trading Scheme (EU ETS). Regarding the principle of rectification at source, it was found that the EU ETS does not constitute an adequate implementation of the principle, although this is justified by environmental and economic reasons. Nevertheless, there remain important tensions, particularly on the consumption side. The European Environment Agency (EEA) has shown that while EU climate change law has paid a great deal of attention to the production side, the consumption side remains largely unaddressed (EEA 2009: 44). While it is to be expected that price signals may progressively influence, at least to some extent, consumption decisions, the EEA shows that those signals are not enough to generate substantial changes and additional measures will be needed. Eco-labelling has been advocated by the EU for quite some time, but it is not capable of generating significant emission reductions on the consumption side (Mahmoudi 2006). Hence further measures are clearly needed here.

Regarding the specific tests of attribution, subsidiarity and proportionality, the analysis found a number of potential legal tensions stemming from the highly political process through which EU climate change policy—and particularly the 2008 EU climate change and energy package—has evolved. Indeed, in order to strike a deal on relatively ambitious EU mitigation targets, the distribution of the effort has been negotiated at the highest political level. As a consequence, the legal procedure to adopt it—co-decision under Article 175 EC—has had less relevance than in other instances of EU environmental law. While this more political approach has probably been necessary in order to approve the very complex climate change package, it may have limited the possibility to consider in detail potential future consequences for all stakeholders, including competent regulators within Member States. Indeed, the climate change and energy package is so complex, and there are so many interactions among the different instruments, that many of its consequences, including legal ones, will only be discovered over time (Deketelaere 2009). At a minimum, this calls for a constant and careful monitoring of its impacts in order to detect which elements conflict, which are mutually supportive, which can be added and which can be streamlined or eliminated.

The third dimension of the principle of meta-proportionality is concerned with the legal rules governing the access of interested stakeholders to the law-making process and to courts of law. Public participation is generally promoted by the Commission as the initiator of legislation, and this has indeed been the case in the development of climate policy. But a tension exists between promoting participation and expediency, as the Commission has acknowledged explicitly on a number of occasions (Lee 2005: 134 *et seq.*). Climate change is a matter where expediency may be of the essence, and wide public participation may delay the adoption of important measures. The pathway selected has been to push fast for a comprehensive and far-reaching regime, where even the European Parliament has accepted the highly political level of the negotiations and has sought to support the balance developed by Member States at the Council. The EU institutions have justified this expedient approach on the basis of the urgency of the problem and of the broad support given by EU citizens to EU climate change policy. Nevertheless, there are particular areas where public participation might have been restricted to an extent that may cause legal tensions; Directive 2009/29/EC amending the EU ETS is a point in case. Indeed, this Directive requests the Commission to develop a large number of measures to implement it. Those measures will be developed following the applicable comitology procedures, which are not characterised by their transparency. Therefore, private parties that might feel aggrieved by those measures may seek to challenge them before the European courts.

This takes us to the issue of access to justice. As EU law currently stands, private parties lack the standing to challenge general legal measures before European courts. However, given the very high level of harmonisation achieved in the EU ETS, and therefore the relatively reduced margin of discretion afforded to national authorities to take further decisions, private parties could be deprived of their right to seek protection in court. This research has found that European courts might consider broadening their very narrow stance to access to justice in order to allow private parties to challenge decisions made by the Commission in the context of the EU ETS This decision of the court would, *ex post*, compensate to some extent for the limited participation of private parties in the decision-making process, because those parties would be able to challenge them in court and the court would get the chance to examine their legality under EU law.

11.2.5 Policy recommendations

Given the analysis of EU climate policy, the following recommendations can be given with regard to the EU legislator:

- Review the policy regarding biofuels, particularly imported ones, to assess whether this is necessary and, if so, to ensure that no negative social and environmental impacts arise
- Review the approach to burden sharing at EU level, particularly to ensure that old Member States do not profit from past inaction
- Clarify the potential obligations of Member States under Article 10 EC to assist the EU in complying with its mitigation targets
- Consider whether adaptation costs should be subjected to a burden-sharing agreement in the spirit of what has been done in relation to mitigation
- Monitor closely the interactions that may take place between measures adopted within the climate change and energy package, in order to correct undesired effects as soon as possible

- Do not discard the normative power of the 'natural resources' paradigm, and keep searching at international level for a bridge between it and the burden sharing-paradigm

11.3 Is it possible to detect the emergence of new legal principles? A view on the per capita principle

Interestingly, there is a concept that is gaining increasing attention in philosophical and legal literature: per capita allocation of emission rights. Many developing countries have long argued that equal emissions per capita, together with historical per capita emissions, should become the central criterion for allocation at international level. The EU has, however, rejected using a single criterion in burden sharing, with the argument that it disregards the fact that there are significant differences between parties that need to be taken into account (see European Commission 2009b). The EU has even rejected combining this criterion with others in the context of a multi-criteria approach, for two reasons: first, using national per capita emission averages can hide large internal inequalities; second, it might hide significant mitigation potentials due to economic inefficiencies, certainly in the short to mid-term, because in some countries energy is consumed in a much less efficient way than in others (therefore massive reductions can be achieved at a low cost) and this should be taken into account when considering how to share the burden. Per capita allocation cannot lead, in the view of the Commission, to balanced outcomes that would be acceptable for all developing countries. At the same time, the EU has acknowledged that, in the long-term, per capita emissions will converge (as a matter of fact) between developed and developing countries.

The position of the EU makes sense in the context of a 'burden sharing' paradigm, where the overriding concern is achieving emission reductions at the lowest possible cost. However, under the 'right to natural resources' paradigm the second reason provided by the EU looses considerable traction, because equal allocation per capita could be coupled with a global carbon market that would promote allocative efficiency. Of course, this approach would mean that certain countries would receive a large amount of 'hot air'. But it could be argued that this is only fair in the view of the historical imbalances in the appropriation of the absorbing capacity of the atmosphere. Nevertheless, the first reason posed by the EU, that equal per capita allocation hides internal inequalities, remains a real and important one. It does not seem just that wealthy people within poor countries hide behind the low developmental level of their country in order to profit from selling 'hot air' in the global carbon market. On the one hand, one could argue that this is a problem for states to solve under the principle of sovereignty. On the other hand, it could also be argued that this problem justifies taking into account the responsibility of citizens when setting up an allocation mechanism at international level, in order to take into account internal inequalities. There is increasing attention paid to this possibility in philosophical (Page 2006), political (Baer *et al.* 2008; Harris 2008) and economic literature (Chakravarty *et al.* 2009). Hence further legal research is certainly needed to study how such an approach could be implemented and what changes in international law would be necessary. Some of the elements of that approach will be put forward in the next section.

11.4 Personal carbon trading and mitigation: a possible way forward?

If policy-makers focus principally on reducing emissions from domestic industrial facilities, rather than on emissions coming from consumption by all sectors in all major emitting countries, a large share of the global emissions will remain unregulated. The question then is how to make all consumers share the burden of emission reductions in proportion to their responsibilities, capabilities and needs.

There are a number of regulatory approaches to tackling this problem. Babcock has reviewed the types of norm that can be introduced to achieve behavioural changes in the consumption side and which may lead to reductions in emissions (Babcock 2009). These include enforcement and non-enforcement approaches. Among the former he includes sanctioning and naming and shaming. Among the latter he includes providing information, public education and market-based incentives. The latter can include tax-breaks and subsidies, pollution fees and taxes, trading programmes and take-back obligations. After reviewing the pros and cons of each of these instruments, Babcock concludes that there is no 'silver bullet' that will transform 'poor environmental behaviour into good behaviour' (Babcock 2009: 175). Nevertheless, he concludes that providing public education should remain at the centre of efforts. For those who are, given their economic status, insensitive to price increases, naming and shaming may constitute an effective tool. For the rest, price increases and take-back obligations may constitute a good starting point. Price mechanisms have, nevertheless, a number of problems, among which two are particularly salient:[15] first, they do not provide certainty that the target will be achieved; second, unless price changes are made permanent, polluting behaviour will resurge as soon as prices go down again. Babcock is very negative about trading

15 Other problems are the possibility that taxes weaken the personal or communal psychological advantages of carrying out a supportive action and thereby deter the intended behaviour, the risk of commodifying pollution and general public opposition to tax increases. Of course one way of reducing public opposition is by reducing other taxes at the same time so that the tax burden remains at the same level.

programmes, given the problems associated with designing a model to fit so many disparate and small sources of pollution, the costs of gathering data, implementing and enforcing the programme, the need to construct some form of bureaucracy at the local or state level to oversee the workings of the programme, and the loss of individual privacy. This negative view of trading programmes is, however, not shared by others. Starkey and Anderson from the Tyndall Centre for Climate Research have studied the technical feasibility of introducing domestic tradable (carbon) quotas (DTQs) within the UK (Starkey and Anderson 2005). DTQs is a cap and trade scheme for GHG emissions for energy use under which emissions rights are allocated directly to end users of energy, free and on an equal per capita basis to individuals, and through auctions to organisations. (Starkey and Anderson 2004).

DTQs is based on the idea that every citizen has a right to emit an equal amount of carbon (and, one should add here, perhaps also on the idea that every citizen has a responsibility to reduce his or her own emissions to the extent needed to avoid dangerous climate change). A DTQs scheme has four elements: setting the carbon budget, allocating carbon units, surrendering carbon units and trading carbon units. It would be conducted in the following manner: a national emissions cap would be set, and emissions rights (in the form of carbon credits) would be allocated across the population as a whole. Individuals would surrender their carbon credits on the purchase of, for example, electricity, gas or transport fuel. Those who need or want to emit above their quota would have to buy allowances from those who emit less. Over time, the overall emissions cap (and hence individual allocations) could be reduced in line with international or nationally adopted agreements.

Starkey and Anderson examine in particular the technological and administrative feasibility, the public acceptability, the effectiveness and the fairness of DTQs. They conclude, in stark contrast to Babcock, that while much more research needs to be done, DTQs seem to be feasible, affordable and fair, and therefore deserve serious attention as an instrument of public policy. However, they find that DTQs do not appear as equally supported by all approaches to justice. Hence, in order to justify adopting equal per capita emissions DTQs it is necessary to choose an approach to distributive justice that supports it.

Given the particular characteristics of the climate change problem, namely the fact that emissions are the outcome of millions of individual and collective decisions, and that drastic emission reductions need to be achieved within 40 years, personal carbon trading appears as an interesting instrument deserving further exploration because it could be, at least in theory, comprehensive, environmentally effective and economically efficient. Moreover, it would seem to preserve freedom of choice of individuals, because they can cover any excess emissions arising from their personal consumption decisions by purchasing additional allowances in the market.

Most of the research and political action on personal carbon trading (PCT, a synonym for DTQ) has taken place in the UK, but up till now the instrument has not been implemented. In particular the central government (the Department for Environment, Food and Rural Affairs [DEFRA]) has taken the position that there are too many difficulties and costs involved in setting up the system. First, the costs identified are large and outweigh, by many times, the estimated potential benefits of PCT; second, the analysis of distributional impacts indicated that PCT would be a financially progressive policy instrument—meaning that it would disproportionately benefit the poor—and any undesirable distributional impacts falling upon particularly vulnerable groups could be addressed through scheme design, allocation methodology or other measures (DEFRA 2008); third, while no insurmountable technical barriers to the introduction of a PCT scheme were identified, the costs were found to be very significant; fourth, while the public accepted that action should be taken to reduce emissions from individuals, the public was generally against the introduction of PCT on the basis that it would be unfair and too intrusive. However, when other alternatives were provided, such as a carbon tax and upstream trading, both with revenue recycling, the responses were identical. DEFRA stated that:

> while personal carbon trading remains a potentially important way to engage individuals, and there are no insurmountable technical obstacles to its introduction, it would nonetheless seem that it is an idea currently ahead of its time in terms of its public acceptability and the technology to bring down the costs (DEFRA 2008).

Nevertheless the idea has not been abandoned. In May 2008 the Environmental Audit Commission's personal carbon trading inquiry was published calling for further research in a number of areas (House of Commons 2008), after concluding that 'existing initiatives are unlikely to bring about behavioural change on the scale required, with many individuals choosing to disregard the connection between their own emissions and the larger challenge'. The report further stated that PCT is 'the kind of radical measure needed to bring about behavioural change'. Considering all the complexities associated with setting up a comprehensive PCT scheme from the outset, the Environmental Audit Commission proposed to start a 'learning by doing' process whereby a pilot scheme with restricted participation could be introduced. Over time this scheme could be expanded and linked to existing schemes in order to achieve comprehensive coverage. DEFRA responded to the report in detail later that year. It started by acknowledging that 40% of total UK emissions are the result of decisions taken directly by consumers, coming mainly from energy use in the home, which is the largest source of all, travelling and food. Then it proceeded to study PCT from a number of perspectives, including public acceptability, environmental effectiveness, cost-effectiveness, technical feasibility and distributional impacts, and formulated a number of challenges in connection with all those criteria. First, it raises the question of whether a downstream scheme would be more effective than an upstream scheme, particularly taking into account the much larger costs involved in a downstream approach. Second, adopting a very small scheme would preclude cost-effectiveness, which requires the introduction of a scheme that is as broad as possible. Third, there would be overlaps with existing trading schemes, particularly the EU ETS. In a sector covered by both schemes, the carbon price paid by the end user will reflect the sum of the allowance price in the EU ETS and the allowance price in PCT. This would increase the overall costs of achieving the overall cap. Moreover, partially overlapping schemes may lead to several and inconsistent carbon prices, which would also reduce cost-effectiveness by encouraging more reductions in some sectors than the optimal amount.

DEFRA provided two reasons to reject the inclusion of PCT in the UK's climate change policy for the time being: first, 'personal carbon trading is a unique concept which would have such wide reach and potential impact on all levels of society that there would need to be a period of comprehensive public engagement and debate before any concrete proposals could be made'; second, 'the Government must remain committed to identifying and implementing the most cost-effective policies for tackling climate change'.

Following these series of assessments for and against the introduction of a PCT scheme in the near future, Bird and Lockwood (2009) have sought to identify the implicit assumptions underlying those assessments. They organise their study around the core principles that both sides use to either support or reject the introduction of PCT, namely environmental effectiveness, efficiency, political acceptability and equity. Bird and Lockwood conclude that available evidence for the UK indicates that the necessity of introducing PCT depends on the efficacy of the—already quite crowded—current legal and policy framework. Only if this framework would clearly fail to generate the expected mitigation would PCT become a serious alternative. They also suggest that lack of political support for PCT, together with prevailing uncertainties about effects and possible costs, makes its introduction a risky business for government. Hence, two courses of action are recommended: first, to increase financial inclusion and literacy, including on carbon trading, among the general population; second, to explore further the understanding that the public has of fairness in the context of climate change policy (Bird and Lockwood 2009: 44-46).

A number of observations can be made from the preceding discussion. First, it is interesting to see the different positions on PCT held by researchers and by official bodies, which could be due to the fact that it is the latter that is ultimately responsible for introducing such a 'radical' policy innovation. Second, PCT is one alternative among several to engage the public in climate change mitigation. It is therefore distinct from the criterion of per capita allocation as a criterion for deciding on the emission reduction commitments for states. It would be possible to accept per capita allocation as one (or the main) criterion for burden sharing at the level of (Member) States while rejecting PCT in the light of, for example, practical considerations. Third, there seem to be currently four important barriers preventing the implementation of PCT: (1) it requires basing the policy choice on a certain view of distributive justice, and this might be difficult in democracies where legislation may need to be based on consensus among parties with very different views; (2) it is not clear whether the public would accept its introduction. However, given the findings of the UK government that the public also rejects taxation, this would seem to be a general problem rather than one specifically associated with PCT; (3)

there is uncertainty over the degree of engagement of the public, and, if it turned out to be low, the mechanism would become a very high tax, reducing cost-effectiveness and having regressive impacts; and (4) it is not cost-effective for the time being. Clearly, while none of these objections rule out the idea in the long term, they certainly reduce the likelihood of its implementation in the short term.

11.5 General conclusion and outlook

The most relevant conclusion of the research is that the content of distributional choices made in EU climate change policy increasingly embodies—despite significant remaining tensions—the core meaning of the relevant legal principles. This would support the view that legal principles can, and have in fact, provided guidance in structuring and rationalising bargaining processes in EU climate change law and policy. Nevertheless, one important and persistent factor limiting the degree of this contribution of legal principles is the lack of agreement on the core content of some important legal principles, particularly the polluter pays principle, the integration principle and the CBDR principle. Thus, ongoing efforts to explore more in depth the core content of those principles should be promoted.

In an effort to find policy solutions that increase the degree of alignment between distributional choices and the legal principles contained in the meta-principle of proportionality, this research has discussed a concept that is gaining more and more attention in philosophical and legal literature: per capita allocation of GHG emission rights. This concept is distinct from PCT, and it is possible to endorse one while rejecting the other. For instance, it is possible to argue for a more dominant role of per capita allocation in the distribution of the burden among states, without supporting the implementation of PCT at national level. Many different regulatory approaches could be implemented at that level in order to address the consumption side. The choice to implement a PCT scheme depends on the strengths and weaknesses of those other policies. In this vein, the study concludes with the recommendation that the possible role of PCT as a distributional instrument that addresses the consumers should be further assessed in comparison to other regulatory policies. Moreover, further research is needed into the role that the per capita criterion can play with regard to the distribution of climate protection efforts among states.

References

Babcock, H.M. (2009) 'Assuming Personal Responsibility for Improving the Environment: Moving Towards a New Environmental Norm', *Harvard Environmental Law Review* 33.1: 117-76.

Baer, P., G. Fieldman, T. Athanasiou and S. Kartha (2008) 'Greenhouse Development Rights: Towards an Equitable Framework for Global Climate Policy', *Cambridge Review of International Affairs* 21.4: 649-69.

Bird, J., and M. Lockwood (2009) *Plan B? The Prospects for Personal Carbon Trading* (London: Institute for Public Policy Research).

Chakravarty, S., A. Chikkatur, H. de Coninck, S. Pacala, R. Socolow and M. Tavoni (2009) 'Sharing Global CO_2 Emission Reductions among One Billion High Emitters', *Proceedings of the National Academy of Sciences* 106.29: 11,884-88.

DEFRA (UK Department for Environment, Food and Rural Affairs) (2008) *Distributional Impacts of Personal Carbon Trading* (London: DEFRA).

De Sadeleer, N. (2006) 'The Precautionary Principle in EC Health and Environmental Law', *European Law Journal* 12.2: 139-72.

Deketelaere, K. (2009) 'Concluding Remarks', at the *Conference on European Greenhouse Gas Emissions Trading: Lessons to be Learned*, Maastricht, 29–30 January 2009.

ENDS Europe Daily (2009), 'Czechs sell 8.5 million AAUs to Spain, Austria', ENDS Europe Daily, 14 October 2009; www.endseurope.com/22382?referrer=search, accessed 19 March 2010.

European Commission (2007) 'Green Paper from the Commission to the Council, the European Parliament, the European Economic and Social Committee and the Committee of the Regions. Adapting to Climate Change in Europe: Options for EU Action' (COM/2007/0354 final; Brussels: European Commission Publications Office, 29 June 2007).

—— (2009a) 'Adapting to Climate Change: Towards a European Framework for Action' (White Paper COM/2009/147 final; Brussels: European Commission Publications Office, 1 April 2009).

—— (2009b) 'Comparability of Developed Countries' Mitigation Efforts', report presented by the European Commission to the EU Environmental Council on 25 July 2009; www.endseurope.com/docs/90729a.pdf, accessed 19 March 2010.

—— (2009c) 'Communication from the Commission to the European Parliament, the Council, the European Economic and Social Committee and the Committee of the Regions: Towards a Comprehensive Climate Change Agreement in Copenhagen' (COM/2009/39 Final; Brussels: European Commission Publications Office).

European Environment Agency (2009) 'Greenhouse Gas Emission Trends and Projections in Europe 2009: Tracking Progress in Europe Towards Kyoto' (Report No 9/2009; Copenhagen: European Environment Agency).

Harris, P. (2008) 'Climate Change and Global Citizenship', *Law and Policy* 30.4: 481-501.

Hayward, T. (2006) 'Global Justice and the Distribution of Natural Resources', *Political Studies* 54.2: 349-64.

House of Commons, Environmental Audit Committee, (2008) 'Personal Carbon Trading', Fifth Report of Session 2007–2008', London; www.publications.parliament.uk/pa/cm200708/cmselect/cmenvaud/565/565.pdf, accessed 19 March 2010.

Lee, M. (2005) *EU Environmental Law: Challenges, Change, and Decision-making* (Oxford, UK: Hart Publishing).

Müller, B. (2009) 'Additionality in the Clean Development Mechanism: Why and What?', *Oxford Institute for Energy Studies* EV44 (March 2009).

Mahmoudi, S. (2006) 'Integration of Environmental Considerations into Transport', in R. Macrory (ed.), *Reflections on 30 Years of EU Environmental Law: A High Level of Protection?* (Groningen, Netherlands: Europa Law Publishing): 185-95.

Ministry of the Environment of the Czech Republic (2009a), 'Czech Republic and Spain conclude an agreement on the sale of 5 million Kyoto Units', 14 October 2009; www.mzp.cz/en/news_pr091014aau_spain, accessed 19 March 2010.

—— (2009b) 'Austria has purchased an additional 3.5 million Czech emission credits', 13 October 2009; www.mzp.cz/en/news_pr091013aau_Austria, accessed 19 March 2010.

Page, E.A. (2006) *Climate Change, Justice, and Future Generations* (Cheltenham, UK/Northampton, MA: Edward Elgar).

Starkey, R., and K. Anderson (2005) 'Domestic Tradable Quotas: A Policy Instrument for the Reduction of Greenhouse Gas Emissions' (Tyndall Technical Report 39; Manchester, UK: Tyndall Centre for Climate Change Research; www.tyndall.ac.uk/content/investigating-domestic-tradable-quotas-policy-instrument-reducing-greenhouse-gas-emissions-e, accessed 19 March 2010).

—— and K. Anderson (2004) 'Summary of Work to Date on Domestic Tradable Quotas', Tyndall Centre for Climate Research; www.tyndall.ac.uk/content/investigating-domestic-tradable-quotas-policy-instrument-reducing-greenhouse-gas-emissions-e, accessed 19 March 2010.

Thomas, E.W. (2005) *The Judicial Process: Realism, Pragmatism, Practical Reasoning, and Principles* (Cambridge, UK: Cambridge University Press).

Tol, R. (2007) 'Europe's Long-Term Climate Target: A Critical Evaluation', *Energy Policy* 35.1: 424-32.

Tridimas, T. (2006) *The General Principles of EC Law* (Oxford, UK: Oxford University Press).

Vilaça, J.L. da Cruz (2004) 'The Precautionary Principle in EC Law', *European Public Law* 10.2: 369-406.

12

Climate change liability and the application of the precautionary principle

Miriam Haritz
Department of International and European Law and Institute of Transnational Legal Research METRO, Maastricht University, The Netherlands

> Over the course of the last few years, climate change litigation has been transformed from a creative lawyering strategy to a major force in transnational regulatory governance of greenhouse gas emissions (Burns and Osofsky 2009: 1).

The year 2005, when the research project whose findings are summarised in this chapter was started, marked a special year in terms of the climate: flooding and heatwaves struck almost every continent, and the names of hurricanes such as Katrina, Wilma, Rita and Stan, which hit North and Central America and the Caribbean with an unprecedented destructive strength, became familiar to everyone.[1]

These weather catastrophes received wide public and media attention and were more and more associated with climate change, a notion that became increasingly familiar in daily newspaper coverage. Yet the picture is not as clear as it seems at first sight: while there is indeed growing certainty that humans do contribute to

1 Hurricane names in the Atlantic are chosen by representatives of the weather body of the UN, the World Meteorological Organization (WMO), with names being withdrawn and replaced when a storm has caused particularly devastating effects. In 2005, for the first time since 1953, when the names were attributed, all 21 names were used up.

climate change, the exact consequences of this finding are uncertain and lie at the heart of the Vulnerability, Adaptation and Mitigation (VAM) research recapitulated in this chapter.

The main research question that has been guiding the research is to what extent the precautionary principle, which seeks to make uncertainty manageable, is of relevance in the context of climate change liability. In particular, the question of whether the precautionary principle can be applied in making uncertainty judiciable is addressed. Therefore, the influence the precautionary principle may exercise on climate change liability, both in terms of facilitating the establishment of liability as well as potentially creating new uncertainties, and even liabilities through its (erroneous) application, is examined. Starting with an abstract presentation of each aspect (namely risk and uncertainty, the precautionary principle and liability law), each aspect in the particular context of climate change and, where applicable, climate change liability is analysed. The methodological focus is the law and economics perspective in an integrative and problem-oriented manner, looking at exemplary national, European and international jurisdictions.

12.1 A novel legal perspective on the consequences of a meteorological phenomenon

Beyond its terminological origin in meteorological and climate sciences, climate change as a global environmental problem is traditionally regarded as an area confined to the legislative and executive powers that are responsible for issuing relevant regulation in that field, despite these uncertainties. Mostly, it was considered that the judiciary should simply acknowledge its lack of expertise and refrain from handling legal cases involving climate change (Harper 2006: 696).

It is true that politicians, particularly from international politics but increasingly from national politics as well, are the prime addressees for setting the path of climate change policy as determined by scientific research. Additionally, lower layers of governments such as local communities and affected industrial and business sectors down to the individual consumers need to respond in an adequate manner to the manifold challenges stemming from climate change, with corresponding interlinks and overlaps as have been analysed by other contributors to the VAM project.

Policy measures regarding mitigation of and adaptation to climate change remain the main step in fighting climate change. Still, in the light of the magnitude and the cross-border nature of the predicted consequences, there need to be other, complementary, steps accompanying it.

Despite diverging interpretations and lacking commitment, starting with international climate change policy commitments as agreed in the United Nations

Framework Convention on Climate Change (UNFCCC) produced during the United Nations Conference on Environment and Development in Rio de Janeiro in 1992, both the international and different national approaches to combating climate change focused for a long time on the mitigation of climate change. But enhancing climate change mitigation cannot be the sole response, as there is increasing evidence of present and imminent climate change consequences resulting from historic emissions that cannot be altered any more. This makes adaptation to prevent the worst from happening an inevitable task.

What is more, mitigation, adaptation and climate change damage that is already occurring and that is predicted to occur all involve costs in one way or another, so the question of financial distribution and, in this context, the allocation of responsibilities do not only play a role in terms of distributional burden-sharing mechanisms that have been analysed by other VAM contributors, but also in terms of liability, on which this research has focused.

In this context it is important to point at the goals and functions of liability in regard to climate change damage and at the same time its supplementary contribution to other efforts in combating climate change as such.

First, liability law has a corrective function in the sense of *ex post facto* law enforcement. With regard to climate change this function can be recognised in the polluter pays principle.[2]

Second, from the viewpoint of law and economics in particular, liability law also has a preventive function in the sense of *ex ante facto* law enforcement and enhancing the deterrent effect.[3] In the context of climate change, even those critical of this type of preventive effect recognise the potential of climate change liability in forcing decision-makers in the political as well as in the economic arena to enhance efforts to mitigate and adapt to climate change (Spier 2006: 5).

Lastly, with regard to compensation of the victim, liability law also has a reparative function of a distributional nature and serves as an instrument for upholding the rule of law in both the national and the international legal order (Lefeber 1996: 1,313-14). As far as climate change is concerned, the compensatory aspect is relevant given the numerous victims affected by the damaging consequences of climate change and the predictions of a significant increase in the number of victims. Currently, there is no direct possibility for victims of climate change damage to receive compensation, and the indirect means through property insurance and natural disaster funds are per se limited and not available to all victims, especially not to the most vulnerable ones in the developing world who are considered to be the most exposed to the most devastating climate change consequences.[4]

2 This established principle in international environmental law says in essence that the one responsible for pollution shall bear the costs (see Lefeber 1996: 2ff.; Sands 2003: 279).

3 See, for example, the arguments of Van den Bergh and Visscher (2008), in the context of private collective actions in the field of consumer protection.

4 See UNFCCC 2007. The range of the Adaptation Fund that has been established under the UNFCCC is, like other funds aimed at supporting mitigation in developing countries, per

Climate change liability has been deemed to be 'the best tool for addressing climate change in the foreseeable future'.[5] Yet liability can neither be a goal on its own nor should it be considered the sole or even best way to compensate victims of climate change or to prevent climate change in the future. Liability for the consequences of climate change is but one means of compensating for ongoing and predicted damage, along with other proposals that deserve separate analysis in future research such as the idea of an international compensation fund (see e.g. the analysis of the general feasibility and value of such funds in Van Langendonck 2007) or obligatory or voluntary climate change insurance.[6] The development of a coherent liability law system for climate change consequences is complementary to these measures in forcing worldwide actors and stakeholders to enhance their efforts and should form part of the necessary global response.

It is in this light that Gupta's convincing statement has to be understood: 'Within the legal community, there is a growing conviction that ultimately local, if not global, environmental justice will require courts to play a critical role in putting pressure on the legislature and executive' (Gupta 2007: 78).

In fact, the pressure to enact climate change regulation, as part of the consequences of climate change litigation and liability claims in particular, has already been considered with regard to the US situation in particular to 'pervasively transform the manner in which American businesses relate to shareholders and consumers' and it is argued that companies should be required to take this into account regardless of whether the scientific predictions on the catastrophic effects of climate change become true or not (Faulk 2009: 59, 64-65).

In view of the global nature of climate change, these conclusions also hold true for other countries. Indeed, one can discern a global gradual development towards obliging states and operators to assume responsibility for climate change consequences. This is also reflected in the fact that major CO_2-emitting companies are increasingly becoming aware of possible financial consequences and seeking remedies by taking out insurances and investing in greener energies (see Ross *et al.* 2007: 316; Zacharias 2009).

With regard to the economic implications, companies will take measures without being legally obliged to do so only where there are economic and financial incentives. In the case of climate change, this behaviour means that companies will only start mitigating activities if the possible costs of the economic consequences and a possible liability for climate change outweigh the costs of taking precautionary or even preventive measures. A major cost factor that plays a role in this decision, next

se limited, since it covers only concrete adaptation projects in developing countries that are parties to the Kyoto Protocol, and because it is financed under the clean development mechanisms enshrined in that Protocol.

5 Quoted by a US lawyer on the prospect of climate change litigation in Grossman 2003: 6.

6 For a discussion of the broader context see Richardson 2002; Sugarman 2006. More specifically on climate change see Benoist 2007; Hawker 2007. Faure and Skogh (2003: 262-86) discuss the limits to liability and the broader insurance possibilities in the context of international risk sharing in cases such as climate change.

to consumer choices under risk, is the role of financial markets and capital investments. This is the case both in light of the danger of economic losses as well as the potential profitability of emerging trading markets, but also concerning the risk of liability (see Kunreuther and Michel-Kerjan 2007). The ongoing activities in this field show that even from an economic perspective, climate change is perceived as both a real threat and a field for economic expansion when it comes to combating measures.

Among lawyers the conviction is spreading that some of their clients, namely the actual operators that contribute to CO_2 emissions, will increasingly face litigation, drawing a comparison to the beginning of legal cases surrounding health damage inflicted by tobacco consumption in previous decades, where uncertainty particularly in regard to causation also posed a major obstacle that was overcome (see Grossman 2003; Jaap Spier, Advocate-General of the Supreme Court of the Netherlands, in Spier 2006).

Not surprisingly, climate change has thus also become a major topic for insurance and reinsurance companies, not just with regard to property insurance against damage resulting from climate change, but increasingly with regard to the prospect of liability claims over climate change damage and insurance coverage sought by the companies viewed as potential wrongdoers, i.e. legally speaking a tortfeasor, for their contribution to climate change (Munich Re 2009).

Similarly, decision-makers are facing increasing pressure to consider CO_2 emissions, even if their governments do not participate in any of the Kyoto mechanisms within the international efforts to achieve emissions reductions, as environmental groups and affected individuals are using the indirect means offered by existing legislation through court cases.

We thus currently observe various cases of climate change litigation pending before different courts (for a recent overview of such cases see Burns and Osofsky 2009). Most of the ongoing cases deal with the failure to take into account climate change effects when allowing certain conduct, for example, the operation or funding of power plants based on fossil energy or when issuing environmental legislation. These cases can be considered as paving the way for actual liability claims against companies for CO_2 emissions.

A growing number of such cases (be it actual claims or petitions) are putting direct and indirect pressure on both decision-makers and operators. Regulators and agencies are being increasingly considered as possible targets of such claims when disregarding climate change considerations in their decision-making.

For instance, in 2007, the US Environmental Protection Agency was successfully sued by a coalition of federal states for the failure to regulate CO_2 in the context of its obligations relating to pollution control.[7]

Other relevant cases relate to the disclosure of information over greenhouse gas (GHG) emissions activities, the conduct of environmental impact assessments,

7 *Commonwealth of Massachussetts et al. v. EPA*, 415 F.3d 50, 05-1120, 127 S. Ct. 1438 (2007).

the failure to comply with emissions reductions requirements as laid down by the Kyoto Protocol or alleged human rights violations as a result of climate change consequences (such as the petition of the Inuit Circumpolar Conference against the USA[8] for the loss of their natural habitat due to decreasing Arctic ice, or the possible lawsuit of the low-lying Pacific island states threatened by rising sea levels (as discussed, e.g. in Culley 2002; Jacobs 2005).

Actual liability claims until now have been brought only in the US, be it as public or private nuisance claims or property damage claims, such as *California v. General Motors*,[9] *Kivalina v. ExxonMobil*[10] or *Comer v. Murphy Oil*[11] against companies directly responsible for GHG emissions and their contribution to climate change, or *Korsinsky v. EPA*[12] as an example of a claim against a governmental authority for failure to protect against the consequences of GHG emissions. On the plaintiff side, either affected individuals or states have brought claims. Clearly, the issue of standing has in such cases always been a controversially discussed complex issue and has until recently been denied. Yet, in September 2009 the US Court of Appeals was the first court ever to grant plaintiffs standing and to consider climate change

8 *Inuit Circumpolar Conference v. USA, Inter-American Commission on Human Rights* (IACHR), Petition to the Inter-American Commission on Human Rights Seeking Relief From Violations Resulting from Global Warming Caused by Acts and Omissions of the United States, Submitted by Sheila Watt-Cloutier, With the Support of the Inuit Circumpolar Conference, On Behalf of all Inuit of the Arctic Regions of the United States and Canada, 7 December 2005. The Petition was dismissed, but the case is considered as serving as a basis for other potential cases and in academic discussion (see Osofsky 2009).

9 *State of California on behalf of the People of California v. General Motors Corporation, Toyota Motor North America Inc., Ford Motor Company, Honda North America Inc., Chrysler Motors Corporation and Nissan North America Inc.*, Case No. C 06-05755, US District Court for the Northern District of California, 20 September 2006. The case was dismissed on grounds of the political question doctrine in 2007 and an appeal was filed at the US Court of Appeals for the 9th Circuit. The State of California voluntarily dismissed its appeal in June 2009, alleging that under a changed climate change approach under the administration of US President Obama the concerns addressed by their claim would be respected by legislative and executive means.

10 *Kivalina v. BP PLC, BP American, BP Products North America, Chevron, Chevron USA, ConocoPhillips, Royal Dutch Shell PLC and Shell Oil, Peabody Energy, AES, American Electric Power, American Electric Power Services, DTE Energy, Duke Energy, Dynegy Holdings, Edison International, MidAmerican Energy Holdings, Mirant Corp., NRG Energy, Pinnacle West Capital, Reliant Energy, The Southern Co. and Xcel Energy*, Case No. 4:2008CV01138, filed at the US District Court San Francisco on 26 February 2008. The case was dismissed on 30 September 2009 for a lack of standing and on grounds of the political question doctrine, but it is likely that it will be appealed.

11 *Comer v. Murphy Oil USA*, Case No. 05-CV-436, US Southern District Court of Mississippi, 2006 WL 1066645/No. 07-60756, Fifth Circuit Court of Appeals. Case was appealed, yet vacated for a lack of quorum and may thus be decided by the US Supreme Court only. See footnote 13.

12 *Korsinsky v. US EPA*, Case 1:05-CV-00859-NRB, US District Court for the Southern District of New York, 2005. Case on appeal to the US Court of Appeals for the 2nd Circuit.

as justiciable despite its political implications on behalf of the executive and legislative.[13]

Nevertheless, no liability claim so far has granted any of the remedies that were applied for, and the most difficult legal hurdles in a climate change liability claim, such as causation and the proof thereof that this research addresses among others, have not yet been judged on in substance.

12.2 Climate change and uncertainty

It is evident that climate change is surrounded by numerous scientific uncertainties that make decision-making on this issue problematic. Stakeholders will therefore be affected not only by the factual but also by the legal consequences of climate change long before there might be scientific certainty as to the exact scope and causal distribution of climate change damage (Faulk and Gray 2009: 16). For this reason, the research starts with clarifying the terms of risk and uncertainty in general before moving on to what these notions entail in the context of climate change. The introduction of the broad concept of incertitude encompassing uncertainty, ambiguity and ignorance in the context of assessing risks in general, is followed by an overview of the current state of affairs as far as climate change science and the continuing uncertainties and controversies over the topic relevant to this research are concerned. To that purpose, the terminology of climate change as a phenomenon beyond the meteorological implications as such and methods of combating its causes and effects are clarified, and an overview of the past, present and future of climate change science and policy is presented as well as an analysis of the concepts of risk and uncertainty relevant to climate change.

Concluding from the scientific consensus view as presented, for example, with growing urgency and increasing evidential conviction in the consecutive reports of the Intergovernmental Panel on Climate Change (IPCC)[14] gathering what is to be considered as authoritative views from a broad range of scientists, climate change

13 *State of Connecticut, et al. v. American Electric Power Company Inc., et al.*, Docket Nos. 05-5104-cv, 05-5119-cv, US Court of Appeals for the 2nd Circuit, Decision of 21 September 2009. On 16 October 2009, the same reasoning and conclusion was followed in the appeal case of *Comer v. Murphy Oil*, see *Comer* v. *Murphy Oil USA*, No. 07-60756, US Court of Appeals for the 5th Circuit, Decision of 16 October 2009.

The current state of affairs is thus that lower courts in public nuisance cases over climate change damage have rejected judging on the content of the cases by refusing to grant plaintiffs standing and qualifying such cases as non-justiciable for the political implications. On appeal, so far some of these cases have been decided differently and have been referred back to the lower courts for a decision in substance or might have to be decided by the US Supreme Court in the future.

14 The reports were released in 1990, 1995, 2001 and 2007 and can be found at the IPCC website www.ipcc.ch and see IPCC 2007 for the latest, Fourth Report.

damage has to be considered real and a fundamental threat for the future. The measured and predicted factual effects of a general temperature rise lead to rising sea levels, an increased likelihood of extreme weather events, severe ecological harm and numerous effects on human health, property and welfare having drastic socioeconomic and financial implications (for a focus on the economic consequences see Stern 2007). Still, many of these predictions as well as the exact nature and degree of causal distribution and interrelations remain entangled in a web of numerous uncertainties.

Generally, the IPCC reports reflect a qualitative and quantitative assessment of uncertainty by the choice of words such as the level of agreement in relation to the level of evidence, the degree of confidence in relation to a punctuation of X out of 10 or the probability of occurrence.[15]

It can be seen that more uncertainty exists with regards to the phenomena that stem not from Global Warming as such but from the consequences induced by climate change, such as extreme weather events like the frequency and strength of storms, flooding from precipitation or sea-level rise and drought periods.

Starting with the uncertainty about future global emission scenarios, down to the exact response of nature's capacity to absorb CO_2 in relation to the range of possible reaction of the climate system, the regional changes in the climate, and, more particularly, the exact scope, timing and likelihood of possibly damaging impacts become more and more difficult to predict. This has been referred to as an 'uncertainty explosion' or 'cascade of uncertainties' (Schneider 2002: 443; Schneider and Kuntz-Duriseti 2002: 67-68).

These climate change uncertainties mainly result from the fact that it is not clear how a system of such complexity and subject to natural variability and randomness will evolve per se in the future. In addition, uncertainty is increased by the varying extents of human interference in the form of GHG emissions as a result of changing energy policy choices and consumption behaviour, determined by different risk perceptions, changes in population patterns and technological innovations (with their benefits and down-sides, e.g. for biomass production in the field of renewable energies) (Nordhaus 2007). For a detailed analysis of the scientific climate change uncertainties see Van Asselt and Rotmans 2002.

Moreover, despite improved assessment methods, there remains an uncertainty in measurements and predictions that 'ranges from inexactness to irreducible ignorance' (Van Asselt and Rotmans 2002: 80) or 'from unreliability to more fundamental uncertainty' (Van Asselt and Rotmans: 81).

In fact, climate change confronts scientists and decision-makers alike with novel uncertainty features that make it different from and more complex than any other environmental problem experienced before (Farber 2003: 149). An illustrative dif-

15 Virtually certain >99%; extremely likely >95%; very likely >90%; likely >66%; more likely than not >50%; about as likely as not 33% to 66%; unlikely <33%; very unlikely <10%; extremely unlikely <5%; exceptionally unlikely <1%. (See the explanation as to the overall treatment of uncertainty in IPCC 2007: 27.)

ferentiation of the most relevant uncertainty factors can be found in Wibisana 2008: 360).

Even though certain uncertainties have decreased over time owing to the availability of new scientific evidence and scientific prediction methods, such as the question of anthropogenic contribution, others have increased, such as the quantity of future emissions of developing countries or countries in transition, or have been discovered anew, such as the likelihood of abrupt climate change. Generally, because of the diverse (and possibly varying over time) factors that contribute to how and when predicted damage will manifest itself and as a result of the non-linearity of climate change per se, it is unfeasible that all uncertainties will ever be overcome, while consequences arise and require regulatory action and also—owing to resulting damage—a compensatory response.

At the same time, an overview of the problems regarding scientific proof of climate change shows that evidence indicating the harmful effect of GHG on the climate existed long before international action to combat climate change started.[16]

For instance, the Keeling Curve commenced in 1956, documenting the growing, continuing CO_2 accumulation in the atmosphere based on samples measured in a Hawaiian observatory,[17] showed a striking resemblance to the so-called Hockey-stick Curve[18] (referred to in the Third IPCC report of 2001) as far as the post-industrialised time period is concerned.

Moreover, despite the increasing evidence of the harmful effects, actual measures to reduce emissions only started with the coming into force of the Kyoto Protocol in 2005, whereas damage is and will be resulting from historic emissions and unabated or insufficiently reduced emissions on a global average.

12.3 The precautionary principle as a regulatory tool to handle uncertainty

Today, a vast field of literature exists that analyses the precautionary principle in a remarkably extensive way in all different jurisdictions, from different academic angles and regardless of praise, criticism or even condemnation of its existence

16 An overview of the growing evidence of the anthropogenic nature of climate change over the past 150 years can be found e.g. in Christianson 1999 and Weart 2008.

17 The Keeling Curve was established by the oceanographer and climatologist Charles Keeling and is still in use today; see the latest data from Scripps CO_2 Program, Scripps Institution of Oceanography, May 2009; scrippsco2.ucsd.edu/graphics_gallery/mauna_loa_record/mauna_loa_record.html, accessed 19 March 2010.

18 Based on the MBH98 reconstruction by Mann *et al.* (1998), as included in Folland *et al.* 2001. The Hockeystick Curve is still considered to reflect the calculated temperature rise due to CO_2 over several centuries correctly, despite disputes concerning its calculation methods.

and application. This research does not repeat those kinds of analysis. Instead, it explores the existing literature to give an overview of the principle's content and application as an instrument enabling decision-making in the face of uncertainty, in particular with regard to climate change policy and ultimately to facilitate or to obstruct climate change liability.

Without going into detailed discussion of the pros and cons of applying the precautionary principle, it is to be concluded that inconsistencies resulting from the precautionary principle are more the result of the lack of a coherent approach or even contradictory application than of its diverse interpretations at national, European and international levels.[19]

Furthermore, situations that require the use of the precautionary principle inevitably involve risk trade-offs and a due balance between errors of so-called false negative nature (i.e. too little precaution, as later on the issue proves to have harmful consequences) and false positive nature (i.e. too much precaution, as later on the issue proves to be harmless), regardless of the degree of precaution set by the different interpretations of the principle.

In addition, it should be underlined that the precautionary principle itself does not mandate a specific type of action in any of its interpretations. Generally, a wide array of measures can be justified on the grounds of the precautionary principle, while the exact choice of measure is to be determined on a case-by-case basis, weighing considerations that are beyond any definition under the precautionary principle (such as cost–benefit and cost-effectiveness considerations). On this basis, any regulatory decision, from banning a product to requiring additional information, might be justified, while the precautionary principle only mandates the taking into account of possible errors in decision-making with an environmental and public health bias in the sense of 'erring on the side of precaution' (Arcuri 2006).

For the purpose of this research, three main interpretative versions can be singled out from the many formulations of the precautionary principle in numerous legal documents, judgements, treaty texts and literature, varying in strength and intensity on a worldwide level. Even though primarily addressed at governmental decision-makers, they are also applicable to private decision-makers.

In version 1, a modest interpretation of the precautionary principle allows for action in the face of uncertainty, granting those who invoke it a right to act. This version is, for example, reflected in the phrasing chosen in Article 3 of the UNFCCC.

In the more proactive version 2, the precautionary principle can be regarded as an obligation to actively counteract in the face of a possible threat, both granting decision-makers a right to act and also, equally, placing them under a duty to act.

In its strictest formulation to be found in version 3, the precautionary principle can require a reversal of the burden of proof on top of the obligatory regulatory action to be derived from version 2. Here, the activity involving the risk in question

19 For a detailed comparison of these practical inconsistencies see the case studies on the precautionary principle's application in Zander 2009; see also Trudeau 2003.

can be prohibited in a risk-minimising manner unless the surrounding uncertainty is proven to be as close to a certainty of non-harmfulness as possible. Here, one has to distinguish between the burden of proof regarding the regulation of the activity in question[20] and the burden of proof in a liability claim.[21]

If one compares the three versions it becomes evident that they share as common elements the requirements of uncertainty, the threat of serious and irreversible damage and some form of regulatory consideration and precautionary response. The differences mainly consist in allowing or prescribing action in the

Figure 12.1 **The procedural precautionary principle in interaction with the content-created version 1–3 (based on Arcuri's Three-dimensional Framework**

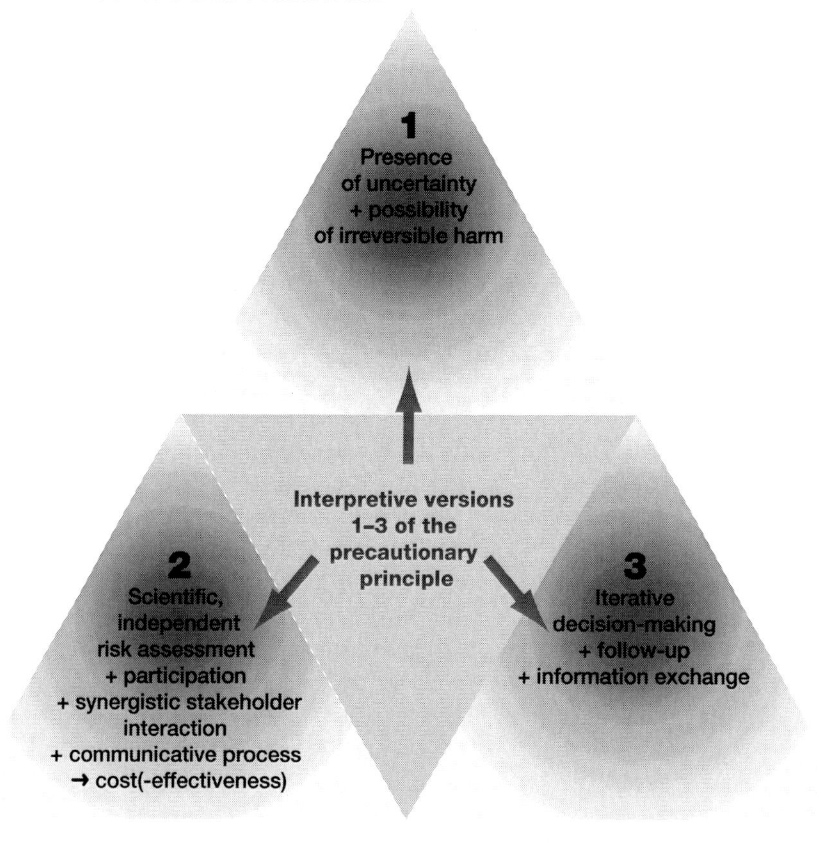

Source: Arcuri 2006: 53

20 This is the sense in which Zander (2009) has analysed the burden of proof with regards to the different versions of the precautionary principle in his case studies on Sweden, the UK and the US.

21 This is the sense relevant to climate change liability in view of the *ex post* application of this version of the precautionary principle.

Table 12.1 **The main 'versions' of the precautionary principle**

Version	**Triggering threshold**	**Enabling action**	**Degree of rigidity**	**Affecting the regulatory burden of proof**	**To be found for example in:**
1 (modest)	Lack of full scientific certainty	Allowing for regulatory action	Right to act	Modification	• Principle 15 Rio Declaration Art. 3 of the UNFCCC
2 (proactive)	Higher degree of uncertainty (scientific minority views)	Obliging to take regulatory action	Right to act + duty to act	Modification up to shifting the onus	• Swedish Environmental Code • French Environmental Charter
3 (risk-minimising)	Anything above a mere suspicion	Obliging to take more stringent regulatory action	Right to act + duty to act + necessity to prove harmlessness of a conduct upon indication of harmfulness	Reversal	• Final Ministerial Declaration. 3rd International Conference on the Protection of the North Sea
Procedural	n.a.	Setting the framework for precautionary decision-making	n.a.	Setting the procedure depending on the content of version 1–3	n.a.

face of uncertainty and the question of proof of the safety or danger of or harmlessness of conduct, in terms of both the degree of threat that requires precaution and the question of who bears the burden of such proof. Despite these differences, the boundaries between the three versions are fluid, sometimes overlapping and even changing over time.[22]

In addition, the procedural precautionary principle is singled out, not as an alternative in content-related interpretation but as an additional feature of the principle in setting a specific precautionary procedure to be followed in regulatory decision-making at the public level. Even if there are cultural diversities in defining the respective procedural requirements of the precautionary principle, there is a common understanding that makes it a pluralistic principle in interdepend-

22 As can be seen, for example, with regard to the EU, where the precautionary principle can be said to have developed over the years from the interpretation of version 1 to the stricter interpretations of versions 2 and 3 (see Löfstedt 2002).

ence with the given institutional circumstances (Fisher 2009: 23ff.). The procedural precautionary principle is aimed at contributing to a 'democratisation of science' (Arcuri 2006: 23, 53, 59) and thereby plays a role in balancing what has been termed the de- and re- 'politicisation of science' (Everson and Vos 2009) in the context of decision-making, in the face of risk regulation under conditions of uncertainty (Scott 2004: 66) (see Fig. 12.1 and Table 12.1).

12.4 Liability law faced with climate change uncertainties

In a next step, it was necessary to analyse different exemplary liability systems in an integrative manner, singling out the main problems and commonalities of liability law within different legal traditions of national law and on European and international levels when faced with the uncertainties presented by climate change. With this aim, examples from national legislations are given by looking at individual countries that are representative of the identified legal systems in handling scientific uncertainty. Subsequently, specific European and international examples of legislation are analysed to draw conclusions as far as possible for the legal uncertainties posed by climate change. For Europe, the Product Liability Directive 1985/374/EEC, the Environmental Liability Directive 2004/35/EC and the recommendations from the principles of European tort law are taken into consideration.

Next, existing treaties in the field of climate change (the UNFCCC and the attached Kyoto Protocol) and a range of general rules and obligations of international law relevant to climate change liability (such as the no-harm-rule, the duty to prevent trans-boundary harm and the principle of common but differentiated responsibility) are examined.

Finally, an overview of potential liability scenarios and actual litigation with relevance to climate change liability from different jurisdictions is presented.

The research cannot address all the aspects that are to be considered in a possible climate change liability claim.[23] Instead, the most relevant elements in the light of uncertainty and the application of the precautionary principle are scrutinised: the choice of the basis of liability (strict or fault liability), determining wrongfulness and the required standard of care; establishing the chain of causation and the different levels of causal uncertainty (effective causation in terms of general and specific causation in connection with normative causation theories in different legal systems); distributing causal responsibility among multiple actors (individual liability, joint and several liability or proportional liability); and defining the temporal and remedial extension of liability (retroactivity and available remedies) and the scope of possible defences.

23 Such as applicable forums and the optimal place for bringing a climate change liability claim.

With regard to a liability claim for the consequences of climate change, arguments for or against either basis of liability are given. Regarding causation, the exclusion of the background risk (given that climate variations and extreme weather events might also occur without human interference) and the adequate threshold for probability (referring to the degree of likelihood required to establish legal certainty despite scientific uncertainty) are found to be particularly problematic. As to the choice of the appropriate model for apportioning liability among multiple tortfeasors, different solutions are elaborated on. Retroactivity, the evolution of the standard of care and *ex post* liability, as far as historic emissions are concerned, are considered as well as the available remedies, namely damages and injunctions in consideration of present and future emissions and resulting damage. Finally, under the heading of possible defences, the element of foreseeability is analysed in terms of the 'state-of-the-art' defence (the argument that the state of the art in climate change science does not require a different conduct), the defence of *force majeure* (the argument that climate change damage was unforeseeable and unavoidable at the time of action) and the 'regulatory compliance' defence' (the claim that as long as industry follows the standards set by the government no liability can be incurred for later damage). Moreover, the argument that climate change mitigation requirements need to be balanced with cost–benefit considerations and consumer behaviour (e.g. considering energy efficiency patterns) is scrutinised in regard to the 'risk–cost' defence and the equitable defence of 'unclean hands' or contributory negligence.

The main conclusion on these elements is the finding that the scientific uncertainties identified in the available evidence and predictions of climate change causes and consequences translate into legal uncertainties in a liability claim over climate change damage. These uncertainties in turn make it extremely difficult to adjudicate such a claim, whereby not only the compensatory function of liability law but also the foremost preventive goal of liability law is inhibited.

These findings are in line with the verdict of the sociologist Ulrich Beck, who considers climate change as a prime example for increasing 'manufactured risks' (Beck 1986) in a world that is growing more complex (Beck 2007), namely that the scientific uncertainties surrounding these risks also expand in a legal sense. It is in this context that Beck perceives a shift in paradigms from compensation towards more precaution in order to handle newly emerging risks that are increasingly surrounded by several factors of uncertainty and defined by him as anticipated catastrophes (Beck 2007). Consequently, the question of due precaution and how to handle these uncertainties gains more relevance to liability questions as well. The precautionary principle's essence in ethical terms is reflected in the sentence '*mieux vaut prévenir que guérir*' (Radé 2000: 75)—it is better to prevent than to cure. This essence needs to be remembered when it comes to adjudicating the consequences of not having stuck to this guidance translating that failure into a feasible liability claim.

12.5 The precautionary principle and climate change liability

In the field of environmental damage in particular, liability law is affected by uncertainty, and its effectiveness in compensation, reparation and prevention is inhibited. When applying the precautionary principle the certainty requirements set by traditional concepts of liability law are replaced with sufficiently grounded probabilities so that the goals of liability law can be attained despite involved uncertainties (de Sadeleer 2002: 211-12).

The precautionary principle is to be considered in a triangle of interacting principles of environmental law with operational interdependency, namely the precautionary principle, the prevention principle and the polluter pays principle. These all modify the traditional structures of civil liability in legal systems that are increasingly required to anticipate environmental risks and distribute the responsibility for resulting damage accordingly for the purpose of strengthening both the compensative and the preventive goals of liability.[24]

In discussing the exact consequences of the precautionary principle's application to liability law in general as it is recognised in the academic literature,[25] diverging opinions on the disputed role of the precautionary principle are presented in the research and applied to the climate change scenario with the goal of showing how the precautionary principle in its three interpretations singled out above alleviates or increases uncertainty within climate change liability claims.

In the regulatory sphere, even the most basic and least stringent version of the precautionary principle implies that scientific uncertainty is not to be used as a justification for inaction. Therefore it is only consequent to apply this requirement of enhanced foresight retrospectively, when a defendant is at fault for not taking precautionary measures against risks that were uncertain and not usually foreseeable at the time (Boutonnet 2005: 438; de Sadeleer 2002: 212; see also Baudouin and Deslauriers 2003: paragraph 152; Khoury and Smyth 2007: 228). It can thus be observed that the precautionary principle links naturally to uncertainty in all of its versions also in this *ex post* sense, leading to a natural partnership embracing public and private law at national, European and international levels alike.

24 See de Sadeleer 2002: 225, 369; stressing that the polluter pays principle would establish responsibility for GHG emissions and thereby facilitating the finding of causation, see Baer 2006 131-32; Cullet 2007 119-20; see also the thorough analysis of the contribution of the precautionary principle and the prevention principle to liability law in particular in Tapinos 2008.

25 A growing number of legal scholars in different jurisdictions have started to attach increased importance to the preventive function of liability in particular stemming from the application of the precautionary principle to liability law (see the conclusions and overview in Van Boom Pinna 2008: 275; see also Baghestani-Perrey 1999: 461; Lascoumes 1997). 'The days of absolute certainty are over; in the future greater importance will necessarily be attached to doubt and consequently to the precautionary principle within the mechanisms of liability' (de Sadeleer 2002: 211).

The so-called 'toxic tort' cases, for example over exposure to asbestos, have provided a prime example of where tortfeasors can be tried for a lack of precaution despite indications of the harmfulness of the substance over a long time-period.[26] The example of climate change liability shows that the precautionary principle can even be used as a judicial tool in claims against direct emitters of GHG responsible for climate change as well as in shaping the liability of public authorities for insufficient or excessive precaution. In regard to the latter aspect, the procedural precautionary principle is to be considered as freeing governmental institutions from further liability as long as the correct procedure for decision-making requirements is followed.

Looking at the main elements clouded by uncertainty in a climate change liability claim, it is concluded that the precautionary principle does not aid in advocating for either fault or strict liability as the most appropriate basis of a liability claim. Thus there is no effect of the precautionary principle in the choice of the basis of liability.

Yet, within fault liability, it can be observed that the precautionary principle does affect the required standard of care. It imposes precaution over inaction in its stricter interpretations, thereby facilitating the finding of negligent behaviour and thus fault. But also in its modest version 1, the so-called bonus pater familias standard—i.e. the requirement to act prudently and protectively like a good father would do when looking after his family—which is used as the traditional standard of care in most legal systems, is extended to include additional information requirements. These would oblige stakeholders to consider climate change evidence prior to 1990 and to enforce climate change mitigation with more commitment.

Consequently, as far as the finding of fault in climate change liability under the precautionary principle is concerned, it is argued that both governments and entrepreneurs have been under obligation to question their conduct over CO_2 emissions from the moment that the risks in question amounted to more than a suspicion from the 1950s onwards. Under all three interpretations of the precautionary principle this questioning would require proactive gathering of information and investment into follow-up research instead of continuing business as usual until further notice of certainty.[27] Moreover, the active and biased support of scientists suggest-

26 See the 12 late lessons case studies in Harremoës *et al.* 2001. See also the case *Janssen v. Nefabas*, Judgment of the Hoge Raad, 6 April 1990, NJ 1990 573. In that case, the court ruled that a stricter standard of care and a partial reversal of the burden of proof on behalf of the plaintiff in proving fault was justified, facilitating and thus paving the way for subsequent asbestos claims. The fact that despite absence of regulation an employer was placed *ex post* under the obligation to investigate potential dangers for his employees even beyond the information available in the Netherlands is explicitly regarded as an emanation of the precautionary principle in de Sadeleer 2002: 212-13, in particular footnote 477.

27 Today, proposed strategies for companies to evade liability include proactive research, monitoring and reporting, engagement in the quest for alternative technologies and development of a climate change action strategy (see Healy and Tapick 2004: 101ff.).

ing that climate change is less of a problem than, for instance, the IPCC reports claimed would have to be considered as wrongful behaviour under the precautionary principle.[28]

As regards the conduct of governmental authorities, one could argue that participation in international climate change policy, including the setting up of a body to accumulate and assess research such as the IPCC, and the active effort to reduce emissions following the Kyoto Protocol or similar reductions approaches, partly reflects states' efforts to comply with the requirements of the modest version of the precautionary principle. Uncertainty was not used as an excuse for inaction. Still, this would cover only the period from 1990 (the year the first IPCC report was released)/1992 (the year the UNFCCC was agreed on), whereas scientific evidence beyond the stage of mere suspicion was already available before that, when governments did in fact remain inactive. Because past emissions stemming from a time without emissions regulations show their effects now and in the near future, the time before 1990/1992 has to be taken into account to some extent under the modest version of the precautionary principle for the finding of fault.

Moreover, the proactive version of the precautionary principle requires more action than that taken in international climate change policy by means of the UNFCCC and its attached Kyoto Protocol, because the efforts in mitigation undertaken so far will not lead to a substantial reduction in temperature rise and will therefore not suffice to prevent a large part of the predicted damage. Furthermore, the precautionary principle generally facilitates alleviating or shifting the burden of proof of such negligence. In addition, the burden of proof is affected by the application of the precautionary principle in establishing the chain of causation.

While causation is already an issue when deciding on the plaintiffs' standing, the real difficulty lies in establishing causation on the merits. Here, a distinction needs to be made between proof of general causation (where the overall causality of emissions-favouring conduct is to be acknowledged under the precautionary principle), specific causation (where the causation of a specific incident is facilitated by the precautionary principle as well) and multi-party causation (where the effect needs to be different in view of opting for proportional or joint-and-several liability).

More specifically, the effect of the three versions of the precautionary principle ranges from mere facilitation of taking into account causal evidence in alleviating the burden of proof requirements (version 1) to shifting the burden of proof, whereby the onus shifts onto the defendant once the plaintiff has been able to prove an initial fault on the side of the tortfeasor (version 2). For the strictest version 3, an even stronger modification of the burden of proof is to be assumed, commonly referred to as a reversal of the burden of proof. Yet, a full reversal without any

28 It has been asserted, with reference to internal documents that were made public, that companies have funded one-sided scientific research on climate change with the goal of suppressing evidence of the harmful effects in order to delay governmental regulation obliging them to reduce emissions, similar to the conduct of tobacco producers (see Müller and Plehwe 2008; Union of Concerned Scientists 2007).

prior indication of harm would amount to the juridical translation of a zero-risk approach, which is to be considered as over-interpretation of the precautionary principle.

Moreover, it is concluded that under the precautionary principle liability could be established from at least 1990, the date of the release of the first IPCC Assessment Report, in its stricter interpretations with regards to previous times, considering that there was scientific evidence of climate change available long before international action on the topic really started. Thereby taking the contribution of historic emissions into account would be facilitated.

In addition, the precautionary principle is found to have a deep impact on the decisive criterion of foreseeability. In light of the principle's obligation to conduct more extensive and continuous risk assessment, it can no longer be argued that it was unforeseeable that the risk stemming from ongoing, unabated GHG emissions could lead to the predicted or manifested damage. Therefore, recourse to possible defences relating to a claim of *force majeure*, the reliance on regulatory compliance and the consideration of development risk and state-of the-art arguments would be substantially limited.

Moreover, applying the precautionary principle as a principle that by nature attaches a higher value to environmental considerations would inhibit recourse to the risk–cost defence (which allows balancing the costs of risk reduction in favour of the defendant) regardless of arguing that climate change mitigation involves substantial economic cutbacks (as argued by Goklany 2000: 221-22) or provides economically and socially beneficial incentives for technological innovation and a more effective energy consumption behaviour (see Wibisana 2008: 378-79). Here, the application of the precautionary principle is particularly justified because conventional cost–benefit analysis is unable to take into account the factor of uncertainty to the full extent—completely unmitigated emissions will increase climate change beyond existing knowledge or estimates, and that, coupled with the non-linearity of the earth's climate in general, is a factor too risky not to be assessed properly; in addition, the inherent uncertainties make this assessment even more difficult, thereby rendering a sole cost–benefit analysis inefficient in these terms (Ackerman and Heinzerling 2004; Dasgupta 2007: 143-44).

The climatologist Schneider in an interview at Stanford University in 2007[29] explicitly referred to this inaptitude and agreed that despite the risk of erroneously balancing false positives against false negatives, the precautionary principle should be mandatory to climate change, because the potential costs of climate change clearly outweigh the costs of choosing the wrong measures to counter the risk, especially in view of worldwide growing GDPs. Within that precautionary framework, justice and equity considerations can then play a role in balancing, for instance, the effect of CO_2 reductions on the developing world against the improvement of living con-

29 Author's interview with Stephen Schneider, Stanford University, 23 February 2007. Schneider is a climatologist who has researched and published extensively for many years in the field of climate change and contributes to the IPCC reports.

ditions such as access to electricity and other technologies precisely responsible for GHG emissions, without, however, limiting the combating of climate change to a dichotomy between lowering emissions and economic prosperity (Cullet 2008: 114ff.). Overall, in determining which regulatory path to follow, other implications and considerations have to be balanced against the scope of possible damage, but not against the use of the precautionary principle as such (Peel 2005: 153).

Finally, it is to be concluded that the precautionary principle defines the extent of the damage to be taken into consideration, including expenses incurred for adaptation against present and future damage, albeit with limitations as to the effect on compensation for future uncertain harm (unless concerning adaptation costs). In its stricter interpretations of versions 2 and 3, however, this could entail a justiciable obligation to stop or modify the GHG-emitting activity in question (thereby giving rise to injunctions).

It is also found that the precautionary principle can be invoked against governmental authorities, both for the failure to apply precautionary measures (false negative error leading to under-regulation) and for the comparatively less likely and less damaging event of unnecessary and thereby harmful over-precaution leading to unwanted over-deterrence (false positive error leading to over-regulation). This in turn creates new uncertainties that need to be balanced cautiously and prudently in inherent risk trade-offs (Kourilsky and Viney 2000: 141).

It is, however, also concluded that relying on the precautionary principle in the sense of deriving an individual right from its breach is to be considered as non-justiciable for a lack of substantiation. The precautionary principle can be relied on for protecting fundamental rights, but it is not itself a right in the sense of a vague right to precaution without being connected to a concrete failure to protect such fundamental rights. Moreover, as far as the procedural precautionary principle is concerned, authorities may be able to rely on the justificative effect of regulatory norms.

Nonetheless, the precautionary principle can guide the behaviour of the potential tortfeasor in the form of an incentive or stimulus both *ex ante* and *ex post* in line with law and economics considerations.

Despite the precautionary principle's capacity to function as a judicial tool in solving many climate change liability uncertainties, it cannot be disregarded that the precautionary principle also generates new sources of uncertainty. These stem primarily from its diverse interpretations, which make it difficult for judges to apply one definition as best fitted to decide on the case in question. Yet, this interpretational variety is to be considered as an indispensable feature in reflecting cultural diversities and different approaches towards risk and risk perception (see also the conclusions on the role of the precautionary principle in civil liability in de Sadeleer 2002: 221-22). In fact, the resulting flexibility in interpreting the precautionary principle when addressing problems generated by uncertainty has to be considered an asset of the principle.[30] Yet adjudicating its content requires

30 See also COMEST 2005: 21: 'Arguably, a strength of the PP being a principle is thus its open-endedness and flexibility, which creates a possibility and an incentive for social learning.'

cautious handling so as not to be a possible source for flaws when being used as pretext for an inconsistent application of the principle. Nonetheless, this is not to be considered an insurmountable obstacle to its application to liability law (as can deduced from Tinker 1996: 53). Rather, it is precisely because of its flexibility that the precautionary principle's potential to serve as a judicial tool to determine the outcome of liability cases is to be recognised (Leone 1999: 12-16; Trudeau 2003).

In addition, the precautionary principle may generate a new source of uncertainty because of possible false positive errors; yet practice shows that both the likelihood and the extent of the damage that may arise out of such an erroneous application are comparatively small when balanced against the outcomes of false negative errors that the precautionary principle seeks to avoid (Sandin *et al.* 2002; Underwood 1999).

Therefore, even if the precautionary principle at least potentially entails new uncertainty, this effect is outweighed by the benefits from reducing the legal uncertainties that stem from scientific uncertainties. Otherwise the courts would be left with even less guidance, inhibiting the preventive and compensative goals of climate change liability as explained.

The practical relevance of the different versions of the precautionary principle applied to climate change liability has also been tested on actual climate change liability cases as analysed in the research, showing that—with varying degrees of intensity—also in practice all versions of the precautionary principle have the effect of substantially altering the outcome of these cases.

Drawing parallels with 'toxic tort' cases, in particular asbestos and tobacco litigation, it is found that looking at climate change liability through the lens of the precautionary principle helps to structure the debate over the role of the precautionary principle in liability law in general and the prospect of climate change liability in particular, beyond what tort law has so far been capable of. In practice, the precautionary principle needs to be formulated on a case-by-case basis and depending on the scale of potential damage (Spier 2008: 2523). See, too, the effects of the precautionary principle on climate change liability in Table 12.2.

12.6 Conclusion

The analysis has revealed that the precautionary principle plays an important role in addition to conventional cost–benefit and cost-effectiveness considerations in *ex ante* risk analysis and everything that follows *ex post* in a liability claim whenever a risk develops into damage, as is the case with climate change. It is this *ex ante* and *ex post* dimension of the precautionary principle taken together that imposes the obligation of anticipating risks that in the end provide a new framework for liability. In turn, it is only when applied in the context of liability law that the precautionary principle can fully attain its true objective, namely to avoid uncertain risks developing into damage, however uncertain.

Table 12.2 **The effects of the precautionary principle(s) on climate change liability**

Version	Choice of liability basis	Determining the standard of care	Time perspective	Available remedies	Limiting recourse to defences	Affecting the legal burden of proof*
1 (modest)	(–)	Due diligence = active information duties	1990/1992	Damages for present damage from historic emissions as of 1992 + injunction to reduce emissions	Development risk. Regulatory compliance, *force majeure*, risk–cost (+)	Modification: alleviation of evidence standards
2 (proactive)	(–)	Due diligence = take into account minority views and act accordingly	> 1980	Damages for present damage from historic emissions as of 1980 + adaptation costs + injunction for stricter emissions reductions	(+)	Shift: upon initial onus on the claimant, the burden shifts on the defendant
3 (risk-minimising)	(–)	Due diligence = follow minority views and strict counter-action	> 1960	Damages for present damage from historic emissions as of 1960 + adaptation costs + injunction for strictest emissions reductions	(+)	Reversal: full onus on the defendant
Procedural	n.a.	Due diligence = obligation to follow the procedural steps in decision-making	n.a.	Damages can be awarded for damage resulting from a breach of the procedural requirements	Regulatory compliance defence is allowed if procedural steps have been followed	(-) the claimant must prove breach of procedure

* Note that the burden of proof is to be distinguished into proof of negligence and proof of different aspects of causation (see text).

(+) effect; (–) no effect.

The precautionary principle requires decision-makers to consider all possible alternatives as well as all uncertainties resulting from an activity in question, in particular in view of inter- and intra-generational equity and justice considerations. The example of climate change has shown that if the precautionary principle has not been applied correctly in the decision-making phase this will trigger an application of the precautionary principle to questions of legal responsibility. In that respect, it facilitates liability claims based on a failure to apply the principle or on an incorrect application of the principle in the determined policy area.

To some extent, parallels to toxic tort cases can be drawn, where courts have started judging in a way that implicitly relates to requiring more precaution from tortfeasors. Yet, this implicit demand of precautionary behaviour is not to be equated with a consistent application of the precautionary principle. Essentially, there are three important arguments against such a simplification of the precautionary principle's contribution to liability law, which are of particular relevance to the case of climate change:

First, there is a difference in applying precautionary considerations on a case-by-case basis at the judge's discretion and a systematic inclusion of the precautionary principle as a judicial tool that prescribes these considerations. Only the latter will provide some certainty in cases entangled in uncertainty, for the benefit of the compensation-seeking victim, the future liability-evading behaviour of potential tortfeasors and the judicial guidance of judges in charge of such cases. Climate change, being a problem of unprecedented scale that requires international, European and national regulation in a comprehensive and cooperative effort, would certainly benefit from such a systematic application.

Second, most of the toxic tort cases arose after many years of inactivity, during which the uncertainty grew into such a degree of certainty that it was, if anything, beyond pure reparation, prevention rather than precaution that really triggered judicial action. By that time, immense damage had been incurred, and the inherent precautionary considerations led to compensation, but did not provide a real incentive in terms of the preventive and deterring effect as far as the behaviour of potential tortfeasors was concerned. In contrast to that, the application of the precautionary principle would allow courts to step in at a prior stage and would particularly enhance the preventive effect of liability law, as argued before. In the case of climate change this is particularly relevant since, contrary to toxic torts, which were tried retrospectively, most of the damage resulting from climate change is future-related.

Third, the chain of causation is longer in climate change than in any of the cases described before, thus making climate change liability more difficult to adjudicate than any of the other cases. Tobacco consumption may lead directly to specific types of disease such as lung cancer and cardiovascular diseases. These may have other causes as well, which makes it difficult to establish the causal link between the disease that has developed and tobacco consumption. Alternative causes could be genetic predisposition, exposure to alternative toxic substances, or, in view of all the potential dynamics that contribute to the human immune system that scientists do not know about, simple coincidence. Aside from asbestos (where mesothelioma and asbestosis can have almost no other origin), in most other toxic torts exposure to the substance is a major criterion for suffering from the illness that needs to be singled out from other potential causes. Consequently, the exclusion of background risk and other interfering factors is a difficult feature that is common to both toxic torts and climate change claims in some instances. Yet in the climate change scenario it is not the emissions that directly cause the damage in question; instead it is emissions that cause global warming, which is in turn responsible for climate change, leading to an increased likelihood of changed climatic and weather-related patterns, be it long term or in individual extremes. Thus, the chain of causation from the defendant's conduct to the individual damage is much longer and therefore more susceptible to the uncertain contribution of other factors. In view of this enhanced scientific uncertainty, the legal uncertainty courts are facing is much greater, which is why the application of the precautionary principle would bring about judicial solutions beyond what has been feasible and practised until now. This is not opening the floodgates of liability law beyond the reasonable but, instead, an adequate response to the challenge posed by a new type of litigation that has become necessary in view of the environmental, economic and societal consequences and threats on a previously unknown scale resulting from climate change.

The research has shown that applying the precautionary principle in its main interpretations has a varying effect on aspects of liability for climate change. In the most modest version, the precautionary principle contributes to putting the polluter pays principle into practice by providing the necessary framework for finding faulty behaviour and causality as well as the requirement of damage, albeit in the future. In the regulatory sphere, the most essential interpretation prohibits using scientific uncertainty as an excuse for inaction. In liability law, this requirement of enhanced foresight is to be applied retrospectively, when a defendant is at fault for not taking precautionary measures despite risks being uncertain. Moreover, when applied in its more stringent interpretations entailing an obligation, the precautionary principle could become a truly innovative normative principle of liability law obliging preventive and collective action in the field of uncertain environmental risks.

Judicial systems around the world seem to be characterised by the search for certainty and are marked by a concern to avoid uncertainty; that said, our risk society in general, and more particularly in view of climate change, should rather mandate judges to seek an improved approach in judicially handling uncertainties, since these will most likely never cease to exist in one way or another. It is particularly in the field of environmental damage that liability law is affected by such uncertainty. Its effectiveness in terms of compensation, reparation and prevention is enhanced when applying the precautionary principle, thereby replacing the certainty requirements set by traditional concepts of liability law with sufficiently grounded probabilities. These probabilities could then be constructed in a climate change liability claim along the lines of assessing uncertainty in the IPCC reports.

Here, the precautionary principle in its interpretative versions functions as a judicial tool in claims against direct emitters of GHGs responsible for climate change. It also functions partly as a normative principle whose violation can give rise to climate change liability claims against public authorities, should a lack or excess of precaution occur so far as the procedural precautionary principle is concerned.

In practice, the precautionary principle should be formulated on a case-by-case basis and depending on the scale of potential damage. Thus, following the distinction between the three interpretations of the precautionary principle, if the expected costs are much larger than the benefits, the precautionary principle should be interpreted in line with the stricter versions 2 and 3. In any case, the consequences of these findings are equally relevant to countries using a modest version of the principle as well as to explicit proponents of a stricter version.

Even though courts have not yet explicitly embraced uncertainty with the precautionary principle as a judicial tool, the precautionary principle's contribution to climate change liability can be considered of significant relevance to the path liability law is to take in the future at national, European and international levels.

Climate change is a prime example of an environmental problem with severe effects on human health and property burdened by uncertainties, which for a long time have been used, and often continue to be used, as an excuse to substantially combat the causes and consequences on behalf of governments and operators worldwide. The effects of the precautionary principle on climate change liability manifest themselves at national, European and international levels; therefore a cautious handling of the precautionary principle should be taken into account by decision-makers in government and private companies alike when addressing the question of financial distribution and the allocation of responsibility in view of actual and potential damage when deciding their future climate policy options for the post-Kyoto phase.

References

Ackerman, F., and L. Heinzerling (2004) *Priceless: On Knowing the Price of Everything and the Value of Nothing* (New York: New Press).

Arcuri, A. (2006) 'The Case for a Procedural Version of the Precautionary Principle Erring on the Side of Environmental Preservation', in M. Boyer, Y. Hiriart and D. Martimort (eds.), *Frontiers in the Economics of Environmental Regulation and Liability* (Aldershot, UK: Ashgate Publishing): 19-63.

Baer, P. (2006) 'Adaptation: Who Pays Whom?', in W. Adger (ed.), *Fairness in Adaptation to Climate Change* (Cambridge, MA: MIT Press): 131-54.

Baghestani-Perrey, L. (1999) 'Le principe de précaution: nouveau principe fundamental régissant les rapports entre le droit et la science', *Recueil Dalloz*, 29.10: 457-62.

Baudouin, J.-L., and P. Deslauriers (2003) *La responsabilité civile* (Montreal: Éditions Yvon Blais, 6th edn).

Beck, U. (1986) *Risikogesellschaft: Auf dem Weg in eine andere Moderne* (Bielefeld, Germany: Suhrkamp).

—— (2007) *Weltrisikogesellschaft* (Bielefeld, Germany: Suhrkamp).

Benoist, G. (2007) 'Climate Change Impacts on Personal Insurance', *Geneva Papers* 32: 16-21.

Boutonnet, M. (2005) *Le principe de précaution en droit de la responsabilité civile, collection bibliothèque de droit privé* (tome 444; Paris: Librairie Générale de Droit et de Jurisprudence).

Burns, W., and H. Osofsky (2009) 'Overview: The Exigencies That Drive Potential Causes of Action for Climate Change', in W. Burns and H. Osofsky (eds.), *Adjudicating Climate Change: State, National, and International Approaches* (Cambridge, UK: Cambridge University Press).

Christianson, G. (1999) *Greenhouse: The 200-Year Story of Global Warming* (London: Constable).

COMEST (World Commission on the Ethics of Scientific Knowledge and Technology) (2005) *The Precautionary Principle* (Paris: United Nations Educational, Scientific and Cultural Organization).

Cullet, P. (2007) 'Liability and Redress for Human-Induced Global Warming: Towards an International Regime', *Stanford Journal of International Law* 43A: 99-121.

—— (2008) 'The Global Warming Regime after 2012: Towards a New Focus on Equity, Vulnerability and Human Rights', *Economic and Political Weekly* 43.28: 109-17.

Culley, D. (2002) 'Global Warming, Sea Level Rise and Tort', *Ocean and Coastal Law Journal* 8.1: 91-125.

Dasgupta, P. (2007) 'A Challenge to Kyoto: Standard Cost–Benefit Analysis May Not Apply to the Economics of Climate Change', *Nature* 449 (13 September 2007).

De Sadeleer, N. (2002) *Environmental Principles* (Oxford, UK: Oxford University Press).

Everson, M., and E. Vos.(2009) 'The Scientification of Politics and the Politicisation of Science', in E. Vos and M. Everson (eds.), *Uncertain Risks Regulated* (Abingdon, UK: Routledge Cavendish): 1-18.

Farber, D. (2003) 'Probabilities Behaving Badly: Complexity Theory and the Environmental Uncertainty', *UC Davis Law Review* 37: 145-73.

Faulk, R. (2009) 'Lifting the Veil: Pressures Mount for Climate Change Disclosures', *Electricity Journal* 22.6: 59-67.

——, and J. Gray. (2009) 'A Lawyer's Look at the Science of Global Climate Change', *World Climate Change Report* (Bureau of National Affairs) 44: 2-17.

Faure, M., and G. Skogh (2003) *The Economic Analysis of Environmental Policy and Law: An Introduction* (Cheltenham, UK: Edward Elgar).

Fisher, E. (2009) 'Opening Pandora's Box: Contextualising the Precautionary Principle in the European Union', in E. Vos and M. Everson (eds.), *Uncertain Risks Regulated* (Abingdon, UK: Routledge Cavendish): 21-46.

Folland, C., T. Karl, J. Christy, R. Clarke, G. Gruza, J. Jouzel, M. Mann, J. Oerlemans, M. Salinger and S.-W. Wang (2001) 'Observed Climate Variability and Change', in J. Houghton, Y. Ding, D. Griggs, M. Noguer, P. van der Linden, X. Dai, K. Maskell and C. Johnson (eds.), *Climate Change 2001: The Scientific Basis. Contribution of Working Group I to the Third Assessment Report of the Intergovernmental Panel on Climate Change* (Cambridge, UK: Cambridge University Press): 99-181.

Goklany, I. (2000) 'Applying the Precautionary Principle in a Broader Context', in J. Morris. (ed.), *Rethinking Risk and the Precautionary Principle* (Oxford, UK: Butterworth Heinemann): 189-228.

Grossman, D. (2003) 'Warming up to a Not-So-Radical Idea: Tort-Based Climate Change Litigation', *Columbia Journal of Environmental Law* 28: 1-61.

Gupta, J. (2007) 'Legal Steps Outside the Climate Convention: Litigation as a Tool to Address Climate Change', *Review of European Community and International Environmental Law* 16.1: 76-86.

Harper, B. (2006) 'Climate Change Litigation: The Federal Common Law of Interstate Nuisance and Federalism Concerns', *Georgia Law Review* 40: 661-98.

Harremoës, P., D. Gee, M. MacGarvin, A. Stirling, J. Keys, B. Wynne and S. Guedes Vaz (2001) 'Late Lessons from Early Warnings: The Precautionary Principle 1896–2000 '(Environmental Issue Report 22; Copenhagen: European Environment Agency).

Hawker, M. (2007) 'Climate Change and the Global Insurance Industry', *Geneva Papers* 32: 22–28.

Healy, K., and J. Tapick (2004) 'Climate Change: It's Not Just a Policy Issue for Corporate Counsel—It's a Legal Problem', *Columbia Journal of Environmental Law* 29: 89-118.

IPCC (Intergovernmental Panel on Climate Change) (2007) *Climate Change 2007: Synthesis Report. Contribution of Working Groups I, II and III to the Fourth Assessment Report of the Intergovernmental Panel on Climate Change* (Geneva: IPCC).

Jacobs, R (2005) 'Treading Deep Waters: Substantive Law Issues in Tuvalu's Threat to Sue the Unites States in the International Court of Justice', *Pacific Rim Law and Policy Journal* 14.1: 103-28.

Khoury, L., and S. Smyth (2007) 'Reasonable Foreseeability and Liability In Relation to Genetically Modified Organisms', *Bulletin of Science, Technology and Society* 27.3: 215-32.

Kourilsky, P., and G. Viney (2000) *Le principe de précaution, rapport au premier ministre, la documentation française* (Paris: Editions Odile Jacob).

Kunreuther, H., and E. Michel-Kerjan (2007) 'Climate Change, Insurability of Large-Scale Disasters, and the Emerging Liability Challenge', *University of Pennsylvania Law Review* 155: 1,795-842.

Lascoumes, P. (1997) 'La précaution, un nouveau standard de jugement', *Esprit*, November 1997: 129-40.

Lefeber, R. (1996) *Transboundary Environmental Interference and the Origin of State Liability* (The Hague: Kluwer Law International).

Leone, J. (1999) 'Les O.G.M. à l'épreuve du principe de precaution', *Petites Affiches* 164: 12-16.

Löfstedt, R. (2002) 'The Precautionary Principle: Risk, Regulation and Politics', introductory paper for *21st Century Trust Conference*, Merton College, Oxford University, 5–13 April 2002; www.21stcenturytrust.org/precprin.htm, accessed 19 March 2010.

Mann, M., R. Bradley and M. Hughes (1998) 'Global-Scale Temperature Patterns and Climate Forcing over the Past Six Centuries', *Nature* 392 (23 April 1998): 779-87.

Müller, U., and D. Plehwe (2008) 'Nicht öffentlichkeitsfähig: Wissenschaft als Lobby-Instrument, Mythos Wissensgesellschaft: Verklärung oder Aufklärung?', *Forum Wissenschaft* 2 (15 May 2008).

Munich Re (2009) '13th International Liability Forum: Climate Change Litigation and Environmental Liability: Commonalities and Differences'; www.munichre.com/publications/302-06161_en.pdf, 23 April 2009, accessed 19 March 2010.

Nordhaus, W. (2007) 'The Challenge of Global Warming: Economic Models and Environmental Policy', Yale University, 24 July 2007; nordhaus.econ.yale.edu/dice_mss_072407_all.pdf, accessed 19 March 2010.

Osofsky, H. (2009) 'The Inuit Petition as a Bridge? Beyond Dialectics of Climate Change and Indigenous Peoples' Rights', in W. Burns and H. Osofsky (eds.), *Adjudicating Climate Change: State, National, and International Approaches* (Cambridge, UK: Cambridge University Press): 272-91.

Peel, J. (2005) *The Precautionary Principle in Practice: Environmental Decision-Making and Scientific Uncertainty* (Annandale, VA: Federation Press).

Radé, C. (2000) 'Le principe de précaution, une nouvelle éthique de la responsabilité?', *Revue Juridique de l'Environnement*, numéro special: 75-89.

Richardson, B. (2002) 'Mandating Environmental Liability Insurance', *Duke Environmental Law and Policy Forum* 12.2: 293-329.

Ross, C., E. Mills and S. Hecht (2007) 'Limiting Liability in the Greenhouse: Insurance Risk-Management Strategies in the Context of Global Climate Change', *Stanford Environmental Law Journal* 26A and *Stanford Journal of International Law* 43A.1: 251-317.

Sandin, P., M. Peterson, S. Hansson, C. Rudén and A. Juthe (2002) 'Five Charges against the Precautionary Principle', *Journal of Risk Research* 5.4: 287-99.

Sands, P. (2003) *Principles of International Environmental Law* (Cambridge, UK: Cambridge University Press, 2nd edn).

Schneider, S. (2002) 'Can We Estimate the Likelihood of Climatic Changes at 2100?', *Climatic Change* 52: 441-51.

—— and K. Kuntz-Duriseti (2002) 'Uncertainty and Climate Change Policy', in S. Schneider, A. Rosencranz and J. Niles (eds.), *Climate Change Policy: A Survey* (Washington, DC: Island Press): 53-87.

Scott, J. (2004) 'The Precautionary Principle before the European Courts', in R. Macrory (ed.), *Principles of European Environmental Law* (Groningen, Netherlands: Europa Law Publishing): 50-72.

Spier, J. (2006) 'Legal Aspects of Global Climate Change and Sustainable Development', *Revista Para el Análisis del Derecho* 2: 2-24.

—— (2008) 'Het WRR-rapport Onzekere veiligheid: een welkome stap voorwaarts', *Nederlands Juristenblad* 83 (14 November 2008): 2,521-25.

Stern, N. (2007) *The Economics of Climate Change: The Stern Review* (Cambridge, UK: Cambridge University Press).

Sugarman, S. (2006) 'Roles of Government in Compensating Disaster Victims', Berkeley Electronic Press; www.bepress.com/ils/iss10/art1, accessed 19 March 2010.

Tapinos, D. (2008) *Prévention, précaution et responsabilité civile: risque avéré, risque suspecté et transformation du paradigme de la responsabilté civile* (Paris: L'Harmattan).

Tinker, C. (1996) 'State Responsibility and the Precautionary Principle', in D. Freestone and E. Hey (eds.), *The Precautionary Principle in International Law* (The Hague: Kluwer Law International): 53-72.

Trudeau, H. (2003) 'Du droit international au droit interne: l'émergence du principe de precaution en droit de l'environnement', *Queen's Law Journal* 28: 455-527.

Underwood, A. (1999) 'Precautionary Principles Require Changes in Thinking about and Planning Environmental Sampling', in R. Harding and E. Fisher (eds.), *Perspectives on the Precautionary Principle* (Sydney: Federation Press): 254-66.

UNFCCC (United Nations Framework Convention on Climate Change) (2007) *Climate Change, Impacts, Vulnerabilities and Adaptation in Developing Countries* (Bonn: UNFCCC Secretariat; unfccc.int/resource/docs/publications/impacts.pdf, accessed 20 March 2010).

Union of Concerned Scientists (2007) 'Smoke, Mirrors and Hot Air: How ExxonMobil Uses Big Tobacco's Tactics to Manufacture Uncertainty on Climate Science'; www.ucsusa.org/assets/documents/global_warming/exxon_report.pdf, accessed 19 March 2010.

Van Asselt, M., and J. Rotmans (2002) 'Uncertainty in Integrated Assessment Modelling', *Climatic Change* 54.1–2: 75-105.
Van Boom, W., and A. Pinna (2008) 'Le droit de la responsabilité civile de demain en europe: questions choisies', in B. Winiger (ed.), *La responsabilité civile européenne de demain: Projets de révision nationaux et principes européens, Colloque international à l'Université de Genève* (Zürich: Schulthess): 261-77.
Van den Bergh, R., and L. Visscher (2008) 'The Preventive Function of Collective Actions for Damages in Consumer Law', *Erasmus Law Review* 1.2: 5-30.
Van Langendonck, J. (2007) 'International Social Insurance for Natural Disasters', in W. van Boom and M. Faure (eds.), *Shifts in Compensation between Private and Public Systems* (Vienna/New York: Springer): 181-98.
Weart, S. (2008) *The Discovery of Global Warming* (Cambridge, MA: Harvard University Press, rev. edn).
Wibisana, A. (2008) 'Law and Economics Analysis of the Precautionary Principle' (Doctoral thesis defended 25 April 2008, Maastricht University, Netherlands).
Zacharias, C. (2009) 'Climate Change Is Heating Up D&O Liability', *John Liner Review* 23.1 (Spring 2009): 7-14.
Zander, J. (2009) 'Different Kinds of Precaution: A Comparative Analysis of the Application of the Precautionary Principle in Five Different Legal Orders' (Doctoral thesis defended 17 April 2009, Maastricht University, Netherlands).

13

Incentives for international cooperation on adaptation and mitigation

Rob Dellink, Kelly de Bruin and Ekko van Ierland
Environmental Economics and Natural Resources Group,
Wageningen University and Research Centre (WUR), The Netherlands

Global warming poses one of the biggest global environmental threats for current and, to a greater extent, future generations. Greenhouse gas (GHG) emissions increase the temperature and the variability of the climate, causing damage (IPCC 2007). Both GHG mitigation and adaptation to the impacts of climate change are essential for effective and efficient climate policy. From a biophysical perspective the near-term impacts of climate change are already 'locked in', irrespective of the stringency of mitigation efforts, thus making adaptation inevitable. Meanwhile, the magnitude and rate of climate change will probably exceed the capacity of many systems and societies to adapt, making mitigation inevitable (IPCC 2007: ch. 18). Though adaptation and mitigation are substitutes, from an economic perspective, implementing both adaptation and mitigation will minimise the total social costs of climate change. This is based on the assumption that while initial levels of both mitigation and adaptation can be achieved at low cost in relation to the avoided climate damage, both sets of responses will face progressively rising marginal costs. Therefore, an optimal climate policy would require a mix of both mitigation and adaptation measures, as opposed to solely one or the other.

To understand the full effects of climate change and the interactions between climate change and the economy, integrated assessment models (IAMs) with their comprehensive and internally consistent representation of dynamic interactions of the climate, environment and the economy can be particularly useful to policy-makers. Typically, they provide insights on costs and benefits of climate change policies, on their optimal timing and the optimal sharing of their burden. In general they can guide policy decisions through the multiple trade-offs triggered by climate change damages and policies: for example, more action against climate change today may decrease economic growth in the short run, but stimulate it in the future. The right balance needs to be found between mitigating climate change, adapting to climate change and accepting (future) climate change damages. IAMs provide a consistent framework to investigate these issues.

Virtually all existing IAMs focus on the trade-off between damages due to climate change and the costs of mitigation. Adaptation, however, is either ignored or only treated implicitly as part of the damage estimate (Tol and Fankhauser 1998).[1] This means that adaptation is not modelled as a decision variable that can be controlled exogenously, but a certain level of adaptation is assumed when calculating damages. Adaptation has not been explicitly examined in IAMs for several reasons. First, adaptation is often seen as primarily a private choice (Tol 2005). Second, adaptation options are difficult to quantify and compare with each other. Adaptation takes many forms and does not have a clear common performance indicator, such as mitigation has (Lecocq and Shalizi 2007). Finally, while there is extensive literature on both top-down and bottom-up estimates of the costs of mitigation, the literature on adaptation costs and benefits is scarce (Agrawala and Fankhauser 2008).

Systematic inclusion of adaptation in IAMs can further strengthen the policy relevance of IAMs. The inclusion of an explicit adaptation function can allow the formulation of different policy scenarios where adaptation is in fact a decision variable, in combination with emission reduction. Furthermore, the effect of adaptation restrictions, that is, barriers or constraints resulting in suboptimal levels of adaptation, can be identified. Moreover, as mentioned above, both adaptation and mitigation can be used to limit the final effects of climate change. The substitution between the two is, however, limited as adaptation can limit damage that mitigation no longer can and mitigation can limit damage that is too large to adapt to. An IAM can give insights into both the theoretical and empirical optimal levels of both these controls and their substitutability.

Recently, adaptation has been included in an IAM framework (de Bruin *et al.* 2009a, 2009b, 2009c) using the DICE and RICE models (Nordhaus 2007; Nordhaus and Boyer 2000) to create the AD-DICE and AD-RICE models. In these models the adaptation costs and benefits are made explicit and represented by an adaptation cost curve. This curve shows the costs of various levels of adaptation, where adaptation is measured as the fraction of initial gross damages reduced. These models have been used to understand the role of adaptation in various settings. In this chapter we utilise the AD-RICE model to study the issue of cooperation in climate change. We look at different forms of cooperation and the incentives of various

1 The situation has not evolved much since this review.

regions to cooperate. We study the role of an adaptation fund in cooperation and consider how suboptimal adaptation may affect incentives to cooperate.

This chapter is structured as follows. Section 13.1 describes the AD-RICE model we use in this chapter. Section 13.2 describes the calibration of the adaptation cost curves. The third section describes the results and the final section concludes.

13.1 AD-RICE model

The AD-RICE model incorporates adaptation into the RICE model,[2] using the method as introduced in de Bruin *et al.* (2009a, 2009b, 2009c): that is, by calibrating an adaptation cost curve that describes the marginal costs of adaptation efforts, analogous to a marginal abatement cost curve as often used for mitigation efforts. RICE is a regional version of the Dynamic Integrated Climate and Economy (DICE) model. A full model description is given in Nordhaus (2007); here we present only a brief overview. The RICE model is a Ramsey-type growth model and includes a simplified system of equations to describe economic activity and economic growth as well as a system of equations that describes the climate system; both systems are linked: economic activities influence the climate, and climate change influences the economy. In this model, utility, calculated as the discounted natural logarithm of consumption, is maximised. Consumer goods are produced by combining inputs of capital, labour and energy and cause emissions of CO_2. In each time-period, consumption and savings/investment are endogenously chosen subject to available income reduced by the costs of climate change (residual damage, adaptation costs and mitigation costs). Climate change damage is represented by a damage function that depends on the temperature increase compared to 1900 levels. The RICE model does not have an explicit mitigation variable but mitigation is incorporated implicitly in specification of the carbon energy input. AD-RICE adds adaptation as a control variable to RICE as described in detail in de Bruin *et al.* (2009b, 2009c) and briefly explained here. We define gross damage as the initial damage by climate change if no changes were to be made in social and economic systems. If these systems were to adapt to limit climate change damage, the damage would be lower. This 'left-over' damage is referred to here as residual damage. Reducing gross damage, however, comes at a cost: that is, the investment of resources in adaptation. These costs are referred to as adaptation costs. In the RICE model some regions experience net benefits of low levels of climate change. This does not mean, however, that adaptation becomes irrelevant or ineffective: one can adapt to improving circumstances as well, in order to maximise the benefits from climate change. For example, in cold regions new land can be cultivated for agriculture. To incorporate this phenomenon, adapting to net benefits of climate change is also introduced.

2 We use the RICE-99 model available online: nordhaus.econ.yale.edu.

Adaptation costs are assumed as strictly positive, whether it is to amplify climate benefits or to reduce damage.

Furthermore, the net damage in the RICE model is assumed to be the optimal mix of adaptation costs and residual damage. Thus the net damage function given in RICE is unravelled into residual damage and adaptation costs in AD-RICE, yielding the level of adaptation as a new policy variable. Adaptation (P) is given on a scale from 0 to 1, where 0 represents no adaptation: none of the gross climate change damage is decreased through adaptation. A value of 1 would mean that all gross climate change damage is avoided through adaptation. If, however, damage is negative, that is, there are gross benefits from climate change, adaptation can be used to increase these benefits. This type of adaptation (PR) will thus increase the gross benefits into residual benefits. Here a value of 1 would mean that gross benefits double and a value of 0 would mean gross benefits do not increase. Naturally a region either has gross benefits or gross damage from climate change or will apply only PR or P, respectively.

Thus, the AD-RICE model contains all equations of the original RICE model, except the equation for damage. This single equation is replaced by a system of three equations that disentangle damage and adaptation. First, adaptation (denoted by P when damage is positive and by PR when damage is negative) is expressed as the fraction by which gross damage (GD) is reduced to the level of residual damage (RD) or gross benefits are increased to residual benefits. Second, adaptation costs (PC) depend on the level of adaptation. Third, gross damage as a percentage of output (Y) is a function of temperature increase (TE), as in the original RICE model. Using t to denote the time-period and j to denote the region, we have the following revised climate impact module:

$$RD_{t,j} = GD_{t,j} \cdot (1 - P_{t,j}) + PR_{t,j} \cdot GD_{t,j} \qquad (1)$$

$$PC_{t,j} = \gamma_{1,j} P_{t,j}^{\gamma_{2,j}} + \gamma_{1,j} PR_{t,j}^{\gamma_{2,j}} \qquad (2)$$

$$\frac{GD_{j,t}}{Y_t} = \alpha_{j,1} TE_t + \alpha_{j,2} TE_t^{\alpha_{j,3}} \qquad (3)$$

Where $0 \leq P_{t,j} \leq 1$ and either or equals zero in each period for each region.

The AD-RICE model is calibrated such that it best replicates the results of the optimal control scenario of the original RICE model when adaptation is assumed to be at its optimal level throughout the model horizon. To this end regional adaptation cost curves are constructed so that the discounted squared difference between net damage (D_{tj}) in the original RICE and net damage ($RD_{tj} + PC_{tj}$) in AD-RICE is minimised.

13.2 Data and calibration of adaptation costs curve in AD-RICE

In this section we describe the empirical foundation and calibration of the adaptation cost curves in the AD-RICE model. For a full description of the data and calibration method as applied to AD-DICE and AD-RICE see de Bruin and Dellink (forthcoming); here we only provide a summary. We use the regional and sectoral breakdown of damage as used in the RICE model (Nordhaus and Boyer 2000; Nordhaus 2007) to provide a more detailed estimation of the adaptation variables. The world is disaggregated into 13 regions: Japan, USA, Europe,[3] Other High Income countries (OHI),[4] High Income Oil exporting regions (HIO),[5] Middle Income countries (MI),[6] Russia, Low-Middle income countries (LMI),[7] Eastern Europe (EE), Low Income countries (LI),[8] China, India and Africa.[9]

In this calibration procedure we separate the AD-RICE-99 estimates of damage into residual damage and adaptation costs to create the AD-RICE model. To calibrate our model we assess each impact category described in the RICE model and use the relevant literature, supplemented with expert judgement where necessary, to estimate the (optimal) levels of the relevant adaptation variables (adaptation, adaptation costs, residual damage and gross damage). We thus assess each impact category, consider the main adaptation possibilities in that category and assess the related costs and benefits for each region. The RICE damage function is calibrated at the point where global atmospheric temperature has increased by 2.5 degrees compared to the 1900 level. We calibrated our damage function to replicate the RICE-99 damage function over the model horizon in the optimal case.

Very few empirical studies have focused on estimating the costs and benefits of adaptation. Furthermore, the few studies that do exist often focus on specific local adaptation options, because the costs and benefits of adaptation are often very location specific. Agrawala and Fankhauser (2008) give an excellent overview of the current literature available on adaptation costs and benefits. We have drawn strongly on this literature in our analysis and gathered other literature where possible.

The Nordhaus and Boyer damage function contains seven damage categories: Agriculture, Coastal, Health, Settlements, Other vulnerable markets, Non-market

3 Includes Austria, Belgium, Denmark, Finland, France, Germany, Greece, Greenland, Iceland, Ireland, Italy, Liechtenstein, Luxembourg, the Netherlands, Norway, Portugal, Spain, Sweden, Switzerland and the United Kingdom.
4 Includes Australia, Canada, New Zealand, Singapore, Israel and rich island states.
5 Includes Bahrain, Brunei, Kuwait, Libya, Oman, Qatar, Saudi Arabia and UAE.
6 Includes Argentina, Brazil, Korea and Malaysia.
7 Includes Mexico, South Africa, Thailand, most Latin American states and many Caribbean states.
8 Includes Egypt, Indonesia, Iraq, Pakistan and many Asian states.
9 Includes all sub-Saharan African countries, except Namibia and South Africa.

time use and Catastrophic. The regional damage estimates for these seven categories are reproduced in Table 13.1. The empirically estimated levels of optimal adaptation (P) and the ratio of residual damages to protection costs (RD/PC) are also given in Table 13.1.

The 'Agriculture' category refers to damage in the agricultural sector due to climate change. The damage estimates are based on studies done on crop yield variation under different temperatures and precipitation. The damage is assessed assuming that crop production will be adjusted to the new climate. To estimate the adaptation in this sector we use regional estimates by Tan and Shibasaki (2003) and Rosenzweig and Parry (1994). Tan and Shibasaki estimate damage/benefits with and without adaptation for several world regions, but they only consider low-cost adaptation measures. Rosenzweig also looks at more substantial adaptation options in developing and developed regions.

The 'Coastal' category refers to damage due to sea-level rise. As the climate warms, the level of the sea rises. Adaptation options considered consist of either building sea walls to protect against sea-level rise (incurring protection costs) or accepting the land loss (incurring residual damage). Nordhaus and Boyer (2000) use US estimates and extrapolate them based on a coastal vulnerability index (the coastal area to total land area ratio). To estimate the adaptation in this sector we use the FUND model (Tol 2007), which directly gives both the optimal protection level as costs and benefits of adaptation for more than 200 countries in the world. As can be seen in Table 13.1, the adaptation potential in this sector is high.

The 'Health' category refers to all damage incurred owing to malaria, dengue, tropical diseases and pollution. Although heat- and cold-related deaths are also affected by climate change, they are not included here. In general, regions that are already vulnerable to such diseases have substantial damage, and thus developing regions tend to have the greatest damage in this category. To estimate adaptation in this sector we use the study by Murray and Lopez (1996) on which Nordhaus and Boyer base their estimates. This study assumes a level of adaptation based on general improvements in healthcare and related domains. We also use data from the malaria report 2008 (WHO 2008), which estimates the use of mosquito nets in various vulnerable regions.

The next category is 'Settlements'. This is a willingness to pay (WTP) analysis. They estimate the WTP to climate proof, that is fully protect against climate change, certain highly climate sensitive settlements. They estimate that countries are on average willing to pay 2% of the output of a climate-sensitive region to protect them from the effects of climate change. Furthermore, they estimate the willingness to pay to protect vulnerable ecosystems. Estimates of adaptation in this sector are based on expert judgment and Nordhaus and Boyer (2000).

The 'Other vulnerable markets' (OVM) category refers to the effect of climate change on other markets. Nordhaus and Boyer conclude that the only significantly affected markets are energy and water. More energy will be needed in some regions for air conditioning whereas colder regions will need less energy for heating. Water use is also expected to increase, for example, owing to increased irrigation needs.

Table 13.1 **RICE impacts per sector (net damage in % of GDP at 2.5°C climate change), estimates of optimal adaptation (P: fraction of gross damage reduced), and ratio of residual damage to adaptation costs (RD/PC) from empirical data**

	USA	China	Japan	EU	India	HIO	MI	LMI	Africa	LI
Agriculture										
Net damage	0.06	-0.37	-0.46	0.49	1.08	0	1.13	0.04	0.05	0.04
P	0.58	0.63	0.43	0.48	0.63	0.48	0.58	0.655	0.53	0.58
RD/PC	0.89	1.14	1.14	0.89	0.89	n.a.	0.89	0.89	0.89	0.89
Coastal										
Net damage	0.11	0.07	0.56	0.6	0.09	0.06	0.04	0.09	0.02	0.09
P	0.97	0.95	0.95	0.59	0.95	0.98	0.90	0.95	0.99	0.95
RD/PC	1.86	0.02	0.40	0.43	0.00	0.66	0.45	0.28	0.02	0.02
Health										
Net damage	0.02	0.09	0.02	0.02	0.69	0.23	0.32	0.32	3	0.66
P	0.60	0.41	0.42	0.60	0.36	0.82	0.78	0.50	0.35	0.40
RD/PC	0.70	0.72	0.73	0.70	0.77	0.85	0.89	0.77	0.79	0.76
Settlements										
Net damage	0.1	0.05	0.25	0.25	0.1	0.05	0.1	0.1	0.1	0.1
P	0.75	0.75	0.75	0.75	0.75	0.75	0.75	0.75	0.75	0.75
RD/PC	0.43	0.43	0.43	0.43	0.43	0.43	0.43	0.43	0.43	0.43
Non-market time use										
Net damage	-0.28	-0.26	-0.31	-0.43	0.3	0.24	-0.04	-0.04	0.25	0.2
P	0.9	0.7	0.9	0.9	0.3	0.3	0.7	0.7	0.3	0.3
RD/PC	9	9	9	9	9	9	9	9	9	9
Other vulnerable markets										
Net damage	0	0.13	0	0	0.4	0.91	0.41	0.29	0.09	0.46
P	0.8	0.6	0.8	0.8	0.6	0.6	0.7	0.6	0.5	0.5
RD/PC	0.24	0.25	0.24	0.24	0.25	0.24	0.24	0.25	0.25	0.25
Catastrophic										
Net damage	0.44	0.52	0.45	1.91	2.27	0.46	0.47	1.01	0.39	1.09
P	0.2	0.2	0.2	0.2	0.2	0.2	0.2	0.2	0.2	0.2
RD/PC	0.95	0.95	0.95	0.95	0.95	0.95	0.95	0.95	0.95	0.95

Nordhaus and Boyer estimate this damage based on US data that is then extrapolated using the average temperature effects in the other regions. To estimate our adaptation variables in this sector we use expert judgement. We assume that dryer, hotter regions will have more trouble adapting. Furthermore, developing regions may lack the infrastructure to adapt.

'Non-market time use' is a more abstract category and refers to the change in leisure activities. Owing to a change in climate, people's leisure hours will be affected. In colder regions, a warmer climate will lead to extra enjoyment of outdoor leisure activities. In warmer climates, however, leisure activities will be more restricted if the amount of extremely hot days increases. Table 13.1 shows that most regions have benefits in this category. In this sector we rely on expert judgement to estimate the adaptation variables. Most of the impacts in this category will be adaptation costs, as people will adapt their leisure activities to fit the new climate. Thus the net costs or benefits in this category are mostly changes in adaptation costs and not residual damage or benefits.

The 'Catastrophic' category refers to the WTP to avoid catastrophic events. Nordhaus and Boyer define a catastrophic event as an event that destroys 30% or more of a region's GDP. They quantify the associated expected damage by estimating a risk premium, that is, they do not actually quantify the damage of catastrophic events but rather use the concept of insurance premium to value these effects. This does not include 'minor' catastrophic events such as extreme weather events. Estimates of adaptation in this sector are based on expert judgement and Nordhaus and Boyer (2000). As the events considered in this category are large catastrophes, the potential to reduce their effects through adaptation is minimal.

Table 13.2 shows the resulting calibrated values of the gross damage function and adaptation cost function in AD-RICE.10 Three regions (Eastern Europe, Other High Income regions and Russia) have been excluded from the calibration procedure for the adaptation cost curves as they have very low, near zero, net benefits from climate change. For these regions we assume that no adaptation will take place as the impacts are so close to zero.

Adaptation costs curves are drawn for the remaining ten regions in Figure 13.1, where the x-axis shows the level of adaptation as fraction of gross damage reduced and the y-axis shows the associated costs as a fraction of output. As can be seen, the adaptation costs in the different regions vary widely. India, Africa and Low Income countries in particular have high adaptation costs, whereas Japan, China and the USA have relatively low adaptation costs.

10 Note that the form of the adaptation cost curve is also influenced by the form of the DICE damage function.

Table 13.2 **Calibrated parameter values for AD-RICE model**

	Parameters for gross damage			Parameters for adaptation costs	
	α_1	α_2	α_3	γ_1	γ_2
USA	-0.0015	0.0008	2.9216	0.0267	6.1853
CHINA	-0.0033	0.0012	2.3863	0.0038	1.5023
JAPAN	-0.0026	0.0009	2.9804	0.0080	2.9831
EUROPE		0.0038	2.4405	0.0534	2.2105
INDIA	-0.0518	0.0625	1.2229	0.0895	2.1360
HIO		0.0051	1.7011	0.0229	1.5000
MI	-0.6814	0.6872	1.0100	0.0334	3.3891
LMI		0.0044	1.9903	0.0322	1.8540
AFRICA		0.0182	1.1937	0.1427	2.4492
LI		0.0082	1.6391	0.0419	1.5905
GLOBAL	0.0012	0.0015	2.6787	0.0565	3.3869

Figure 13.1 **Calibrated adaptation cost curves: adaptation costs (expressed as a fraction of output) as a function of adaptation (expressed as fraction of gross damage reduced) for AD-RICE regions**

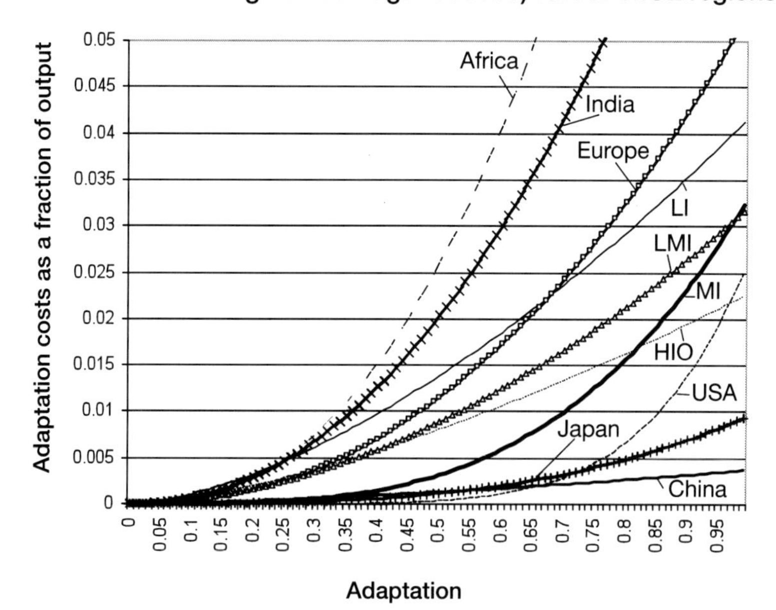

13.3 Results

In this section we discuss the results obtained in our analysis. We first introduce and explain the policy scenarios we will use in the analyses. We then discuss several results, namely the optimal emissions paths, the composition of climate change costs and the gains from cooperation for the different policy scenarios.

13.3.1 Policy scenarios

The focus of this chapter is to study the incentives of regions to cooperate on climate change policies. To this end, we use the AD-RICE model to study the three cooperation scenarios. To understand the differences between these scenarios we first define the social welfare function.

In our model regional utility (U_j) is derived from the discounted sum of consumption per capita

$$c_{jt} = \frac{C_{jt}}{L_{jt}}$$

in each period over the planning horizon T. Here L_{jt} denotes the population size. In our model, output (Y) equals (national) income and regions' consumption (C_{jt}) is equal to their output minus all costs related to climate change (residual damage, RD) and climate policy (adaptation costs, PC, and mitigation costs, MC) as follows:

$$C_{j,t} = Y_{j,t} - RD_{j,t} - PC_{j,t} - MC_{j,t}. \tag{4}$$

Consumption per capita is discounted over time using a discount factor (Ξ_t):

$$U_j = \sum_{t=1}^{T} \rho_t \cdot \left(c_{j,t}\right). \tag{5}$$

The regional utilities are weighted to create a global 'social welfare' (SW):

$$SW = \nu_j \sum_{j=1}^{J} U_j. \tag{6}$$

Maximising SW involves choosing optimal values for the mitigation and adaptation levels and adaptation transfers. This will be explained in more detail further on. The three cooperation scenarios are as follows:

1. **Nash: no cooperation**. In the uncooperative Nash case each region optimises its own utility taking the emissions of the other regions as given. The social welfare function weights are then the inverse of marginal utility of consumption per capita, that is,

$$\nu_j = \frac{1}{C_j}$$

This implies that the existing income distribution is taken as given, and thus yields the competitive solution as optimum (Negishi 1960). In this case the shadow prices of capital in all regions are equalised and monetary transfers will, other things being equal, not increase social welfare: any welfare increase from shifting one unit of income to one (host) region will be matched exactly by an equivalent welfare loss in the donor region.

2. **Climate cooperation.** Differences in climate change impacts are considered. In this case, the social welfare function weights are given by the inverse marginal utility of income before climate change damage is subtracted, that is,

$$\nu_j = \frac{1}{Y_j}$$

In this case shadow prices are equalised when there is no damage from climate change.[11] Monetary transfers will thus only be desirable from a social welfare perspective if damage among regions is unequal: compensation for damage in a high-impact region by a low-impact region will boost global welfare.

3. **Full cooperation.** Differences in income are considered. In the case of full cooperation, all regions have the same welfare weight, that is,

$$\nu_j = \frac{1}{J}$$

where J denotes the number of regions. In this case summed utility of all regions is maximised, shadow prices will only be equalised across countries if consumption levels are equal across countries and monetary transfers will increase social welfare if they flow from a high-income region to a low-income region (as marginal utility decreases in income levels).

For each of these scenarios, we can implement an adaptation fund, where adaptation in one region can be financed by another region. De Bruin *et al.* (2009c) investigate the role of international financing of adaptation in detail. They show that adaptation transfers will fully crowd out domestic adaptation expenditures when adaptation and mitigation are set at their optimal levels, that is, all extra foreign investments in adaptation will decrease domestic adaptation by the same amount. But this is not necessarily the case when domestic adaptation is suboptimal. Furthermore transfers will run from low-impact to high-impact regions in the case of climate cooperation and from rich to poor regions in the case of full cooperation. We study what effects an adaptation fund can have on cooperation in the different scenarios. When an adaptation fund is in place financial transfers can take place from one region to another. We assume that the receiving host region can only use these funds for additional adaptation purposes, that is, it cannot decrease its level of adaptation expenditures, so there is no crowding out. In our setting the regions first set their optimal domestic level of adaptation expenditures after which transfers can take place. Adaptation transfers give regions a new tool to cooperate. Adaptation transfers can be used to compensate regions with high damage from climate change. Transfers, however, also introduce a means of transferring funds

11 This is because this weight will result in the competitive equilibrium in the case of no damage from climate change.

from one region to another, which was not possible because of the assumption of no trade in the model. We will discuss this further in the next section.

The discount rate is one of the most important and contentious aspects of cost–benefit analysis of climate policies. Our base model assumes a small but positive pure rate of time preference (at an initial level of 1.5%), in line with the DICE and RICE models (Nordhaus 2007). For each form of cooperation we also investigate the effects of using a pure rate of time preference of zero, that is, all generations get equal weights; this is roughly in line with the assumptions used in the Stern review (2005). Changes in the discount rate affect cooperation over time as opposed to across regions, as a higher discount rate gives smaller weight on future generations compared to current generations.

This leads to the following set of scenarios:

- A1a Nash, with positive time preference
- A2a climate cooperation, with positive time preference
- A3a climate cooperation with adaptation transfers, with positive time preference
- A4a full cooperation, with positive time preference
- A5a full cooperation with adaptation transfers, with positive time preference

- A1b Nash with zero time preference
- A2b climate cooperation with zero time preference
- A3b climate cooperation with adaptation transfers and zero time preference
- A4b full cooperation with zero time preference
- A5b full cooperation with adaptation transfers and zero time preference

None of our cooperation scenarios fully reflects what is likely to happen in the real world. They do, however, reflect the four main motivations behind the behaviour of regions when cooperating on climate change. First, regions naturally are concerned about their own well-being: this is reflected in the Nash scenario. Second, when regions cooperate on climate change, they will want to compensate those most affected by climate change: this is reflected in the climate cooperation scenario. Third, regions also consider the level of income in regions when cooperating. Regions generally have some, though perhaps low, incentives to compensate low-income regions. This is reflected in the full cooperation scenario. Finally the pure rate of time preference reflects people's concern for future generations. In a real world context, motivations will lie somewhere in between these extremes and will probably also depend on ethical considerations, including a region's historical responsibility in contributing to the climate change problem (see Dellink *et al.* 2009, for a systematic analysis of international burden-sharing arrangements based on equity considerations).

13.3.2 Optimal emission paths over time

International cooperation on climate policies serves two objectives: increasing the environmental effectiveness and reducing the associated costs. Figure 13.2 shows the impact of cooperation on the optimal emission paths as assessed by AD-RICE. Although all scenarios are represented in the figures, the impact of the financial transfers on emissions in the case of climate cooperation is negligible, and thus largely overlaps with the corresponding scenario without transfers.

Figure 13.2a shows the results for the scenarios with the default discount assumption of the RICE model: that is, with a positive pure rate of time preference. Figrate of time preference. For the base case—the Nash scenario with positive time preference—emission levels are projected to increase over time. Mitigation levels are positive, but as no international cooperation takes place, there is no incentive for countries to jointly stabilise emissions. This clearly illustrates the overwhelming power of the free-rider incentives that underlie climate policy. Mitigation is a public good, and all regions benefit from the mitigation efforts of other regions. Mitigation costs, however, are borne domestically. If no compensation payments are available, it is well known that regional mitigation levels will be quite limited (Barrett 1994; Nagashima *et al.* 2009; Nordhaus 2007). Climate cooperation, either without (scenario A2a) or with financial transfers for adaptation (A3a), induces substantially higher mitigation levels: as it is in the global interest to reduce damage in those regions that are hurt most by climate change, ambitious mitigation policies are employed, and a clear reversal in the emission trend is visible. Allowing financial transfers for adaptation leads to slightly lower levels of mitigation in the case of climate cooperation. This is because funds can be transferred to compensate for climate change damage, as a substitute for mitigation. In particular, regions with low damage and high mitigation costs can benefit from using adaptation transfers as a compensation mechanism.

In the case of full cooperation the level of mitigation is high, especially in the next century. Full cooperation (scenarios A4a and A5a) invokes a different motive for mitigation: here any policy that transfers income from the rich to the poor regions is deemed beneficial. As time passes, however, the difference between incomes in regions decreases (this reflects the usual assumption of convergence). To some extent, income equalisation can be achieved through increases in mitigation levels. But differences in climate costs across regions are only a small element in the income difference, and there is no one-on-one correspondence between income levels and damage from climate change. Mitigation policies are thus a blunt instrument to achieve a redistribution of wealth. When transfers are possible (as assumed in scenario A5a), emissions are much higher as compared to the same specification without transfers (scenario A4a), as adaptation transfers can be used instead of mitigation to compensate income differences. Consequently, global emissions are substantially lower than in the Nash case, but a radical break with past trends is not foreseen.

Figure 13.2a **Emissions per decade for different cooperation scenarios and positive time preference**

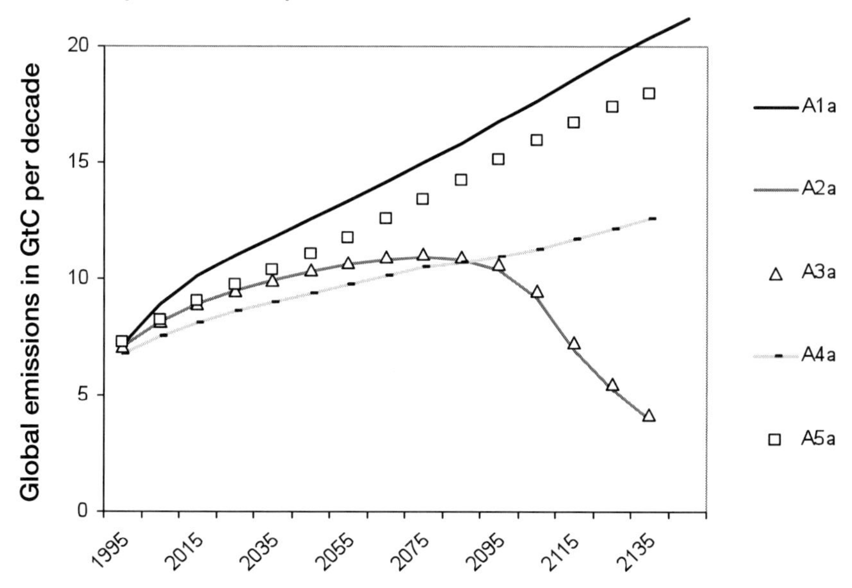

Figure 13.2b **Emissions for different cooperation scenarios with zero time preference**

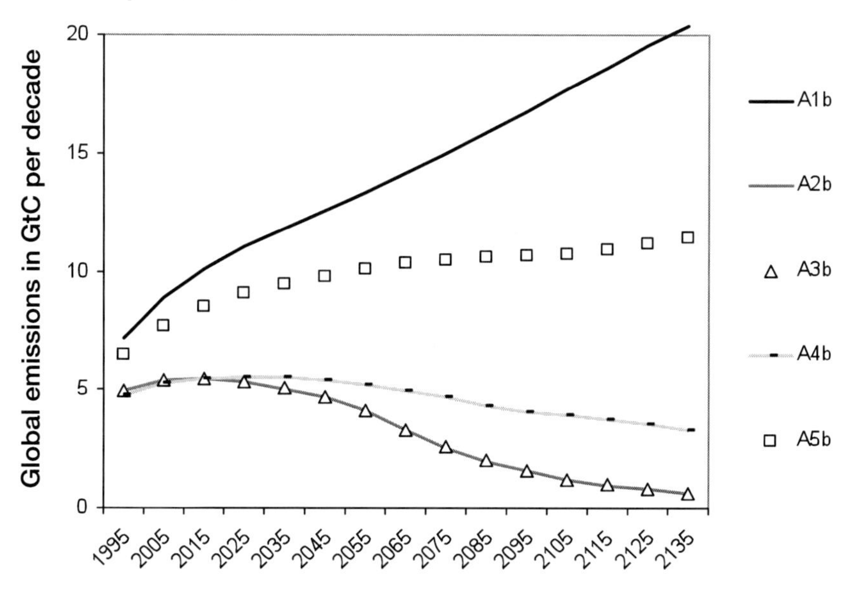

Note: The levels for 1995 are different under the different scenarios as this is the base year of analysis in the model. Climate policy will invoke a re-evaluation of the entire consumption path, including current consumption. Fixing the level of consumption in this starting period will not drastically alter the results.

The situation is somewhat different for the case with zero pure rate of time preference (Fig. 13.2b). The effectiveness of mitigation is larger in this case, as future benefits from mitigation have a larger present-value weight. Optimal emission levels are lower for all scenarios under this discounting assumption, but the relative ordering of the scenarios, with the most ambitious mitigation policies adopted in the case of climate cooperation, remains the same as in the case with positive time preference.

13.3.3 Build-up of climate change costs

In our setting, climate change costs consist of two elements: residual damage and adaptation costs. The model adopts a two-stage cost–benefit analysis to find the optimal climate policies: first, optimal mitigation levels are determined to equate the marginal costs of mitigation with the marginal avoided climate change costs. Second, optimal adaptation levels are set such that the resulting climate change costs are minimal, given the level of GHG concentrations. This involves a trade-off between marginal adaptation costs and marginal avoided damage. As marginal costs of mitigation and adaptation are increasing with the ambition level of the policy, it will not be optimal to avoid all damages. Rather, an optimal mix of mitigation costs, adaptation costs and residual damage emerges.

Figure 13.3a shows the development of adaptation costs over time for the scenarios with positive time preference. As expected, adaptation costs increase over time, as damage from climate change becomes more pronounced. The Nash scenarios (A1a and A1b), which provided the weakest incentives for mitigation, also boost adaptation. The climate cooperation scenarios show much lower levels of adaptation costs. The reason is that the much more ambitious mitigation policy in the long run reduces damage, so much less adaptation is required. The full cooperation scenarios (A4a and A5a) have a similar time profile, again linked to the changes in optimal mitigation policies. This illustrates that while adaptation and mitigation policies may not have very strong direct links, they do interact through the competing claims on income, and through their impact on the social welfare function.

The alternative assumption of zero time preference (see Fig. 13.3b) moves the balance between the effectiveness of adaptation and mitigation policies in favour of more mitigation. Other things being equal, this would imply lower levels of adaptation. But the present value of residual damage is higher, and the Nash scenario (A1b) does not imply an immediate stabilisation of GHG concentrations, so in this setting there is room to expand adaptation as well, although not as much as mitigation. In the cooperation scenarios (A2b–A5b), adaptation levels are lower than in the setting with positive time preference, and the role of adaptation in the optimal climate policy mix remains limited. This does not imply that adaptation is irrelevant in these scenarios. The avoided damage from adaptation is still quite substantial, even though the associated adaptation costs are relatively small.

We now examine how the composition and level of climate change costs differ over the scenarios. In Figure 13.4, the net present value (NPV) of adaptation costs

Figure 13.3a **Global adaptation costs as a percentage of output for the scenarios with positive time preference**

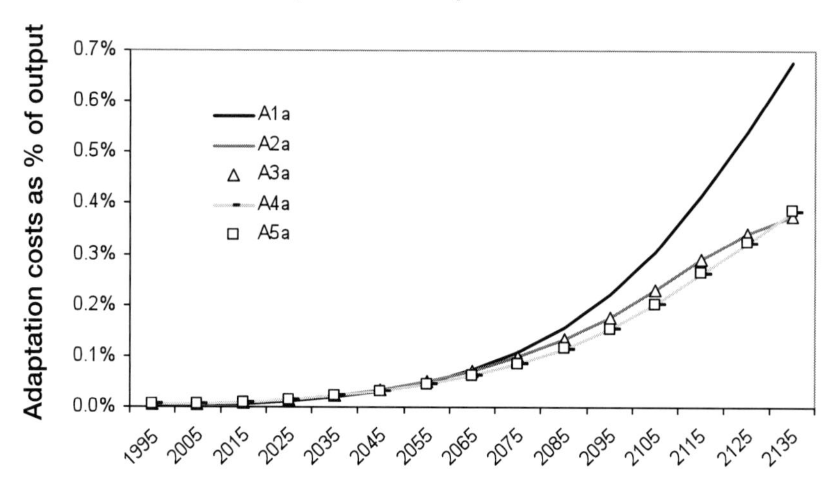

Figure 13.3b **Global adaptation costs as a percentage of output for the scenarios with zero time preference**

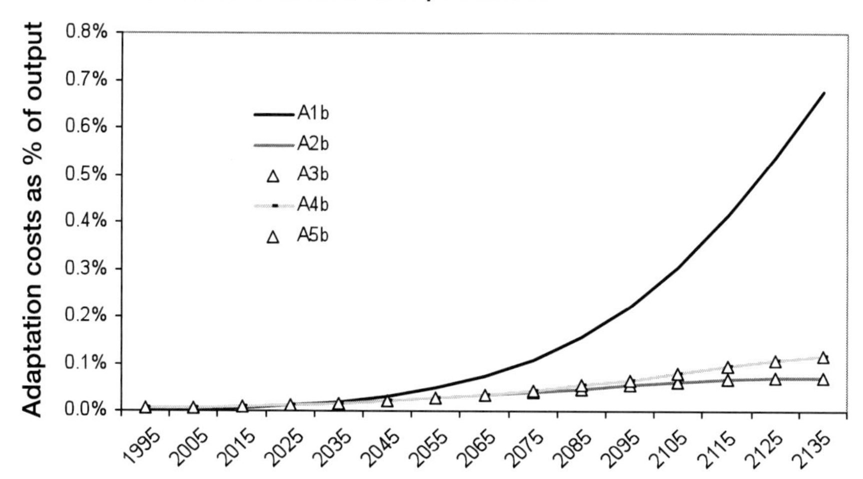

Figure 13.4a **Climate change cost components in NPV US$ trillion for the different scenarios with positive time preference**

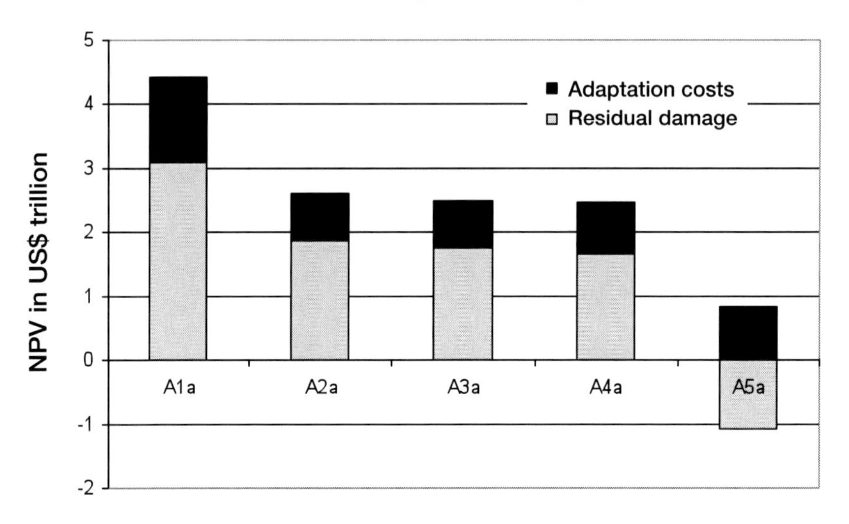

Note: Mitigation costs cannot be calculated as they are not explicit in the RICE model.

Figure 13.4b **Climate change cost components in NPV US$ trillion for the different scenarios with zero time preference**

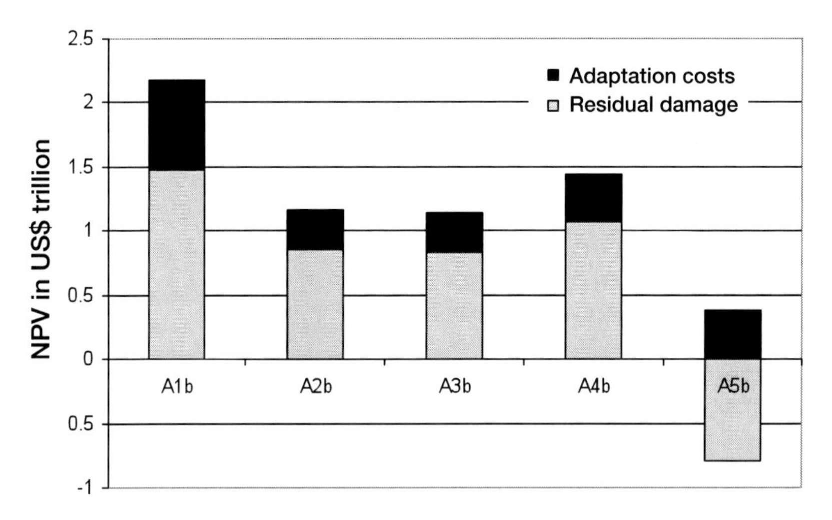

and residual damage are given, where, as usual, Figure 13.4a shows the scenarios with a positive rate of time preference and Figure 13.4b shows the scenarios with a zero rate of time preference. Figure 13.4a, shows that, as expected, climate change costs are highest in the Nash scenario (A1a). Where regions do not cooperate on mitigation, they will set lower levels of mitigation and higher levels of adaptation. The cooperation scenarios both lower residual damage and adaptation costs. Comparing full cooperation to climate cooperation, we see that in the case of full cooperation adaptation costs are slightly higher and residual damage slightly lower. This is because (as seen in Fig. 13.2) full cooperation leads to higher levels of mitigation (and thus lower emissions) in the first century. In later time-periods, however, mitigation decreases. Furthermore, in the full cooperation scenario, income differences among regions are minimised. This has a negative impact on global income, as inefficient methods must be used to decrease the income differences between regions. Richer regions will invest more in mitigation to compensate poorer regions and decrease their damage from climate change. Thus in the first century mitigation is high, resulting in small temperature changes and lower residual damage. In later periods, however, less mitigation is undertaken, and adaptation efforts will be increased.

Another interesting result is that in the case of full cooperation with financial transfers (scenario A5a), on balance, global residual damage becomes negative, that is, the net residual effects are positive. In this scenario regions have two ways to equalise income (mitigation and adaptation), and the optimal level of mitigation will decrease and the level of adaptation will increase compared to the scenario without transfers. According to the calibration on the RICE model, and hence also of our AD-RICE model, damage is negative in some regions for low levels of temperature increase; that is, certain regions experience residual benefits from a limited change in the climate system. For instance, this is assumed to be the case for agriculture in the temperate regional zones. Remember that adaptation can also be used to enhance climate change benefits. Thus, we see that while damage remains in all regions in later time-periods, and in some regions also in the short term, these are outweighed by the NPV of these short-term benefits after adaptation.

Figure 13.4b shows the case for zero social time preference. We see the same general results as in Figure 13.4a concerning the ratio of residual damage to adaptation costs and the relative costs across scenarios. However, the actual level of costs involved is much lower than in the case of positive time preference. This is because of the more stringent mitigation policies in the case of zero time preference: this dampens both damage and adaptation costs.

13.3.4 Gains from cooperation

One can also look at the different scenarios from the perspective of incentives to cooperate. The Nash scenario provides a benchmark where all regions act in their own interest. In contrast, the cooperation scenarios reflect some sort of joint welfare maximisation. We can investigate the regional gains from cooperation to shed light on the incentives of regions to behave cooperatively in the area of international climate policy. A full analysis of the strategic effects of climate policies is clearly beyond the scope of this chapter. We refer the reader to Barrett (1994) and Nagashima *et al.* (2009) for an introduction to this literature, which adopts a game-theoretic perspective, and to Zehaie (2009) and de Bruin *et al.* (forthcoming) for the first few available applications that consider adaptation explicitly. Here, we merely compare consumption levels across the different scenarios, in order to see which regions benefit from cooperation, and which lose. To this end, we calculate the gains from cooperation as the increase in the net present value of regional consumption in the respective scenarios, compared to the Nash scenario. For ease of comparison, these are expressed in percentage change from the value in the Nash scenario.

Table 13.3a, illustrates that climate cooperation can be beneficial for all regions combined (see row 'Global') but may not be beneficial for individual regions. There are sizable gains that can be procured under cooperation that are not attainable in the Nash scenario due to the strong free-rider incentives; these gains are mostly reaped by the non-OECD regions, especially Russia, China and India.

Table 13.3a **Gains from cooperation for scenarios with positive time preference, as a percentage change from the corresponding Nash scenario**

	A2a	A3a	A4a	A5a
JAPAN	-0.1%	0.0%	-0.2%	-8.8%
USA	-0.1%	-0.1%	-0.5%	-1.2%
EUROPE	0.0%	0.0%	-0.1%	0.0%
OHI	0.1%	0.1%	-0.2%	0.1%
HIO	-0.4%	-0.3%	-0.4%	0.3%
MI	0.3%	0.4%	0.0%	0.2%
RUSSIA	0.7%	0.0%	0.4%	2.6%
LMI	0.4%	0.4%	0.1%	0.4%
EE	0.6%	0.6%	0.0%	1.0%
LI	0.3%	0.4%	0.2%	1.6%
CHINA	0.8%	0.8%	0.0%	0.2%
INDIA	0.9%	1.0%	0.4%	2.6%
AFRICA	0.2%	0.3%	0.3%	2.8%
Global	0.2%	0.2%	-0.1%	-0.8%

Table 13.3b **Gains from cooperation for scenarios with zero time preference as a percentage change from the corresponding Nash scenario**

	A2b	A3b	A4b	A5b
JAPAN	0.2%	0.2%	-2.3%	-11.1%
USA	-0.3%	-0.3%	-5.3%	-2.0%
EUROPE	0.5%	0.5%	-1.7%	-0.1%
OHI	-0.4%	-0.4%	-3.6%	-1.0%
HIO	-4.4%	-4.4%	-4.9%	-0.5%
MI	-1.7%	-1.7%	-1.2%	0.1%
RUSSIA	-8.4%	-8.5%	-5.5%	1.0%
LMI	-3.3%	-3.2%	-1.1%	0.5%
EE	-6.3%	-6.3%	-3.6%	0.2%
LI	-4.2%	-4.2%	0.5%	2.0%
CHINA	-6.9%	-6.9%	-1.5%	0.2%
INDIA	-4.3%	-4.3%	1.1%	3.3%
AFRICA	-3.0%	-3.0%	1.3%	3.4%
Global	-1.6%	-1.6%	-2.5%	-1.3%

The role of adaptation transfers is quite limited in the case of climate cooperation. Table 13.3 confirms this (compare scenarios A2a and A3a): the transfers cannot convert any region that loses from cooperation into a net beneficiary. Note however that these adaptation transfers are not designed to maximise the incentives to cooperate; as Nagashima *et al.* (2009) show, specifically designed transfer schemes can increase the incentives to cooperate substantially.

As expected, the full cooperation case generates a more substantial redistribution of wealth: most OECD countries are worse off than under the Nash scenario, whereas most non-OECD regions are better off. The negative global figure also illustrates that there are costs involved in using climate policy as a blunt instrument for redistributing income across regions. The full cooperation with transfers case, finally, implies a sizable shift in consumption away from adaptation donor regions Japan and USA to the less-developed regions. While this is optimal in the setting of this scenario, with a social welfare function that gives equal weight to all countries, it puts a downward pressure on global consumption levels.

Perhaps surprisingly, in the scenarios of climate cooperation with a zero rate of time preference the main gains from cooperation are found in Japan and Europe. This can be explained by looking at the regional incentive structure of these regions: they have relatively high absolute damage levels, partly due to the high level of income, partly due to the coastal nature of large parts of these economies. Furthermore, it is well known that the marginal mitigation costs are high in these regions. Thus, they have a large incentive to stimulate an ambitious climate policy, but they do not have the means to combat climate change alone. The low discount rate makes this phenomenon more pronounced. The other regions have no incentive to cooperate, however, as they would be substantially worse off from cooperating. These results are in line with the observations in Nagashima *et al.* 2009.

13.4 Conclusion

Climate change requires a combination of mitigation and adaptation strategies. So far, many IAMs have focused on mitigation policies under the assumption that adequate adaptation would occur, without making adaptation costs or benefits explicit. In this chapter we describe how adaptation can be made explicit in a model with various regions of the world. To this end we create a model that is an extended version of RICE, namely AD-RICE. We use this model to investigate the effects and incentives to cooperate in different cooperation scenarios. We investigate a Nash scenario, where all regions act solely in their own self-interest. We also include a climate scenario where regions want to minimise the differences in regional climate change costs. Finally we consider a scenario where regions want to minimise the differences in income (full cooperation). The different cooperation scenarios create very different results, though it is clear that substantial welfare improvements can be made by means of both forms of cooperation. Mitigation levels increase and adaptation levels decrease in the case of cooperation, as mitigation yields a global benefit through reduced levels of climate change, whereas adaptation has purely regional benefits.

If international cooperation fails, however, it will be necessary in the long run to adjust to substantial climate change and related impacts. This will demand stringent adaptation measures in both industrialised and developing countries, and the costs of adaptation will be high.

This chapter also investigates the role that discounting plays in cooperation, mitigation and adaptation decisions. Lowering the rate of time preference, a main component of the discount rate, increases the weight of future damage. This significantly affects the levels of mitigation and relative shares of expenditures on adaptation and mitigation. Concurrently a zero rate of time preference increases the level of mitigation and decreases the level of adaptation.

In this chapter an adaptation fund is also considered, that is, regions can transfer funds for the purpose of adaptation in other regions. With both forms of cooperation, adaptation transfers will be used to increase social welfare. In the case of climate cooperation, adding adaptation transfers has little effect. In this case, compensation for climate change damage as the motivation behind cooperation and mitigation plays a more prominent role as a means to cooperate. In the full cooperation scenario, however, adaptation transfers prove to be a powerful new way to transfer income across regions. When compensating for income differences, adaptation transfers are very useful to compensate regions. Adaptation levels increase while mitigation levels decrease. Thus the motivation behind cooperation will also affect the best choice of policy instrument.

Both for mitigation and for adaptation the question arises of who will carry the burden. For mitigation it is useful to apply the polluter pays principle as a starting point, as recognised by the OECD. For adaptation, the issue of international financing is complex, and one of the most controversial topics of discussion in the ongoing climate negotiations. By explicitly considering adaptation costs in IAMs it now becomes possible to analyse in which regions the costs of adaptation need to be made under various mitigation scenarios, and this provides a starting point for the discussion on how to share the burden of adaptation.

References

Agrawala, S., and S. Fankhauser (2008) *Economics Aspects of Adaptation to Climate Change: Costs, Benefits and Policy Instrument* (Paris: Organisation for Economic Cooperation and Development).

Barrett, S. (1994) 'Self-enforcing International Environmental Agreements', *Oxford Economic Papers* 46: 878-94.

De Bruin, K.C., and R.B Dellink (forthcoming) 'The Marginal Costs of Adaptation: Empirical Calibration of Adaptation Cost Curves of the AD-RICE and AD-DICE Models' (Wageningen, Netherlands: Wageningen University, mimeo).

——, R.B Dellink and R.S.J. Tol (2009a) 'AD-DICE: An Implementation of Adaptation in the DICE Model', *Climatic Change* 95: 63-81.

——, R.B Dellink and S. Agrawala (2009b) 'Economic Aspects of Adaptation to Climate Change: Integrated Assessment Modelling of Adaptation Costs and Benefits' (OECD Environment Working Paper 6; Paris: Organisation for Economic Cooperation and Development).

——, R.B Dellink and R.S.J. Tol (2009c) 'International Cooperation on Climate Change Adaptation from an Economic Perspective' (ESRI Working Paper 323; Dublin: Economic and Social Research Institute).

——, R.B Dellink and H.P. Weikard (forthcoming) 'The Role of Adaptation in Coalition Formation', (Wageningen, Netherlands: Wageningen University, mimeo).

Dellink, R.B., M. den Elzen, H. Aiking, E. Bergsma, F. Berkhout, T. Dekker and J. Gupta (2009) 'Sharing the Burden of Financing Adaptation to Climate Change', *Global Environmental Change* 19: 411-21.

IPCC (Intergovernmental Panel on Climate Change) (2007) *Fourth Assessment Report* (Geneva: IPCC).

Lecocq, F., and Z. Shalizi (2007) 'Balancing Expenditures on Mitigation and Adaptation to Climate Change: An Exploration of Issues Relevant for Developing Countries' (World Bank Policy Research Working Paper 4299; Washington, DC: World Bank).

Murray, C.J.L., and A.D. Lopez (1996) *The Global Burden of Disease* (Geneva: World Health Organization).

Nagashima, M., R.B. Dellink, E.C. van Ierland and H.P. Weikard (2009) 'Stability of International Climate Coalitions: A Comparison of Transfer Schemes', *Ecological Economics* 68: 1,476-87.

Negishi, T. (1960) 'Welfare Economics and Existence of an Equilibrium for a Competitive Economy', *Metroeconomics* 12: 92-97.

Nordhaus, W.D., and J. Boyer (2000) *Warming the World: Economic Models of Global Warming* (Cambridge, MA: MIT Press).

—— (2007) *A Question of Balance* (New Haven, CT: Yale University Press).

Rosenzweig, C., and M.L. Parry (1994) 'Potential Impact of Climate Change on World Food Supply', *Nature* 367: 133-38.

Stern, N. (2007) *The Economics of Climate Change: The Stern Review* (Cambridge, UK: Cambridge University Press).

Tan, G., and R. Shibasaki (2003) 'Global Estimation of Crop Productivity and the Impacts of Global Warming by GIS and EPIC Integration', *Ecological Modelling* 168: 357-70.

Tol, R.S.J. (2005) 'Adaptation and Mitigation: Trade-offs in Substance and Methods', *Environmental Science and Policy* 8.6: 572-78.

—— (2007) 'The Double Trade-off between Adaptation and Mitigation for Sea Level Rise: An Application of FUND', *Mitigation and Adaptation Strategies for Global Change* 1: 741-53.

—— and S. Fankhauser (1998) 'On the Representation of Impact in Integrated Assessment Models of Climate Change', *Environmental Modelling and Assessment* 3: 63-74.

WHO (World Health Organization) (2008) *World Malaria Report 2008* (WHO/HTM/GMP/2008.1; Geneva: WHO; www.who.int/malaria/wmr2008/malaria2008.pdf, accessed 22 March 2010).

Zehaie, F. (2009) 'The Timing and Strategic Role of Self-protection', *Environmental and Resource Economics* 44.3: 337-50.

14

Imagining the unimaginable

Synthesis of essays on abrupt and extreme climate change

Darryn McEvoy, Chiung Ting Chang and Pim Martens
ICIS, University of Maastricht, The Netherlands

14.1 Introduction

Climate change is likely to be one of the greatest threats facing societies in the coming decades. In response, considerable international effort—most notably represented by the Kyoto Protocol—has gone into policies that will contribute to the reduction of emissions of greenhouse gases. However, it is increasingly recognised that communities also need to be planning for those climate impacts that are unavoidable even with concerted mitigation efforts.

So what do we, as societies, need to be adapting to? Climate change impacts are commonly understood as long-term changes to weather parameters, for example, the predicted increase in average summer temperatures that will affect European countries later this century, which will undoubtedly have implications for socio-economic systems. However, while these trends are important to understand and plan for, it is the likely pace and intensity of climatic changes that will have the most significant social and economic consequences. Synthesising scientific output since 2007, Steffen (2009) notes that not only is the rapidity of change of concern but evidence suggests long-term feedback processes are starting to develop with potential consequences for abrupt and irreversible changes that will ultimately drive climate-related impacts. The implications of more substantial sideswipes to human activity brought about by such low-probability, uncertain, though potentially high-impact, events have received far less attention to date. Examples of such 'tipping elements' are destabilisation of the Indian and West African monsoon system, major dieback of the Amazonian forest, melting of the Greenland ice sheet, collapse of the West Antarctic ice sheet, and disruption of the Atlantic thermohaline circulation (THC) (Lenton *et al.* 2008).

Recognising the potential importance of these low-probability, high-impact events, the VAM programme committee reserved some money for an explorative study of the societal effects of abrupt forms of climate change. The context for these explorations could have been provided by any of the tipping elements mentioned, but one of them stood out in terms of its immediate interest for Western Europe and the Netherlands, where the VAM programme was funded: the potential disruption of the THC. This THC is driven by differences in temperature and salinity/density and acts as a conveyor belt that transports warmer waters northwards to the maritime regions of Western Europe. Without this naturally occurring phenomenon, parts of Europe would be much colder than is currently the case. In case of a slowdown or shutdown of the THC, society would need to adapt to a period of rapid cooling, with potentially catastrophic environmental, social and economic consequences.

Five essays by academics from across the Netherlands were commissioned to consider the implications of abrupt and extreme climate change (NWO 2008). The diverse collection of essays covers the implications of a shutdown or slowdown of the THC according to a range of different disciplinary perspectives and socio-economic contexts. Faced with high levels of uncertainty, the challenge for each of the authors was to think 'outside the box', drawing on their expert knowledge and opinion in order to explore some of the potential socioeconomic consequences of a low-probability, high-impact event and to highlight some of the likely implications for future society. The five papers covered a selection of legal (Bruggeman and Peeters, Law Faculty, Maastricht University), institutional (Van Koppen, Mol and Van Tatenhove, Environmental Policy Group, Wageningen University), sectoral (Amelung, Huynen and Martens, ICIS, Maastricht University) and economic (Ierland, Environmental Economics and Natural Resources Group, Wageningen University) perspectives of abrupt and extreme climate change, as well as making use of historical and non-weather analogues to explore some of the implications across multiple sectors (Aerts, Smith and Bouwer, IVM, Free University). This chapter reflects on each of the individual essays, synthesises their key findings, before finally elaborating on what their key messages mean for both policy and practice.

A note of caution: as was made clear by several of the essays, a comprehensive understanding of the range of climate risks that communities may face in the future also needs to take account of complex socio-ecological interactions, sometimes referred to as coupled human-environment systems. Different elements at risk can be subject to an array of multiple stressors and feedback processes, not only the effects of a changing climate. For instance, coupling the increase in the occurrence and intensity of extreme events with higher asset values (and greater exposure due to the growing number of people living in areas at risk from weather-related events), it is highly likely that the economic and social costs arising from extreme events will be considerable without some form of planned intervention.

14.2 Legal perspective: exploring governmental and victim responsibilities in view of abrupt climate change in north-western Europe

The opening paper by Bruggeman and Peeters reflects on the complex matter of organisational and individual competencies to adapt to abrupt change, and as a consequence which actors should be politically and legally responsible for the impacts of a climate-related extreme event. At the heart of their argument is a discussion of the extent to which government has a leading role to play, particularly given the uncertainties surrounding future climate risks, either directly through public intervention or indirectly by enacting policy that helps to stimulate action in the private sector; in essence arguing that a public-led approach is likely to prove to be the optimal option. This message resonates with the voices of financial institutions, who have argued that governments have the ultimate responsibility to manage the risks of climate change for society in the long term (Association of British Insurers 2005), especially when risks are beyond normal commercial parameters and there is a need for the provision of a safety net (UNEP Finance Initiative 2006).

Embedding their analysis in a Dutch context, the authors note that while Article 21 of the Dutch Constitution was enacted in 1983 to address social and environmental goals there is no specific article that yet deals with climate change, and while Article 21 could be (re)interpreted as being applicable to the emerging adaptation agenda they also suggest that it has limited 'teeth' when it comes to enforceability. This leads them to the conclusion that the Dutch Constitution implies a requirement to develop and implement an adaptation policy on behalf of government, though with some degree of discretion and a current lack of clarity with regards to enforceability.

Their commentary then turns to the different forms of intervention that are potentially available in the public arsenal. These are differentiated according to whether they are direct or indirect instruments; with planning and fiscal measures cited as examples of direct intervention. However, it is suggested that avoidance of top-down regulation is traditionally preferred by the Dutch Government when effective self-regulation can be stimulated by other non-regulatory measures. The authors also highlight the influential role of supranational legislation in many important areas of spatial planning and environmental protection and a pressing need for multi-level integration of climate policy. Interestingly, they raise the question of how well the current focus on 'climate proofing' development in the Netherlands will perform should there be a collapse in the THC, and as a result a colder climate than is currently being prepared for. These arguments provide stark illustration of the need to ensure flexibility of both policy and practical responses, in order to avoid situations where maladaptation may occur, even if meant with the best of intentions. Against this background, a concept of 'abrupt climate change proof' is introduced.

While the authors place significant emphasis on the public sector for preparing, at least strategically, for an abrupt change in climate, they also argue for adequate accompanying checks and controls to be in place. These checks are recommended not only to reduce the risk of overreaction on the part of any government response, but also to address potential liability. This is clearly an emerging issue as trends would suggest that damage resulting from climate change impacts is a 'legally cognisable injury', particularly if government actors are made aware of the specific risks associated with abrupt climate change. However, the conclusions of this essay would suggest that by following a well-considered adaptation policy public bodies are not likely to be found liable for the damages incurred by affected parties.

The latter part of their paper then went on to address possible financial mechanisms that could potentially be used to support adaptation objectives from the viewpoint of potential victims; again analysing the role of government in supporting this activity. The first mechanism considered was that of liability law. In this case, it was argued that this is likely to prove a very expensive option for compensating for damage incurred, though if imposed it could potentially provide a significant incentive to change undesirable behaviour. However, in their deliberations, the authors raise a very tricky question that permeates across the adaptation agenda—who pays and who benefits? The second mechanism considered was the role of the private sector in providing insurance, that is, compensating for economic damage post event. In this instance the authors lay out some of the main arguments for and against the insurability of climate risks. However, the discussion in this part of the paper focuses predominantly on the provision of insurance by the private sector, and what is less clear in the discussion is that variations affect insurance coverage across Europe at the current time. These differences can be categorised according to three main attributes: geographical differences relating to exposure to climate-related hazards, the role played by public and private actors (the type of set-up influencing whether premiums are provided *ex ante* or *ex post*—the Netherlands operates an institutionalised *ex post* system) and market penetration (for a more detailed discussion of the potential role of insurance, see Bouwer *et al.* 2007). The role of other financial mechanisms, commonly known as alternative risk transfer, was also scoped out for their potential to contribute to the adaptation agenda, with specific attention paid to the examples of catastrophe bonds and weather derivatives.

In the penultimate section on financial mechanisms the discussion returns to the vexed issue of the role of government in not only promoting but also ensuring the long-term sustainability of adaptation measures. It is here that the authors reflect on the tensions between the public and private sectors, with three potential categories of options spotlighted. These include government-run compensation funds (recognising that it is ultimately the taxpayer that will 'foot the bill'), public intervention in private insurance markets (concluding that there is great divergence in opinion on the effectiveness and desirability of such action), and the emergence of new public–private partnerships (seen as offering considerable potential for not only enhancing the effectiveness of insurance and other financial mechanisms but also accessibility to funds).

In their concluding section on observations and recommendations the authors set out a framework for further legal research. This framework include four governmental responsibilities: the need for a clearly defined task regarding monitoring and research; the need for flexible policies, including avoiding maladaptation and malregulation; a right to be informed about climate information; and disaster protection as a governmental task. Three citizen's responsibilities are highlighted: victim responsibility is difficult to encourage through liability, innovative *ex ante* compensation approaches and supporting market solutions as a key governmental adaptation measure.

14.3 Institutional perspective: coping with extreme events—institutional flocking

Drawing on the concept of a 'risk society', the second paper of the NWO portfolio by Van Koppen, Mol and Van Tatenhove considers rapid cooling as experienced by urban and rural areas, as well as for the water sector (from the viewpoint of safety and security). Each of the three examples, taken as being representative of vital areas of Dutch environmental policy and planning, are subject to a structured analysis that considers risk and shifting distributions under changing conditions, as well as its management by existing and potentially new institutional structures.

Their opening discussion on urban infrastructure emphasises the large-scale and centralised nature of much of the infrastructure contained within Dutch cities. Extreme cold events, such as snow and ice, could have significant consequences not only for different forms of physical infrastructure but also the 'organisation, operationalisation and management' of the system in question. Drawing from past weather events across the world the authors suggest that ensuring flexibility in systems can add to their robustness and resilience, as indicated by the comparison of flexible management institutions and good emergency planning versus rigid, highly centralised management and associated emergency responses. In their view what is needed in order to be better prepared for extreme weather events are systems that can be more easily coupled and decoupled, that is, units that not only function in larger-scale network-bound systems but which are also capable of operating in stand-alone mode. This, it is argued, would help to reduce the vulnerability of critical infrastructure in the urban system. In the long term, driven by the increasing need to adapt to extreme events, the authors also propose that a shift in governance processes from a centralised system mode of thinking to a more multi-level and multi-actor provision of utilities in the urban environment will ultimately evolve. For example, while public bodies will continue to have key responsibilities for the provision of infrastructure the authors foresee a future where civil society organisations are likely to play an ever-increasing role.

The paper's introduction to rural land-use planning emphasises the complex governance arrangements that underlie the formal Dutch land-use planning system. Complexity is evidenced by tensions between national and provincial authorities over different land-use issues, moves towards greater horizontal integration with increased involvement of private and civil actors in decision-making processes, and the growing influence of high-level policy emanating from the level of the EU. Turning to the likely impacts of extreme events, it is stressed that farmers and those dependent on the land for their livelihoods are likely to bear the brunt of any rapid change to colder climatic conditions, and increased conflict between different land uses is a likely consequence. However, it is noted that new challenges will be context specific, with uncertain and unpredictable changes to the geographical distribution of climate risk. As such, the authors point towards the need for multi-scalar exchanges of knowledge and innovative experience through new forms of knowledge networks, and new ways of working between different sets of actors involved in the rural landscape, including policy, scientific and wider stakeholder communities, in order to build consensus and contribute to the building of local adaptive capacity.

Protection against sea and riverine flooding provides the substance of the third narrative. The discussion opens by highlighting a shift in the reliance on technological or 'hard engineering' approaches to flood protection, that is, the building of dykes, which have traditionally dominated the Dutch response, to one where there is more importance placed on 'living with water'. This reflects an acknowledgement that it is not always possible, or may not be desirable, to rely solely on engineering solutions to reduce climate risks. This new discourse reflects a change in emphasis, one driven by increasing levels of uncertainty and the new threats posed by heightened frequency and intensity of extreme events, and a movement towards being open to responses more closely aligned with natural processes. Here again, the authors highlight the institutional challenges, multi-actor and multi-scale, which will need to be faced with abrupt climate change, concluding that the 'preparedness' of Dutch society will not only be dependent on the evolution of new flexible institutions but also that these need to invoke trust in the population at large that they are doing 'the right thing'.

14.4 Sectoral perspective: catching a cold or enjoying the breeze? Exploring the linkages between THC collapse, human health and tourism

The essay by Amelung, Huynen and Martens considers the implications of abrupt change for both human health and tourism. In the first instance, they suggest that a decrease in average temperature in the case of a THC collapse will have two distinct effects on temperature-related mortality. Mortality due to heat stress decreases as the number of days with higher temperatures declines. At the same time, however, cold-related mortality increases. It is highly likely that the net effect of these two mechanisms will be negative in the temperate region affected. Here, cold-related mortality is already higher than heat-related mortality. The result of their scenario analysis, using the cooling-then-warming scenario, shows the 3°C cooling is expected to result in about 5.5 thousand additional temperature-related deaths per year up to 2030. However, some vector-borne, air-borne and water-borne health problems may be reduced in this cooling scenario between 2015 and 2030. For instance, the risk of Lyme infection from tick bites will most likely decrease, as their season for survival will become significantly shorter, and outdoor recreation will become less popular as a result of the cooler weather. The pollen season will start considerably later. Effects on the duration of the pollen season, the amount of pollen produced and the geographical distribution of flowering plants are likely, but still poorly understood.

The post-2030 period will see a bounce back in temperatures, hence a reduction in cold-related deaths. As a result of the rapid temperature increase in this period, acclimatisation and adaptation may prove difficult. It may, however, have greater relevance after 2050, when the scenario reconverges with the gradual warming scenario.

Addressing the second case study, tourism, the authors contend that the suitability of north-western Europe for leisure activity will deteriorate as a result of the abrupt drop in temperature. In addition, the length of the summer tourism season will shorten, and even at its height the frequency of very good days will greatly diminish. It is likely that large parts of north-western Europe will enjoy only one month (or even no months) with very good weather. For the traditional Mediterranean holiday destinations, which are thought to have a bleak prospect as a result of climate change, this may be some good news. The period of cooling experienced elsewhere may allow them additional time to adapt to the inevitable prospect of global warming.

For winter sports in the mountains and for skating, a collapse of the THC will open up new opportunities. For example, it will reverse the trend of declining frequencies of Elfstedentocht skating events in the Netherlands. In the period between 2020 and 2040, when the cold period is at its peak, conditions suitable for the Elfstedentocht event are likely to occur in most winters. Perhaps as many as five or six events might actually be organised in that period. After 2040, as winter temperatures quickly rise again, the prospects of frequent Elfstedentocht events quickly disappear again.

14.5 Multi-sectoral perspective: abrupt climate change—analogies in impacts and coping mechanisms

This essay by Aerts, Smith and Bouwer addresses the issue of abrupt climate change from a different angle from others in the NWO portfolio. In addition to a shutdown of the THC, the authors consider three additional weather and non-weather related events that could potentially lead to rapid cooling in Europe, using this wider breadth of knowledge in an attempt to better understand some of the potential socioeconomic implications and to begin to explore what adaptation options are likely to be necessary to cope with a rapid change to the climatic regime. The four events considered were: the Little Ice Age, a nuclear winter, cooling from aerosols and a shutdown of the THC.

The first example considered, the Little Ice Age, was included in their assessment due to it being the only abrupt climate event that has been experienced in industrial times. It is estimated to have lasted from the beginning of the 14th to the middle of the 19th century, with greatest impact occurring between 1550 and 1700. Significantly colder winters and more erratic summers were representative of this period with particularly adverse impacts on the agricultural sector and knock-on implications for food availability. Written records highlighted a fluctuation in the price of grain, for example. Further climate-related impacts such as storms and sea floods also resulted in large numbers of deaths in the Netherlands and Germany during this period. Hypothesising future cooling events, the authors also considered the potential and extreme impact associated with cooling that would occur in the aftermath of nuclear conflict. This cooling would be a direct consequence of the blanketing of the atmosphere with particulate matter, effectively blocking the penetration of sunlight. This represents the most extreme case of cooling with estimates suggesting a possible drop in temperature of between 10° and 15°C in Western Europe for the first year at least. Dealing with the issue of aerosols—as represented by volcanic activity—historical evidence from major events would suggest that a drop in 5°C is a reasonable assumption for the likely scale of impact; though this would be on a more localised basis. Finally, the authors put some figures to the potential reduction in temperature that would accompany a overturning of the THC. For western maritime regions, a drop of 5° to 10°C has been suggested by climate change scientists though the change would most likely be tempered with distance inland.

Evidence from the Little Ice Age gives clear indication that food shortages, and as a direct consequence famine, were a major problem for many societies during this colder period. For instance, the example of Iceland, which lost half its population, is cited. However, the authors also stress that in reality the situation was more complex, with culture and economy paradoxically doing well in some parts of Europe. In fact, new climatic conditions may actually have been an important driver for changing agricultural practice, as witnessed in the Netherlands by innovations such as lay farming, reclamation technologies, the growing of animal fodder and the cultivation of grasslands for cattle. Colonial expansion would have provided a valuable support for these changes in practice by enabling access to resources from other locations. Land productivity and famine are again identified as important impacts of both the nuclear and aerosol scenarios, with vegetative productivity likely to be particularly affected. The aftermath of a nuclear attack is obviously the most extreme event, of which a drop in temperature will be only one of a multitude of threats to human well-being. The impacts of changes to the THC are likened to the Little Ice Age, with impacts rippling through a range of social and economic activity from direct impacts on agricultural production to less direct impacts such as changes to energy demand as households react to the changing climate.

Following identification of some of the likely impacts, the penultimate section of the paper then summarises some of the coping mechanisms that could be called upon by society. Those mechanisms discussed include migration (citing studies that refer to a potential depopulation of the western regions towards more favourable climatic conditions found in southern and eastern reaches), shelter (pointing to the fact that Scandinavian housing has evolved, and is well suited, to existing severe climatic conditions), transport (de-icing is commonplace in many countries already though new management practices may need to be developed under colder conditions), food and water supply (storage and increased trade with other parts of the world may be needed for ensuring food security, with new technologies considered to be adequate to maintain water supply under cooling conditions), energy (changes to generation and supply), health and education (a refocus by the medical services on cold-related injuries and other medical ailments) and governance and economy (financial support and international cooperation may be necessary to address some of the resource and human issues which spill-over beyond the affected regions).

Concluding, the authors signpost the collapse of current agricultural practice as one of the important casualties of rapid cooling, though recognise that numerous other impacts will be felt across all sectors of society. Although consequences are likely to be severe in many cases they argue that Europe has a high level of adaptive capacity with considerable technological, societal and economic resources at its disposal. The ability to learn from the experiences of cold countries, in terms of both infrastructure and behaviour, will be an important component in the adaptation process, and as comes through from the writings of other papers in the NWO portfolio, foresight, planning and flexibility of response are essential if society is to reduce the risk of socioeconomic disorder and adapt to new conditions that abrupt climate change will bring.

14.6 Economic perspective: the slowdown of the North Atlantic Gulf stream—an economic perspective on adaptation

Economic considerations and institutional complexity formed the basis of the last paper in the portfolio. Van Ierland sets out the discussion in a context of decision-making under uncertainty, with explicit consideration of a process of institutional learning over time. Central to the discussion is the issue of timing of response, that is, do we need to take action now and do what can be done or delay the decision until more information is available; essentially a 'wait and see' approach.

For the purposes of this paper, economic implications are sketched out for a range of key sectors. These include agriculture (interestingly the analysis in this case suggests a much smaller impact than is alluded to in other papers), energy demand and supply (likely changes in winter and summer demand are highlighted), water management (safety considered to be of paramount importance with cooling), transportation (cold weather impacting on the demand for all types of transport), tourism and recreation (shift in holidaying patterns with implications for the Dutch visitor economy), health (extreme cold posing particular risks for vulnerable groups such as the elderly or homeless people), immigration/emigration (climate refugees and an internal movement of people within the EU), and ecosystems(change to growing seasons). The decrease in temperature will most likely pose costs to each of these sectors, financially, socially and environmentally.

The author contends that there is sufficient time to adjust to many of the impacts that may result from the slowdown of the North Atlantic Gulf Stream. Financially speaking, there is no need for making excessive outlay now. That said, it is argued that attention should be paid to enhancing robustness to better cope with the extended thresholds induced by an abrupt decrease in temperature. This will ultimately require an integrated approach due to linkages, and potential spill-over effects between sectors. One such example is better insulation of houses, factories and offices. This makes the control of the interior climate more efficient, either by heating or by cooling (possibly much can be learned from construction experience in Scandinavian countries). The challenge is to design buildings for internal human comfort but also to ensure that any mechanical operation is either generated by a minimum of energy requirement or by sustainable energy sources. A comfortable indoor temperature will also reduce the health risk faced by the most vulnerable groups like the elderly.

Although not discussed in any great detail in the paper, the impact on ecosystems is likely to be of great importance, again with cross-sectoral implications (e.g. affecting tourism and agriculture). Identifying and better understanding these linkages will be a significant step in moving towards a more informed adaptation response. Developing and introducing new concepts for spatial planning and the design of infrastructure and cities for instance, highlighted by the author in the introduction, are likely to be critical components of any strategic adaptation strategy.

14.7 Some reflections and implications for policy and practice

By 'imagining the unimaginable' the portfolio of VAM papers has provided multi-disciplinary insights into the potential consequences of a western maritime Europe that cools abruptly as a result of a shutdown of the THC, even at the same time that warming occurs globally. While each of the papers explored different aspects of abrupt change from a range of perspectives there are some key generic messages that can be distilled from across the narratives.

The first common thread is that abrupt climate change will result in significant social and economic impacts for society. The complex nature of impacts will be further complicated by the fact that climate-related risks, and opportunities for adaptation, will be unevenly spread across different communities, sectors and landscape types. Agricultural practice, for example, was identified as particularly vulnerable by several papers. Furthermore, many impacts will not be confined by geographical boundaries, and trans-national issues such as migration and water availability are likely to figure prominently. Improving our understanding of climate risks across sectors and spatial scales will be critical to underpinning a sustainable transition to a system that is well adapted to new climatic conditions (though it is important to note that climate change is only one risk among many that future societies are likely to face, and hence need to adapt to). These risks are not only diffuse and multi-scalar in nature but in many cases adaptation will also need to account for other complexities; for example, in the case of cities, the socio-physical interactions of urban systems which are shaped by many different elements at risk and affected actors with competing priorities, will also need to be accounted for (McEvoy *et al.* 2008; Martens *et al.* 2009).

Second, while the findings suggest considerable disruption to socioeconomic systems, all note that a high level of uncertainty is involved, both in terms of likely impacts and in how best to respond. In urban areas should planners and urban designers be preparing for warming or cooling? What are the implications of this for building regulations and guidance? In rural areas which crops should farmers be planting to best cope with future conditions? Improving decision-making under conditions of uncertainty will be vital if the necessary adaptation is to occur, and mal-adaptation is to be avoided. The challenges for policy-making are considerable, especially bearing in mind the fact that decisions are often framed by short-term election cycles and based on variables that are much more certain than is the case with the risks associated with low-probability/high-impact abrupt change. Even though this complexity makes it challenging for 'non-experts' to have adequate confidence in making informed decisions, knowledge needs to be drawn from all available quarters, including from local experience and on contributions from different parts of civil society. Speaking a common language, translating scientific information into usable knowledge and bridging between different actors, organisations, sectors and regions will all be needed.

The VAM narratives also highlight two interrelated solutions: in the first instance the need for adaptation responses to be 'flexible' and secondly that adaptation should be understood as an iterative learning process on behalf of institutions and individuals, not merely an end-point. Indeed, it has been recognised in the wider academic literature that an iterative learning approach is vital for addressing the uncertainties associated with climate change and variability, representing a shift from traditional technocentric responses to weather-related extreme events to one where flexibility, and responses that are 'fit for purpose', are considered more appropriate for dealing with the complexity of climate change (Pahl-Wostl *et al.* 2007). Others emphasise adaptation as a response to an 'unbounded problem', where: (1) there is no clear agreement about what exactly the problem is, (2) there is uncertainty and ambiguity as to how improvements might be made, and (3) the problem has no limits in terms of the time and resources it could absorb (Chapman 2002). In this sense, learning is needed to coordinate particular collective courses of action, under conditions of uncertainty, in order to achieve desirable outcomes for society. Ultimately, what will be required are adaptation measures that have the flexibility to cope with a wide range of climatic conditions.

As recognised by several of the papers, learning and better informed decision-making is supported by transfer of knowledge. Although a degree of caution needs to be exercised when considering spatial analogues, it was suggested that countries such as the Netherlands could learn from the vernacular architecture and house-building techniques employed by colder Scandinavian countries. It has also been recognised elsewhere that as demands for adaptation are increasingly mainstreamed, cities will increasingly act as 'laboratories' of experimentation as niches develop and innovation occurs. This practical evidence will have considerable added value if shared—raising the critical issue of knowledge transfer platforms. These, along with suitable networking forums will be an essential component of building adaptive capacity. To be effective these need to be targeted and accessible for specific end-users, not only focusing on individual sectors but also acting as a mechanism that helps to bridge across the agendas of different policy and practitioner communities. (McEvoy *et al.* 2008).

Insights from all the essays reinforce the need to understand that adaptation can manifest itself in many forms: technological (including both hard and soft engineering), best practice in management, planning and design, legal responsibility and regulation, insurance, financial arrangement, or other institutional aspects. Adaptation is therefore both process and outcome (McEvoy *et al.* forthcoming). In these narratives particular importance was placed on the institutional dimension when thinking about adaptive responses to abrupt climate change, with issues relating to the roles and responsibilities of governments and citizens, the complexity of multi-actor and multi-level interactions, distributional effects within and across sectors, uncertainty of economic and social implications, to name but a few of the many challenges that will need to faced under abrupt and extreme climate change.

References

Association of British Insurers (2005) *Financial Risks of Climate Change* (London: Association of British Insurers).

Bouwer, L.M., D. Huitema and J. Aerts (2007) 'Adaptive Flood Management: the Role of Insurance and Compensation in Europe', paper presented to the *Conference on the Human Dimensions of Global Environmental Change*, Amsterdam, 24–26 May 2007.

CEA (2007) *Reducing the Social and Economic Impacts of Climate Change and Natural Catastrophes: Insurance Solutions and Public–Private Partnerships* (Brussels: CEA).

Chapman, J. (2002) *System Failure: Why Governments Must Learn to Think Differently* (London: Demos).

Lenton, T., H. Held, E. Kriegler, J. Hall, W. Lucht and S. Rahmstorf (2008) 'Tipping Elements in the Earth's Climate System', *PNAS* (*Proceedings of the National Academy of Sciences of the United States of America*) 105.6: 1,786-93.

Martens, P., D. McEvoy and C. Chang (2009) 'The Climate Change Challenge: Linking Vulnerability, Adaptation, and Mitigation', *Current Opinion in Environmental Sustainability* 1: 1-5.

McEvoy, D., K. Lonsdale and P. Matczak (2008) 'Adaptation and Mainstreaming of EU Climate Change Policy' (CEPS Policy Briefing Note for the European Commission, Brussels).

—— P. Matczak, I. Banaszak and A. Chorynski (forthcoming) 'Framing Adaptation to Climate-Related Extreme Events', *Mitigation and Adaptation Strategies for Global Change.*

NWO (Netherlands Organisation for Scientific Research) (2008) *What if Abrupt and Extreme Climate Change?* (The Hague: NWO).

Pahl-Wostl C., M. Craps, A. Dewulf, E. Mostert, D. Tabara and T. Taillieu (2007) 'Social Learning and Water Resources Management', *Ecology and Society* 12.2: 5.

Steffen, W. (2009) *Climate Change 2009: Faster Change and More Serious Risks* (Canberra: Australian Department of Climate Change).

UNEP Finance Initiative (2006) 'CEO Briefing—Adaptation and Vulnerability to Climate Change: The Role of the Financial Sector' (Geneva: United Nations Environment Programme Finance Initiative).

15
Conclusion

Chiung Ting Chang and Pim Martens
ICIS, Maastricht University, The Netherlands

Bas Amelung
Amelung Advies, The Netherlands

Over the years, the nature of climate change research and policy has changed. Besides the traditional fields of impacts and mitigation, there is now much more attention on vulnerability and adaptation, and on the interaction between adaptation and mitigation. The need for innovative contributions from the social sciences has intensified. In 2004, the Netherlands Organisation for Scientific Research (NWO) responded to these developments, which were emerging at the time, by launching a research programme specifically targeted at climate change research in the social sciences. Under this programme, entitled Vulnerability, Adaptation and Mitigation (VAM), a total of 12 research projects and six essays were funded.

Now that the programme has come to an end, this book takes stock of the results, with this final chapter wrapping up the key insights. It provides a synthesis of the main conclusions from the research projects and the collection of essays. In addition, it presents an outlook on future research, identifying a number of promising avenues. The synthesis of the main conclusions from the various projects and essays is structured around the list of developments and challenges presented in the introductory Chapter 1.

Chapter 1 sets the scene for the book, presenting a scheme for linking the key concepts of adaptation, vulnerability and mitigation, highlighting a number of key developments in climate change research, and identifying the key challenges for integrating the various domains involved in climate change research. These developments and challenges are listed in Box 15.1. They are used as the focal points for the synthesis of the VAM programme's main findings. Rather than summarising the conclusions from the individual chapters, the synthesis aims to reframe these conclusions in order to generate insights with a broader relevance.

Box 15.1 **Developments in climate research, and challenges for integrating climate change research domains**

Development

1. Consensus on climate change
2. From impacts to risk management
3. Acknowledgement of non-climate stressors
4. Need for interdisciplinary work

Challenge

1. Exchanging analogies between mitigation and adaptation domains
2. Bridging differences in temporal and spatial scales
3. Dealing with perceptions of risk and contribution
4. Accounting for distributional effects
5. Analysing actor networks and institutional complexity

15.1 Consensus about climate change

Climate change and the major human contribution to it are accepted as facts in all projects and essays; or at least, these hypotheses are rarely explicitly challenged. One of the few minor exceptions is Chapter 8, in which 'sceptical colleagues' are mentioned as an internal barrier to local adaptation. Overall, the observation in Chapter 1 is supported: that the scientific consensus about the reality of climate change and its causes has increased substantially over the past few years. Climate change is not always perceived as a separate issue, however. In Mozambique, for example, it is only one of the elements local people have to deal with (see Chapter 7).

The reality of climate change as a human-induced phenomenon may be accepted, but there is a lot of uncertainty and confusion around its magnitude and implications. This is a recurrent issue in the VAM programme, and most, if not all, chapters address it. Chapter 12, for example, extensively discusses the uncertainties in the causal chain between emissions and impacts, blurring the responsibility and liability of a particular emitter for a particular impact. It is argued that traditional concepts of liability law, which are based on certainty about the link between cause and effect, may be fundamentally incapable of dealing with liability for climate change. New legal concepts may be required.

15.2 From impacts to risk assessment

The shift in emphasis in climate research from impact assessment to risk assessment is visible in the VAM projects as well. Risk and vulnerability assessments are major components of a majority of projects. Based on projections for the impact of climate change on water discharges of the River Rhine, Chapter 2 concludes that the risks of climate change for inland navigation are greater for transport from the Netherlands to Germany than for transport in the opposite direction. Chapter 3 presents a 'bottom-up' framework for vulnerability assessment for tourist destinations. Vulnerability is shown to be highly dependent on the local mix of tourist activities offered by a destination— an insight that is further strengthened by the (more traditional) 'top-down' assessment of climate change impacts on the climatic attractiveness of coastal areas in Europe for different types of tourist activities. The complementary relationship between impact assessment and risk assessment is explicitly stressed in Chapter 6, which concludes that 'disasters [such as Hurricane Mitch] not only change the asset base of the affected population, but also the nature of their preferences and the weighting of alternative survival strategies'. A key finding from Chapter 9 is that people's risk assessment of flood risk changes when they are exposed to a realistically simulated flooding event: 'Awareness of one's own vulnerability to future flooding due to climate change and insights into the effectiveness of coping actions to deal with these new risks are driven by direct flooding experiences.' On a meta-level of analysis, it can be argued that Chapter 12 recommends that, with respect to climate change, the basis of legal principles move from impact assessment to risk assessment, given the '[uncertain] chain of causation from the defendant's conduct to the individual damage'.

15.3 The consideration of non-climate stressors

As the VAM programme has a strong focus on climate change, non-climate stressors do not play a central role in most chapters, although they are discussed in many of them. The strongest examples are perhaps non-typical as they relate to non-climate factors that influence actors' responses to climate change. Paradoxically, Chapter 4 wonders whether companies' decisions to get involved in partnerships on mitigation and adaptation are mainly driven by motives of adaptation and mitigation or by non-climate factors: 'As reputational and issue management considerations seem to play a clear role for companies, it is a critical question whether and to what extent partnership involvement has real substance (and goes beyond what some might label as "greenwashing").' Reasons of reputation and issue management may also play a role in Chapter 5, which concludes that housing associations concentrate their energy conservation efforts on new houses rather than the existing stock, where large gains can be made. Moreover, energy conservation is relatively low on the priority list of housing associations—a list that is dominated by social issues. Similarly, a key conclusion of Chapter 7 is that climate change is only one of the developments that people in Mozambique have to adapt to, as 'climate change becomes incorporated in the continuity of everyday practice'.

15.4 Connecting the mitigation and adaptation research domains

The most explicit contribution to connecting mitigation and adaptation is made in Chapter 13, where adaptation is added to the Integrated Assessment model RICE as an explicit decision variable. The results clearly show that adaptation and mitigation policies are closely interlinked. The more international cooperation, the stronger the emphasis on mitigation measures, as these have global benefits, whereas the benefits of adaptation are exclusively regional. In the absence of cooperation, climate change mitigation will be limited, so that the costs of adapting to the remaining change will be high. Chapter 3 concludes that many tourist destinations are vulnerable to climate change impacts as well as the impacts of mitigation policies, as many of their visitors use aeroplanes or other high-emission forms of transport to reach them. Transport costs are bound to increase as more stringent mitigation policies come into place.

15.5 Adaptive capacity

Chapter 1 concludes that a common feature of adaptation and mitigation is that they both depend on the capacity of a system to respond. This implies a certain potential for the exchange of concepts and solutions between the adaptation and mitigation research and policy domains. Chapter 8 exploits this potential by turning to the mitigation literature in a search for drivers of climate-change adaptation in Dutch towns. These drivers were found to show 'similarities to key implementation factors identified in the literature on local climate mitigation efforts. By simply replacing the word 'mitigation' with 'adaptation', the list of drivers was found to 'provide a solid baseline for future adaptation initiatives'. In turn, the 'local contextual factors' that were added can perhaps inspire researchers in the mitigation field.

15.6 Temporal and spatial scales

The challenge of connecting the different temporal and spatial scales of adaptation and mitigation is taken up in Chapter 13. This chapter concludes that the different spatial scales at which adaptation and mitigation take place lead to trade-offs that are relevant for negotiations in the international policy arena. In the full cooperation scenario, for example, in which countries strive for low regional income differences, adaptation transfers turn out to be a powerful instrument to compensate regions for lower mitigation levels. The synthesis of the essays (Chapter 14) also touches on the challenge of integrating scales, but does so in the context of adaptation only, concluding that 'Improving our understanding of climate risks across sectors and spatial scales will be critical to underpinning a sustainable transition to a system that is well adapted to new climatic conditions'.

15.7 Risk perception and distributional effects

All projects address risk perception and/or distributional effects, in varying levels of detail. Chapter 13 notices a clear difference between mitigation and adaptation in terms of the principles applied for allocating the burden. For mitigation, the polluter pays principle has become a widely accepted starting point, whereas, for adaptation, 'the issue of international financing is complex, and one of the most controversial topics of discussion in the ongoing climate negotiations'. In Chapter 11, the issue of distributing the burden of mitigation is approached from a different angle, by advocating the principle of equal amounts of emission rights per capita. It is founded on the view that all human beings have an equal right to the limited absorptive capacity of the atmosphere. This principle, which is not necessarily in conflict with the polluter pays principle, connects the challenge of mitigation with environmental equity and effectiveness.

The problems related to linking specific emissions to specific impacts, which were discussed in Chapter 12, are closely related to the observation in Chapter 1 that 'disconnection in space and time can make it difficult for people to link the consequences of their activity with long-term environmental consequences'. In Chapter 1, this observation was mainly made in relation to people's (un)willingness to act on climate change, whereas in Chapter 12 the main focus was on the challenge of holding parties responsible for the consequences of their actions. Chapter 4 concludes that the environmental (and commercial) effectiveness of the climate change-related partnerships that companies get involved in is unclear, partly because partnerships usually do not have a clearly defined target.

The difficulty that people have in understanding how climate change is related to their own lives is not limited to the emissions and mitigation side; it is present on the impact and adaptation side as well. In this context, Chapter 9 concludes that it is difficult for people living in flood-prone areas to imagine how floods might affect them physically and emotionally. Experience is considered vital. Exposing people to simulated floods in virtual environments can compensate for the absence of real experience with floods, although the emotional component is not yet well addressed in current simulators. The important role of experience is also evident in Chapter 6, which concludes that natural shocks such as Mitch make people more risk-averse. The relative importance of actual experience versus calculated risk is unclear, however, as in the case of Nicaragua only *ex post* effects were studied, whereas 'people may refrain from investing in more profitable activities even without a disaster actually occurring'.

Good examples of how subtle the interplay between risk perception and distributional effects can be are given in the Chapters 2 and 3. In Chapter 2, climate change is shown to have a much smaller effect on downstream navigation on the River Rhine than on upstream transport to Germany. As a result, German consumers (hence welfare) have most to gain from infrastructural investments and other adaptation measures. The chapter notes, however, that it may be considered unfair for Germany to pay the bulk of the investment costs, because most barge operators on the Rhine are Dutch. Although irrelevant from a welfare perspective, this may delay or block profitable investments. Major shifts in tourism visitation patterns may result from climate change, as Chapter 3 shows, suggesting that the destinations for which losses are projected might push strongly for stringent mitigation policies. In reality, however, even those destinations tend to oppose mitigation measures, perhaps because they perceive themselves to be vulnerable to mitigation measures too, as many of their visitors use aeroplanes or other high-emission forms of transport to reach them.

15.8 Institutional complexity

Based on the insights from the six essays, Chapter 14 re-emphasises that 'adaptation can manifest itself in many forms: technological . . . , best practice in management, planning and design, legal responsibility and regulation, insurance, financial arrangement, or other institutional aspects'. Often, many of these forms of adaptation are required simultaneously, creating both opportunities and threats for large numbers of actors. The institutional complexity that this entails is further increased by 'unboundedness' of the problem of climate change: 'there is no clear agreement about what exactly the problem is'; 'there is uncertainty and ambiguity as to how improvements might be made'; and 'the problem has no limits in terms of the time and resources it could absorb' (Chapter 14, based on Chapman 2002). This view of climate change as an 'unbounded problem' is reflected in the framework for vulnerability assessment in the tourism sector, which is presented in Chapter 3. The chapter makes clear that, at the end of the day, vulnerability assessment is a result of stakeholder dialogue, i.e. it has a strong element of social construction. Adaptation is both outcome and process, as Chapter 14 concludes from the essays. This is also a major insight from Chapter 7, which emphasises that in Mozambique 'Actors negotiate adaptation in their everyday practice', implying that adaptation policies should be process-oriented rather than top-down and technocratic.

Institutional complexity is a key element in Chapters 4, 5, 8 and 10. In Chapter 4, companies are shown to engage in partnerships with other companies and societal actors for a wide variety of reasons, ranging from trying to shape the climate debate to developing concrete adaptation measures. The effectiveness of these partnerships is unclear and also difficult to assess, partly owing to their elusiveness. Apart from demonstrating institutional complexity in both energy problems and energy policy instruments at the international and EU level, Chapter 10 suggests a framework, a combination of four approaches, for analysing this highly intertwined institutional setting. This attempt at deconstructing complexity, as warned by the author in conclusion, involves a complex and resource-demanding process. A promising insight from Chapter 8 on this point is that for Dutch municipalities engagement in networks and partnerships strongly correlates with a proactive attitude towards adaptation. 'Within such stimulating [local, national and international] networks, local actors are more motivated to explore climate adaptation efforts that would otherwise be too ambitious (resource-demanding) for a single municipality.' The importance of networks is also stressed in Chapter 5, where the success of 'interorganisational collaboration' is one of the main determinants for the success of energy conservation in housing projects. Interestingly, the focus in this latter chapter is on mitigation, where institutional complexity is supposedly less.

15.9 Suggestions for further research

In this early stage of climate research in the social sciences, research opportunities abound. Each chapter mentions a few areas for further research that are either of major scientific or societal interest; often they are both. Rather than repeating these individual recommendations, this chapter aims to make a few suggestions for further research based on the results of the concluding synthesis.

Of the challenges and opportunities for integrated climate research that were formulated in Chapter 1, some were clearly addressed more often than others. A majority of chapters addressed the difficulty of establishing a link between one's own behaviour and climate change, in either direction. The related issues of trade-offs and distribution also featured in many chapters. The VAM programme probably contributed most to addressing these two challenges (risk perception and distributional effects), and ideas of how to move forward on these issues is taking shape. A challenge that was also mentioned in a number of chapters, but for which a research agenda does not really seem to emerge, is institutional complexity. Many adaptation (and mitigation) processes influence and are influenced by a large number and a large variety of actors. What does this mean for agenda-setting, decision-making, implementation, results, the process itself, etc.? Novel approaches may be formulated by bringing together the VAM community and the community of transition researchers, for which other research programmes are available.

Different directions for further research are suggested by Chapter 1's two remaining challenges. A clear finding is that not many projects seem to have exploited the similarities of the mitigation and adaptation agendas. Both are related to a system's adaptive capacity, so there may be relevant analogies to make use of. Almost incidentally, Chapter 8 identifies one of them, suggesting that there may be many more. As the body of scientific knowledge and concepts is much larger for mitigation than for adaptation, searching for analogies may be a fruitful exercise that may prevent researchers from having to reinvent the wheel time and again. It may also considerably speed up the process of catching up in the social sciences.

Climate change affects processes on many different scales, ranging from seconds to millennia, and from the very small to the Earth as a whole. Connecting these temporal and spatial scales is one of the key challenges of climate research. This is the case for the integration of the mitigation and adaptation research domains, but also for many research fields within these domains. Apart from in Chapter 13, which incorporates adaptation in an Integrated Assessment model, the scaling issue is not a major topic in this book. Advances in this field may be difficult to achieve, but are potentially very rewarding if successful.

As the policy debate moves from one of problem framing to one more concerned with implementation, detailed political, ethical, social and normative analyses becomes increasingly important. The production of scientific knowledge alone will not suffice—information will also need to be translated into action 'on the ground'. Hence, a highly organised multi-disciplinary programme of research aimed at improving assessment methodologies, reframing current scientific understanding, and translating new insights into innovative policy options, will be required.

It has also been recognised that there is a need for greater cohesion between climate change and sustainable development objectives. As noted by the IPCC (Klein *et al.* 2007), few plans for promoting sustainability have explicitly included either adaption to climate change impacts, or the enhancement of adaptive capacities. There is considerable added benefit to be gained by ensuring a more integrated approach. Not only will climate change have an adverse impact on progress towards a sustainable future, sustainable development activity can reinforce our responses to climate change by enhancing adaptive capacity and increasing resilience.

The VAM programme was one of the first major platforms in the Netherlands for climate research in the social sciences. Its key results and insights that are brought together in this book will hopefully inspire, guide and inform those committed to climate change research and policy on vulnerability, adaptation and mitigation. It is hoped that the VAM legacy will be part of the foundation of a much larger endeavour to raise the profile of the social sciences in climate change research.

Reference

Klein, R.J.T., S. Huq, F. Denton, T.E. Downing, R.G. Richels, J.B. Robinson and F. L. Toth (2007) 'Inter-relationships between Adaptation and Mitigation', in M.L. Parry, O.F. Canziani, J.P. Palutikof, P.J. van der Linden and C.E. Hanson (eds.), *Climate Change 2007: Impacts, Adaptation and Vulnerability: Contribution of Working Group II to the Fourth Assessment Report of the Intergovernmental Panel on Climate Change* (Cambridge, UK: Cambridge University Press): 745-77.

Acronyms

ADAM	Adaptation and Mitigation Strategies
AG	Advocate General
BCI	Beach Climate Index
C&C	contraction and convergence
CBDR	common but differentiated responsibilities
CCS	carbon capture and storage
CDM	Clean Development Mechanism
CDP	Carbon Disclosure Project
CERs	Certified Emission Reductions
CSIRO	Commonwealth Scientific and Research Organisation
CSTM	Centre for Studies in Technology and Sustainable Development
DEFRA	UK Department for Environment, Food and Rural Affairs
DICE	Dynamic Integrated Climate and Economy
DRR	disaster risk reduction
DTQs	domestic tradable quotas
ECJ	European Court of Justice
ECPI	Energy and Climate Policy Interactions
EEA	European Environment Agency
EM-DAT	International Emergency Disasters Database
EPL	energy performance on location
ESCOs	energy service companies
ETSAP	Energy Technology Systems Analysis Programme
EU ETS	European Union Emissions Trading Scheme
EUAs	European Union Allowances
GCMs	General Circulation Models
GEF	Global Environment Fund
GHG	greenhouse gas emissions
IAMs	integrated assessment models
ICIS	International Centre for Integrated Assessment and Sustainable Development
INGC	National Institute for Disaster Management
IPCC	Intergovernmental Panel on Climate Change
IPQ	Igroup Presence Questionnaire

JI	Joint Implementation
KNMI	Royal Netherlands Meteorological Institute
LSM	Living Standard Measurement Survey
MCA	multi-criteria analysis
MDG-F	Millennium Development Goals Achievement Fund
MICOA	Mozambique Ministry for Coordination of Environmental Action
MRV	monitorable, reportable and verifiable
NAPA	National Action Plan for Adaptation
NAS	National Adaptation Strategy
NPV	net present value
NOW	Netherlands Organisation for Scientific Research
PCT	personal carbon trading
PI	policy interaction
PMT	protection motivation theory
PSEG	Public Service Enterprise Group
RE	renewable energy
RIVM	National Institute for Public Health and the Environment
SEM	structural equation modelling
SRES	Special Report on Emissions Scenarios
TCI	Tourism Climate Index
TGC	tradable green certificates
THC	thermohaline circulation
UNDP	United Nations Development Programme
UNEP	United Nations Environment Programme
UNFCCC	United Nations Framework Convention on Climate Change
UNWTO	United Nations World Tourism Organization
US-CAP	United States Climate Action Partnership
VAM	vulnerability, adaptation and mitigation
VAs	voluntary agreements
VE	virtual environment
VNG	Association of Netherlands Municipalities
VROM	Dutch Ministry of Housing, Spatial Planning and the Environment
VSD	Vulnerability Scoping Diagram
WhC	white certificates
WIMEK	Wageningen Institute for Environment and Climate Research
WMO	World Meteorological Organization
WRR	Scientific Council for Government Policy
WSSD	World Summit on Sustainable Development
WTP	willingness to pay
WTTC	World Travel and Tourism Council

About the contributors

Bas Amelung is director of Amelung Advies, a consultancy firm operating in the fields of climate change, tourism and sustainable development. He obtained a PhD in Integrated Assessment from Maastricht University, and is still loosely connected to ICIS. Bas's research interests include climate change and tourism, polar tourism, and sustainable development.

Luís Artur is a PhD candidate in Disaster Studies at Wageningen University. He is Mozambican, working at Eduardo Mondlane University in Maputo. He first graduated as an agronomist in Mozambique and has a Master's in Rural Development Sociology from Wageningen University in the Netherlands. His research interests include rural development, disaster risk reduction, climate change adaptation and local participation.

Dr **Hans Bressers** is full professor of Policy Studies and Environmental Policy at the Centre for Studies in Technology and Sustainable Development (CSTM). He has published over 200 articles, chapters, reports, papers and books on policy-mapping, -instruments, -implementation and -evaluation and policy networks, mostly applied to environmental policies and water management. He is the founder of the CSTM.

Kees Burger is associate professor of Development Economics at Wageningen University. His research focuses on farm households in Africa, their investments in soil and water conservation and on commodity markets, price formation and stabilisation.

Chiung Ting Chang is a research fellow at the International Centre for Integrated Assessment and Sustainable Development (ICIS), Maastricht University. She is an environmental economist and holds a PhD in environmental economics and management. Her research field is flood management, adaptation for climate change, policy analysis and development studies.

Frans J.H.M. Coenen is currently a senior researcher at the Twente Centre for Studies in Technology and Sustainable Development (CSTM). Some of his recent work includes research on local climate policy, local and regional sustainable development and the knowledge–policy interface.

Kelly de Bruin is a PhD candidate at Wageningen University and Research Centre focusing on the trade-offs between adaptation and mitigation and the stability of international climate agreements. Besides her PhD work she has also been involved with numerous adaptation modelling projects with various research teams. kelly.debruin@wur.nl

Javier de Cendra de Larragán was a former PhD researcher at Maastricht University, and is now senior researcher with the UCL Energy Institute/Faculty of Laws at University College London.

Rob Dellink is an economist and policy analyst at the Organisation for Economic Cooperation and Development (OECD) in Paris and Assistant Professor at Wageningen University and Research Centre. He obtained his PhD in Economics at VU University Amsterdam in 2003. His main research interests include the welfare costs of climate change policies, applied general equilibrium modelling, trade-offs between adaptation and mitigation strategies to climate change, and formation and stability of coalitions for international environmental agreements. He has authored more than 100 scientific publications. rob.dellink@wur.nl

Erhan Demirel obtained his Master's degree in econometrics at Erasmus University, Rotterdam, in 2003. In 2006 he started his PhD project on climate change and inland navigation under the supervision of Piet Rietveld and Jos van Ommeren at the Department of Spatial Economics, VU University, Amsterdam.

Miriam Haritz studied Law at Cologne University, Germany, and at UCL, UK, and European Public Affairs at Maastricht University (UM) and the European Institute of Public Administration (EIPA), the Netherlands. After working as a junior lawyer, she worked at two UM faculties in the field of Institutional European Union Law, European and International Environmental Law and Risk Regulation. She is currently working as a Legal Advisor at the Federal Office of Civil Protection and Disaster Assistance in Bonn, Germany. The chapter in this volume summarises the findings of her PhD thesis.

Dorothea Hilhorst is professor of Humanitarian Aid and Reconstruction at Wageningen University. Her research concerns the aidnography of humanitarian crises and fragile states. Her publications focus on everyday practices of humanitarian aid, disaster risk reduction, climate change adaptation, reconstruction and peace building. She coordinates research in Angola, DRC, Afghanistan, Ethiopia, Sudan, Mozambique and Uganda.

Dr **Thomas Hoppe** is assistant professor of Innovation and Sustainable Governance at the Centre for Studies in Technology and Sustainable Development (CSTM). He specialises in environmental policy, public policy, policy implementation, climate policy, sustainable housing, energy transition and corporate social responsibility. He has also been involved in externally funded projects concerning policy evaluation.

Lia Hull Van Houten is a communications consultant in the San Francisco area and founder of Van Houten Communications. Lia has held positions in corporate social responsibility and communication strategy at marketing consultancies and environmental NGOs in California and the Netherlands. She obtained her MBA degree (cum laude) in 2007.

Ans Kolk is full professor at the University of Amsterdam Business School. Her areas of research and publications are in corporate social responsibility and sustainability, especially in relation to multinational corporations' strategies, and international policy. One of the topics on which she has published extensively in the past decade is business and climate change.

William M. Lafferty is currently professor of Strategic Research for Sustainable Development at the Twente Centre for Studies in Technology and Sustainable Development (CSTM). He is retired professor of Political Science at the University of Oslo, and previous director of the Programme for Research and Documentation for a Sustainable Society (ProSus).

Dr **Kris Lulofs** is senior researcher in Policy Studies and Environmental Policy at the Centre for Studies in Technology and Sustainable Development (CSTM). He specialises in environmental policy strategies and instrumentation, especially in the fields of corporate social responsibility, climate change and water management. He has been a researcher and project leader for numerous externally funded projects, including projects funded by EU research frameworks, the Dutch national science foundation, Dutch ministries, provinces, water boards and international and national knowledge institutes.

Pim Martens is director of the International Centre for Integrated Assessment and Sustainable Development (ICIS), Maastricht University, The Netherlands. He holds the Chair of Sustainable Development at Maastricht University, is a guest professor at Leuphana University Lüneburg, Germany, and an honorary professor at the Institute of Biological, Environmental and Rural Sciences at Aberystwyth University, Wales.

Darryn McEvoy's research expertise covers climate risk assessment and adaptation, innovative adaptation practice in different contextual settings (e.g. climate change and cities), institutional adaptive management, vulnerability assessment, the building of local adaptive capacity, and the synergies and conflicts between the adaptation and mitigation agendas. Of particular personal interest is the translation of theory into practice, and highlighting the implications for climate risk management and decision-making (involving both public and private actors). Following senior research experience in the UK and the Netherlands he now leads the climate change adaptation programme for the Global Cities Institute at RMIT University, Melbourne, and acts as Deputy Director for the newly established Victorian Climate Change Adaptation Research Centre.

Cees Midden is professor of Human Technology Interaction at Eindhoven University of Technology. His research focus is on the social-cognitive factors of human–technology interactions.

Alvaro Moreno studied Forest Engineering at Universidad Politécnica de Madrid (Spain) and obtained his Master's degree in Environmental Sciences at Wageningen University (the Netherlands). He is currently a research fellow at the International Centre for Integrated Assessment and Sustainable Development (ICIS), Maastricht University. He has been actively involved in several other research projects, including the European-funded PESETA project on the economic impacts of climate change in different European sectors and the CLIMAS project on the assessment of impacts in Asturias (Spain).

Vlasis Oikonomou is an economist working in the field of climate change since 2002, focusing on the certificate mechanisms for climate change, energy efficiency and renewable energy. Among other projects, he is appointed as the Dutch expert on the white certificates team of the International Energy Agency/Demand Side Management programme; while finalising his PhD, he has published numerous articles and book contributions on the topic of white certificates and their interactions. Currently he is employed by Joint Implementation Network in the Netherlands on similar issues.

Marjan Peeters is Professor of Environmental Law and Policy, with particular interest in climate change issues, at Maastricht University.

Jonatan Pinkse is assistant professor at the University of Amsterdam Business School. His areas of research, teaching and publications are in strategy and sustainable management.

His PhD thesis, which was awarded the 2006 Organizations and the Natural Environment, Academy of Management Best Dissertation Award, addressed business responses to climate change.

Piet Rietveld studied econometrics at Erasmus University, Rotterdam (degree cum laude), and received his PhD in economics at VU University, Amsterdam. He has worked at the International Institute of Applied Systems Analysis in Austria and was research coordinator at Universitas Kristen Satya Wacana in Salatiga, Indonesia. Since 1990 he has been professor of Transport Economics at the Faculty of Economics, VU University. He is a Fellow of the Tinbergen Institute.

Marrit van den Berg is a lecturer in Development Economics at Wageningen University. Her research is mainly directed at rural household behaviour under risk and market constraints and the impact of policies and programmes on rural livelihoods. She has been involved in projects in Asia, Latin America and Africa.

Maya M. van den Berg is currently a junior researcher at the Twente Centre for Studies in Technology and Sustainable Development (CSTM). She specialises in climate adaptation policies of local governments in the Netherlands. She is pursuing her research interests in a PhD on civil preparedness for climate change in local communities.

Ekko C. van Ierland is full professor of Environmental Economics and Natural Resources at Wageningen University and head of the Environmental Economics and Natural Resources Group. He has been chairman of the board of Wageningen Institute for Environment and Climate Research (WIMEK) and an adviser to the Dutch Scientific Council for Government Policy (WRR) and the 'Algemene Rekenkamer', and has carried out research for the Dutch Ministry of Agriculture (LNV), the Ministry of Physical Planning and the Environment (VROM) and the National Foundation for Scientific Research (NWO). In 1987 he received (with several colleagues) the Education Award of the Stichting Wageningen Fonds. He is author and editor of six books on environmental economics and policy with international scientific publishers and has contributed to many books and international journals. ekko.vanierland@wur.nl

Jos van Ommeren studied econometrics at the London School of Economics, London, and the University of Amsterdam and received his PhD in economics at VU University, Amsterdam, in 1996. He has worked at the Dutch Central Bank, the Netherlands, the European University Institute, Italy, and University College London and Cranfield University, United Kingdom. Since 2001 he has been employed at the VU University, while teaching at University College Utrecht. He is a Fellow of the Tinbergen Institute.

Ruud Zaalberg is a social psychologist. He worked as an assistant professor at Eindhoven University of Technology. His research focuses on the influence of disaster experiences on risk perception and coping behaviour.

Index